Lecture Notes in Computer Science 15268

Founding Editors

Gerhard Goos
Juris Hartmanis

Editorial Board Members

Elisa Bertino, *Purdue University, West Lafayette, USA*
Wen Gao, *Peking University, Beijing, China*
Bernhard Steffen, *TU Dortmund University, Dortmund, Germany*
Moti Yung, *Columbia University, New York, USA*

The series Lecture Notes in Computer Science (LNCS), including its subseries Lecture Notes in Artificial Intelligence (LNAI) and Lecture Notes in Bioinformatics (LNBI), has established itself as a medium for the publication of new developments in computer science and information technology research, teaching, and education.

LNCS enjoys close cooperation with the computer science R & D community, the series counts many renowned academics among its volume editors and paper authors, and collaborates with prestigious societies. Its mission is to serve this international community by providing an invaluable service, mainly focused on the publication of conference and workshop proceedings and postproceedings. LNCS commenced publication in 1973.

Edgar Chávez · Benjamin Kimia · Jakub Lokoč ·
Marco Patella · Jan Sedmidubsky
Editors

Similarity Search and Applications

17th International Conference, SISAP 2024
Providence, RI, USA, November 4–6, 2024
Proceedings

Editors
Edgar Chávez
Ensenada Center for Scientific Research
and Higher Education
Ensenada, Mexico

Jakub Lokoč
Charles University
Prague, Czech Republic

Jan Sedmidubsky
Masaryk University
Brno, Czech Republic

Benjamin Kimia
Brown University
Providence, RI, USA

Marco Patella
University of Bologna
Bologna, Italy

ISSN 0302-9743   ISSN 1611-3349 (electronic)
Lecture Notes in Computer Science
ISBN 978-3-031-75822-5   ISBN 978-3-031-75823-2 (eBook)
https://doi.org/10.1007/978-3-031-75823-2

© The Editor(s) (if applicable) and The Author(s), under exclusive license to Springer Nature Switzerland AG 2025, corrected publication 2025

This work is subject to copyright. All rights are solely and exclusively licensed by the Publisher, whether the whole or part of the material is concerned, specifically the rights of translation, reprinting, reuse of illustrations, recitation, broadcasting, reproduction on microfilms or in any other physical way, and transmission or information storage and retrieval, electronic adaptation, computer software, or by similar or dissimilar methodology now known or hereafter developed.
The use of general descriptive names, registered names, trademarks, service marks, etc. in this publication does not imply, even in the absence of a specific statement, that such names are exempt from the relevant protective laws and regulations and therefore free for general use.
The publisher, the authors and the editors are safe to assume that the advice and information in this book are believed to be true and accurate at the date of publication. Neither the publisher nor the authors or the editors give a warranty, expressed or implied, with respect to the material contained herein or for any errors or omissions that may have been made. The publisher remains neutral with regard to jurisdictional claims in published maps and institutional affiliations.

This Springer imprint is published by the registered company Springer Nature Switzerland AG
The registered company address is: Gewerbestrasse 11, 6330 Cham, Switzerland

If disposing of this product, please recycle the paper.

Preface

This volume contains the papers presented at the 17th International Conference on Similarity Search and Applications (SISAP 2024), held in Providence, Rhode Island, USA, from November 4 to 6, 2024. This year, Brown University had the honor of hosting SISAP 2024, continuing the tradition of providing an annual platform for researchers, practitioners, and developers who are advancing the field of similarity data management.

As our digital landscape continues to expand, the need for efficient similarity search methods becomes increasingly critical. The growing complexity and diversity of digital information demand innovative approaches that extend beyond basic exact query retrieval. Addressing the challenges of exploring similar items and managing vast machine-learning datasets efficiently are central to our field. SISAP has firmly established itself as the leading international forum where scholars and practitioners come together to advance the frontiers of similarity search, continuously striving for faster and more effective solutions. Despite the significant progress achieved, the challenges posed by larger datasets, novel data modalities, and emerging applications continue to inspire the community to pursue further innovation.

This year's conference featured the second edition of the SISAP Indexing Challenge, chaired by Martin Aumüller, Vladimir Mic, and Eric S. Tellez, which builds on the success of the previous edition. This challenge marked a significant milestone in the evolution of the technology and fostered the dissemination of best practices throughout the community. Participants worked with a shared open dataset, offering a benchmark for reproducible results. The proceedings include a detailed description of the challenge by its organizers, along with reports from the participating teams.

A special session on the first day of the conference was dedicated to the presentation of the challenge, giving each team the opportunity to showcase their approach and results. SISAP 2024 also featured three distinguished keynote speakers (Piotr Indyk, Bradley C. Love, and Sanjiv Kumar), who provided a rich diversity of theoretical, practical, and applied insights. Each day began with a keynote presentation.

This year, we received 28 submissions for the SISAP main track and 4 submissions for the SISAP Indexing Challenge from authors representing 15 countries. The Program Committee, composed of 47 members from 18 countries, played a crucial role in ensuring the scientific rigor and quality of the accepted papers. Each submission underwent a thorough review process, with an average of 2.8 double-blind reviews per paper for the SISAP main track. After careful deliberation by the chairs, 24 papers were accepted in total, with 13 being full papers, 7 accepted as short/demo papers, and 4 accepted for the SISAP Indexing Challenge. This resulted in an acceptance rate of 46% for full papers. In addition, the proceedings contain one overview paper of the Indexing Challenge.

The proceedings of SISAP 2024 are published by Springer in the Lecture Notes in Computer Science (LNCS) series. We extend our gratitude to Springer for their continued support and for sponsoring SISAP Awards. As is tradition, authors of selected outstanding papers were invited to submit extended versions of their work for publication in a special issue of the Elsevier journal *Information Systems*.

Finally, we extend our deepest appreciation to all members of the Program Committee for their diligent and timely reviews, which were instrumental in maintaining the high scientific and academic standards of the conference. Their constructive feedback has been invaluable in helping authors refine their work and enhance the quality of these proceedings.

Providence, Rhode Island, USA
November 2024

Edgar Chávez
Benjamin Kimia
General Co-chairs

Jakub Lokoč
Marco Patella
Program Committee Co-chairs

Jan Sedmidubsky
Publication Chair

Organization

General Chairs

Edgar Chávez   CICESE, Mexico
Benjamin Kimia   Brown University, USA

Program Committee Chairs

Jakub Lokoč   Charles University, Czechia
Marco Patella   University of Bologna, Italy

Steering Committee

Giuseppe Amato   ISTI-CNR, Italy
Edgar Chávez   CICESE, Mexico
Stéphane Marchand-Maillet   University of Geneva, Switzerland
Marco Patella   University of Bologna, Italy
Ilaria Bartolini   University of Bologna, Italy
Oscar Pedreira   University of A Coruña, Spain

Publication Chair

Jan Sedmidubsky   Masaryk University, Czechia

Program Committee

Giuseppe Amato   ISTI-CNR, Italy
Laurent Amsaleg   CNRS-IRISA, France
Fabrizio Angiulli   University of Calabria, Italy
Martin Aumüller   University of Copenhagen, Denmark
Ilaria Bartolini   University of Bologna, Italy

K. Selçuk Candan	Arizona State University, USA
Panagiotis Bouros	Johannes Gutenberg University Mainz, Germany
Mario G. C. A. Cimino	University of Pisa, Italy
Edgar Chávez	CICESE, Mexico
Richard Connor	University of St. Andrews, UK
Alan Dearle	University of St. Andrews, UK
Vlastislav Dohnal	Masaryk University, Czechia
Amalia Duch	Universitat Politècnica de Catalunya, Spain
Vlad Estivill-Castro	Universitat Pompeu Fabra, Spain
Rolf Fagerberg	University of Southern Denmark, Denmark
Fabrizio Falchi	ISTI-CNR, Italy
Qiang Gao	Southwestern University of Finance and Economics, China
Claudio Gennaro	ISTI-CNR, Italy
Magnus Lie Hetland	Norwegian University of Science and Technology, Norway
Michael E. Houle	New Jersey Institute of Technology, USA
Björn Þór Jónsson	Reykjavik University, Iceland
Peer Kröger	Christian Albrechts University Kiel, Germany
Jakub Lokoč	Charles University, Czechia
Stéphane Marchand-Maillet	University of Geneva, Switzerland
Conrado Martínez	Universitat Politècnica de Catalunya, Spain
Vladimir Mic	Aarhus University, Denmark
Lia Morra	Politecnico di Torino, Italy
Marco Patella	University of Bologna, Italy
Oscar Pedreira	University of A Coruña, Spain
Miloš Radovanović	University of Novi Sad, Serbia
Marcela Ribeiro	Federal University of São Carlos, Brazil
Kunihiko Sadakane	University of Tokyo, Japan
Maria Luisa Sapino	Università di Torino, Italy
Erich Schubert	TU Dortmund University, Germany
Jan Sedmidubsky	Masaryk University, Czechia
Tetsuo Shibuya	University of Tokyo, Japan
Tomas Skopal	Charles University, Czechia
Eric S. Téllez	CONACyT—INFOTEC, Mexico
Caetano Traina-Junior	University of São Paulo, Brazil
Lucia Vadicamo	ISTI-CNR, Italy
Takashi Washio	Osaka University, Japan
Jan Zahálka	Czech Technical University in Prague, Czechia
Pascal Welke	TU Wien, Austria
Pavel Zezula	Masaryk University, Czechia
Kaiping Zheng	National University of Singapore, Singapore

Arthur Zimek   University of Southern Denmark, Denmark
Andreas Züfle   Emory University, USA

Additional Reviewers

Félix Iglesias
Muhammad Rajabinasab

Keynotes

Talk: *Graph-based algorithms for similarity search: challenges and opportunities*
Speaker: Piotr Indyk
Piotr Indyk is the Thomas D. and Virginia W. Cabot Professor of Electrical Engineering and Computer Science at the Massachusetts Institute of Technology, where he has been on the faculty since 2000. He graduated from the University of Warsaw in 1995 and received his Ph.D. in Computer Science from Stanford University in 2001. He received the Packard Fellowship in 2003 and the Simons Investigator Award in 2013. He is also a co-winner of the 2012 Paris Kanellakis Theory and Practice Award for his work on Locality-Sensitive Hashing. Piotr Indyk is a fellow of the Association for Computing Machinery and a member of the American Academy of Arts and Sciences and the National Academy of Sciences.

Talk: *Embeddings of and for the mind*
Speaker: Bradley C. Love
Bradley C. Love is a Professor of Cognitive and Decision Sciences in Experimental Psychology at University College London (UCL). He is also a distinguished fellow at The Alan Turing Institute for data science and AI, as well as the European Lab for Learning and Intelligent Systems (ELLIS). His research lab is dedicated to advancing the understanding of human learning and decision-making by integrating behavioral, computational, and neuroscience perspectives. Currently, his team is pioneering efforts in large-scale modeling of brain and behavior using deep learning techniques. Additionally, they are developing BrainGPT, an innovative tool designed to assist neuroscience researchers by leveraging large language models.

Talk: *New learning objectives for massive-scale similarity search*
Speaker: Sanjiv Kumar
Sanjiv Kumar is a Google Fellow and VP at Google Research, where he leads a team focused on the theory and applications of large ML Foundational Models and Generative AI. His recent research interests include rethinking existing modeling and computing paradigms in LLMs, with a focus on developing alternative techniques that allow fast training and inference. He also leads the development of massive-scale similarity search techniques, which are widely adopted in Google and the open-source community. He has published more than 100 papers and holds 60+ patents in the area of ML and Computer Vision. His work on the convergence of Adam received the Best Paper Award at ICLR 2018. He is an action editor of *JMLR* and holds a Ph.D. from the School of Computer Science at Carnegie Mellon University.

Contents

Research Track

An Efficient Framework for Approximate Nearest Neighbor Search
on High-Dimensional Multi-metric Data 3
   Reon Uemura, Daichi Amagata, and Takahiro Hara

REHAB24-6: Physical Therapy Dataset for Analyzing Pose Estimation
Methods 18
   Andrej Černek, Jan Sedmidubsky, and Petra Budikova

ETDD70: Eye-Tracking Dataset for Classification of Dyslexia Using
AI-Based Methods 34
   Jan Sedmidubsky, Nicol Dostalova, Roman Svaricek, and Wolf Culemann

Demonstrating the Efficacy of Polyadic Queries 49
   Ben Claydon, Richard Connor, Alan Dearle, and Lucia Vadicamo

Scalable Polyadic Queries 57
   Richard Connor, Alan Dearle, and Ben Claydon

A Dynamic Evaluation Metric for Feature Selection 65
   *Muhammad Rajabinasab, Anton D. Lautrup, Tobias Hyrup,
and Arthur Zimek*

Personalized Similarity Models for Evaluating Rehabilitation Exercises
from Monocular Videos 73
   Miriama Jánošová, Petra Budikova, and Jan Sedmidubsky

Impact of the Neighborhood Parameter on Outlier Detection Algorithms 88
   Félix Iglesias, Conrado Martínez, and Tanja Zseby

Optimizing CLIP Models for Image Retrieval with Maintained
Joint-Embedding Alignment 97
   Konstantin Schall, Kai Uwe Barthel, Nico Hezel, and Klaus Jung

Bayesian Estimation Approaches for Local Intrinsic Dimensionality 111
   *Zaher Joukhadar, Hanxun Huang, Sarah Monazam Erfani,
Ricardo J. G. B. Campello, Michael E. Houle, and James Bailey*

Towards Personalized Similarity Search for Vector Databases 126
  Marek Mahrík, Matúš Šikyňa, Vladimir Mic, and Pavel Zezula

Information Dissimilarity Measures in Decentralized Knowledge
Distillation: A Comparative Analysis 140
  Mbasa Joaquim Molo, Lucia Vadicamo, Emanuele Carlini,
  Claudio Gennaro, and Richard Connor

An Empirical Evaluation of Search Strategies for Locality-Sensitive
Hashing: Lookup, Voting, and Natural Classifier Search 155
  Malte Helin Johnsen and Martin Aumüller

On the Design of Scalable Outlier Detection Methods Using Approximate
Nearest Neighbor Graphs 170
  Camilla Birch Okkels, Martin Aumüller, and Arthur Zimek

A Topological Evaluation Model for Manifold Learning and Embedding
Techniques 185
  Victor Reyes, Margarita Liarou, and Stephane Marchand-Maillet

Local Intrinsic Dimensionality and the Convergence Order of Fixed-Point
Iteration 193
  Michael E. Houle, Vincent Oria, and Hamideh Sabaei

Identifying Propagating Signals with Spatio-Temporal Clustering
in Multivariate Time Series 207
  Jan David Hüwel, Georg Stefan Schlake, Kevin Albrechts,
  and Christian Beecks

Robust Statistical Scaling of Outlier Scores: Improving the Quality
of Outlier Probabilities for Outliers 215
  Philipp Röchner, Henrique O. Marques, Ricardo J. G. B. Campello,
  Arthur Zimek, and Franz Rothlauf

Advancing the PAM Algorithm to Semi-supervised k-Medoids Clustering 223
  Miriama Jánošová, Andreas Lang, Petra Budikova, Erich Schubert,
  and Vlastislav Dohnal

Hierarchical Clustering Without Pairwise Distances by Incremental
Similarity Search 238
  Erich Schubert

Indexing Challenge

Overview of the SISAP 2024 Indexing Challenge 255
    Eric S. Tellez, Martin Aumüller, and Vladimir Mic

Scaling Learned Metric Index to 100M Datasets 266
    David Procházka, Terézia Slanináková, Jozef Čerňanský,
    Jaroslav Olha, Matej Antol, and Vlastislav Dohnal

Grouping Sketches to Index High-Dimensional Data in a Resource-Limited
Setting 274
    Erik Thordsen and Erich Schubert

Adapting the Exploration Graph for High Throughput in Low Recall
Regimes 283
    Nico Hezel, Bruno Schilling, Kai Uwe Barthel, Konstantin Schall,
    and Klaus Jung

Top-Down Construction of Locally Monotonic Graphs for Similarity
Search 291
    Cole Foster, Edgar Chávez, and Benjamin Kimia

Correction to: A Dynamic Evaluation Metric for Feature Selection C1
    Muhammad Rajabinasab, Anton D. Lautrup, Tobias Hyrup,
    and Arthur Zimek

Author Index 301

Research Track

An Efficient Framework for Approximate Nearest Neighbor Search on High-Dimensional Multi-metric Data

Reon Uemura, Daichi Amagata, and Takahiro Hara

Osaka University, Osaka, Japan
{uemura.reon,amagata.daichi,hara}@ist.osaka-u.ac.jp

Abstract. Many objects are often multi-modal data consisting of images, videos, documents, etc., where each model exists in a different metric space. Similarity searches on multi-modal data are widely employed in information retrieval and machine learning. In these applications, each modal is usually represented as dense high-dimensional data (e.g., an embedding vector). This paper addresses the problem of nearest neighbor search (NNS) on high-dimensional multi-metric data. Although some techniques for NNS on multi-metric data exist, they do not consider high-dimensional data and query-dependent weights in multi-metric spaces. A straightforward yet fast algorithm for "exact" NNS on high-dimensional multi-metric data is linear scan, but obtaining exact results is slow on large datasets. We therefore propose an efficient framework for approximate NNS on high-dimensional multi-metric data. For each metric space, we build the same type of proximity graph that allows any search-start node. An approximate NN is found by traversing the proximity graphs while carefully selecting start nodes to avoid redundant node accesses. We conduct experiments on real-world high-dimensional multi-metric data, and the results demonstrate that our framework outperforms state-of-the-art algorithms.

Keywords: multi-metric data · approximate kNN search · high-dimensions

1 Introduction

Given a query object q and a set O of objects, the k nearest neighbor search (k-NNS) problem finds k most similar objects $\in O$ to q based on a given distance function. This problem has many applications, such as computer vision [21], recommender systems [4], geographic information systems [3], and machine learning [22], to name a few.

Motivation. The number of objects has been still increasing, and, at the same time, objects are often represented as *multi-modal* data consisting of images, videos, documents, etc., where each modal exists in a different metric space

[10–12,16,19,23,27,35]. The importance of these objects has been accelerated because of the proliferation of multi-modal searches. For example, VLM (Vision Language Model) [32] accurately deals with text-to-image processing, and Sora[1] effectively handles text-to-video processing. The high accuracy of these models is supported by training sets consisting of huge numbers of multi-modal data. Notice that the similarity between multi-modal data cannot be measured by a single metric, and it needs to focus on multi-metric spaces. That is, similarity searches on multi-modal data need to aggregate multiple metrics.

Motivated by the above observation, this paper addresses the k-NNS problem in multi-metric spaces. Given O, q, k, and $W = \langle w_1, ..., w_m \rangle$ (a set of weights for the metric spaces), this problem finds the k objects $\in O$ that have the minimum aggregate distance to q. (A formal definition appears in Sect. 2.) Azure AI Search[2] is a concrete implementation of this problem.

Challenge. There are two approaches to solving our problem. One is the integrated approach which integrates the multi-metric spaces into a single index [11,12,16,34]. The other is the separate approach, which runs a k-NNS for each metric [10,19,27,35]. The integrated approach does not take W into account. For example, even if a given query ignores some metric spaces, this approach cannot ignore them. The separate approach, which is shown in Fig. 1, does not have this issue, so this paper considers this approach. Next, we need to consider how to accelerate k-NNS in each metric space.

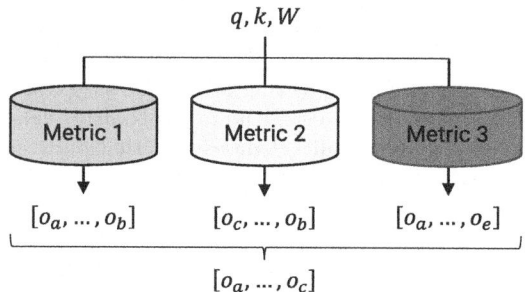

Fig. 1. Example of the separate approach. For each metric space, a k-NNS is done, and the final result is obtained by merging the result in each metric space.

Notice that each element of multi-modal data is represented as a dense high-dimensional vector due to recent embedding techniques. Therefore, each object $\in O$ has multiple high-dimensional vectors. In this setting, it is well-known that the fastest "exact" algorithm for our problem is a linear scan of O [24]. However, this approach is slow when $|O|$ is large. To address this issue, existing works proposed

---

[1] https://openai.com/sora.
[2] https://azure.microsoft.com/en-us/products/ai-services/ai-search.

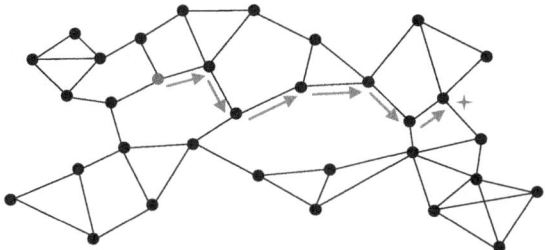

Fig. 2. Example of a proximity graph and greedy search. The red point is a start node, arrows represent graph traversal, and the blue cross is a query.

approximation algorithms based on locality-sensitive hashing [28], quantization [31], and proximity graphs [25]. Among these approaches, many empirical studies show that proximity graphs provide the best speed-accuracy trade-off.

The above discussion suggests that a promising approach is to employ a proximity graph for each metric space. In a proximity graph, each node is an object $\in O$, and there are edges between similar objects (nodes). A k-NNS is done by a greedy search, which, from a (random) start node, iteratively traverses nodes closer to a given query, as shown in Fig. 2. Note that a straightforward implementation of this approach (i.e., the separate approach with proximity graphs) has two drawbacks. One is about the *start node*. As seen in Fig. 2, if a start node is far from a given query, we need to traverse many nodes (i.e., incur many distance computations). The other is about *full separation between metric spaces*. If we run a k-NNS for each metric space *independently*, we may access the same nodes at most m times, where m is the number of metrics, because the proximity graph in each metric space consists of the same nodes (objects).

Contribution. The above drawbacks incur redundant distance computations, so we overcome these by proposing an efficient framework for our problem. For each metric space, we introduce *representative nodes* that are picked so that they exist uniformly in the space, aiming at finding close start nodes for arbitrary queries. Furthermore, they are used to determine the access order of metric spaces. Our framework avoids the full separation between metric spaces and exploits an intermediate k-NNS result to reduce the search space when we move to the next metric space.

To summarize, our contributions are as follows. (i) We propose a new framework for the k-NNS problem in high-dimensional multi-metric spaces. (ii) We conduct experiments on real-world datasets, and the results demonstrate that our framework outperforms state-of-the-art baselines.

2 Preliminary

Let O be a set of n multi-modal data. Each object $o_i \in O$ has m modals and is represented as $o_i = \{o_i^1, ..., o_i^m\}$, where o_i^j is o_i in the j-th metric space with high dimensionality. Let O^i be $\{o_1^i, ..., o_n^i\}$. In our setting, the i-th metric space is represented by $(O^i, dist^i)$, where $dist^i$ is the distance function in this metric space. This distance function satisfies symmetry, non-negativity, identity, and triangle inequality.

2.1 Problem Definition

Let $\mathcal{D}$ and W be $\{dist^1, ..., dist^m\}$ and $\{w_1, ..., w_m\}$ ($\sum_m w_i = 1$), respectively. We first define multi-metric distance.

Definition 1 (Multi-metric distance). *The multi-metric distance between o_x and o_y is*

$$dist^W(o_x, o_y) = \sum_{dist^i \in \mathcal{D}} w_i \cdot \bar{dist}^i(o_x, o_y),$$

where $\bar{dist}^i(o_x, o_y) \in [0, 1]$ is the normalized $dist^i(o_x, o_y)$.

Now we are ready to define our problem.

Definition 2. *(Multi-metric k nearest neighbor search)* *Given O, a query object $q = \{q_i^1, ..., q_i^m\}$, a result size k, and a weighting set W, a multi-metric k nearest neighbor search retrieves $R \subseteq O$ such that $|R| = k$ and $\forall o_x \in R$ and $\forall o_y \in O - R$, $dist^W(o_x, q) \leq dist^W(o_y, q)$. (Ties are broken arbitrarily.)*

As introduced in Sect. 1, it is practically hard to efficiently obtain the exact result when n is large and each modal has high dimensionality. This paper therefore considers an approximation approach, i.e., approximate k-NNS.

2.2 Baseline Algorithm

Before introducing a state-of-the-art baseline algorithm, we introduce notions for proximity graphs. Let G_i be a proximity graph in the i-th metric space, i.e., a proximity graph of O^i. A node set of G_i is hence O^i, and an edge set E^i has node (object) pairs. Essentially, there is an edge between o_x^i and o_y^i if they are similar w.r.t. $dist^i$. We use $E^i(o_x)$ to denote a set of edges $\in E^i$ held by o_x. Also, let $\mathcal{G} = \{G_1, ..., G_m\}$.

Algorithm description. We here introduce a baseline algorithm that employs the separate approach and a proximity graph in each metric space. (This proximity graph is built offline.) This algorithm is implemented in the state-of-the-art vector management system Milvus [29] and Azure AI Search.[3] Its idea is fairly simple. For each metric space, it runs an approximate k-NNS with the greedy algorithm [26] and accumulates the result in this space. At last, it picks k objects with the minimum multi-metric distance in the accumulated result. Algorithm 1 details this algorithm.

---
[3] Milvus and Azure AI Search are not limited to proximity graphs for indices, but the essential approach is the same.

Algorithm 1: BASELINE

Input: O, $\mathcal{G}$, q, k, W, and $\epsilon \geq 1$

1 $R \leftarrow \emptyset$
2 **foreach** $i \in [1, m]$ such that $w_i > 0$ **do**
3      $o_s \leftarrow$ a random start node
4      $C, C' \leftarrow \langle o_s, dist^W(o_s, q) \rangle$
5      $F \leftarrow \{o_s\}$
6   **while** $C' \neq \emptyset$ **do**
7        $o_x \leftarrow$ pop C'
8   **foreach** $o_y \in E_i(o_x)$ such that $o_y \notin F$ **do**
9          $F \leftarrow F \cup \{o_y\}$
10         $\tau \leftarrow$ the $\epsilon \cdot k$-th distance in C
11   **if** $dist^W(o_y, q) < \tau$ **then**
12           $C \leftarrow C \cup \{\langle o_y, dist^W(o_y, q)\rangle\}$
13   Sort the tuples in C in ascending order of the second element
14   **if** $|C| > \epsilon \cdot k_s$ **then**
15   Erase the last tuple in C
16           $C' \leftarrow C' \cup \{\langle o_y, dist^W(o_y, q)\rangle\}$
17   Sort the tuples in C' in ascending order of the second element
18     $R \leftarrow R \cup C$
19 Sort the tuples in R in ascending order of the second element
20 **return** the first k objects in R

2.3 Related Work

NNS in a metric space. As the problem of NNS can be defined based on an arbitrary distance function, many studies proposed efficient NNS algorithms that can be employed in any metric space [13,14]. Recent experimental papers compare existing indexing techniques [15,33]. Note that these works consider only low-dimensional data, so they are inappropriate for our problem.

NNS in multi-metric spaces. Existing works devised some data structures for k-NNS in multi-metric spaces to deal with multi-modal data. In [16], M2-tree was proposed, and this data structure combines all metrics. This integrated approach cannot consider W and is not robust against new metric insertions and existing metric deletions. To deal with any weights, [11,12,19] proposed data structures based on M-tree [17] and employ the separate approach. DESIRE [35] employs pivot-based distance indexing with B$^+$-trees, different from M-tree. Again, these works essentially assume low dimensions.

As introduced in Sect. 2.2, Milvus [29] is a state-of-the-art vector management system and can deal with objects consisting of multiple vectors. Its framework for multi-metric k-NNS is to run a k-NNS for each vector space, which is the same as the separate approach. Milvus can employ any indexing technique to accelerate k-NNS in a given space, and Algorithm 1 is an example when Milvus uses a proximity graph. We use this algorithm as a competitor in Sect. 4. Another state-of-the-art algorithm for our problem is HJG [34]. Essentially, HJG is a HNSW [25] that considers all edges generated from all metric spaces. This algorithm is also used as a competitor in our experiments.

The above separate approach typically runs a k-NNS on each metric space, like Algorithm 1. Let T be the search time of this approach, and T is obtained by

$$T = \sum_{i=1}^{m} T_i + km, \tag{1}$$

where T_i is the k-NNS time in the i-th metric space. As m cannot be reduced, we need to reduce T_i to reduce T. The above works do not consider how to minimize T_i by using the intermediate results obtained in the other metric spaces. Our framework incorporates this idea.

Approximate NNS in high-dimensional spaces. Recent experimental studies [9,24] demonstrate that proximity graphs yield the best trade-off between speed and recall. Therefore, many works developed proximity graphs [30]. Roughly, state-of-the-art proximity graphs are based on either NSW [26] or NSG [20]. NSW is an approximation of the Delaunay graph, whereas NSG is an approximation of a monotonic search graph (i.e., a graph with paths that monotonically reduce distances to a target object for any two objects).

Our framework allows arbitrary proximity graphs that can (i) use any node as a start node and (ii) be employed in any metric spaces. The first requirement is important to implement our idea for reducing search spaces. The second requirement is derived from our problem definition.

3 Proposed Framework

3.1 Main Ideas

The main bottleneck of k-NN is distance computation, so we need to minimize the number of distance computations to reduce the latency of multi-metric k-NNS. Simply employing a proximity graph for each metric space (like Algorithm 1) cannot achieve this well. Recall Fig. 2. The running time of proximity graph-based k-NNS depends on a given start node. If the start node is near a given query object q, its running time will be quite fast. On the other hand, if the start node is far from q, its performance degrades. It is hence important to select a start node to optimize the performance of (multi-metric) k-NNS. Nevertheless, this approach has not been considered in metric spaces.

Assume that we access the metric spaces in order of $O^1, O^2, ..., O^m$, for ease of explanation. When we start to run an approximate k-NNS in O^i ($i \in [2, m]$), we can use the intermediate k-NNS result obtained in the metric spaces $O^1, ..., O^{i-1}$. More specifically, we use the intermediate nearest neighbor node of q w.r.t. $dist^i$ as the start node. This idea brings the following merit:

There are correlations between multi-modal data. For example, similar images have similar captions (documents). This observation suggests that the intermediate nearest neighbor node of q in O^i is similar to q w.r.t. $dist^W$, so we need to traverse only nodes existing around q (and most of them may have been traversed in $O^1, ..., O^{i-1}$).

Clearly, the above merit removes one of the drawbacks of Algorithm 1 (or of running k-NNS in each metric independently). The remaining concern is how to find a start node close to a given query q for O^1 because we have no intermediate k-NNS result for the first accessed metric space.

We overcome this concern by introducing *representative nodes* for the proximity graph in each metric space. Representative nodes are candidates for the start node. Although query objects are unknown in advance, start nodes should be close to a given query. To satisfy this, *representative nodes should be distributed uniformly in the metric space*, as shown in Fig. 3. We pick representative nodes based on this idea.

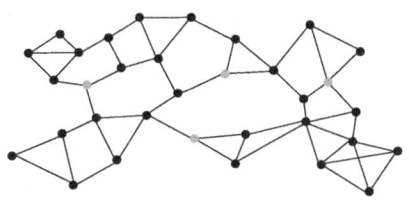 

(a) Representative nodes in a proximity graph of O^1

(b) Representative nodes in a proximity graph of O^2

Fig. 3. Representative nodes (green nodes) in each metric space ($m = 2$)

3.2 Overview

Our framework has two phases: the offline phase (indexing phase) and the k-NNS phase (query phase). The offline phase builds a proximity graph and selects representative nodes for each metric space. In a k-NNS phase, we first determine a start node and the access order of the metric spaces by using the data structure built in the offline phase, and then run the greedy algorithm [26] for each metric space.

3.3 Offline Phase

This phase is done only once if O does not change. (Dynamic data is not the scope of this paper.)

Building proximity graphs. First, we build a proximity graph G_i for each O^i ($i \in [1, m]$). As stated in Sect. 2.3, our framework can employ any proximity graph that can employ an arbitrary node as the start node and distance function. Therefore, the performance of our framework can be enhanced by a future state-of-the-art proximity graph that satisfies the above requirements.

Selecting representative nodes. To set an effective start node at the first accessed metric space, we select $r = \mathcal{O}(1)$ representative nodes in each metric space. To make them distributed uniformly in the metric space, we employ K-means++ algorithm [8] ($K = r$ in our setting) because it (i) yields useful cluster centers that tend to exist uniformly in a given data space, (ii) is robust against outliers, and (iii) runs in a reasonable time, i.e., $\mathcal{O}(Kn)$ time.[4]

Next, for each representative node o_i^j, we compute the maximum distance to its neighbor node in the proximity graph of O^j, i.e.,

$$mdist_i^j = \max_{o_l \in E^j(o_i)} \frac{dist^j(o_i, o_l) - mean^j}{std^j},$$

where $mean^j$ and std^j are the average distance and standard deviation of the distance in the j-the metric space.[5] That is, $mdist_i$ is z-normalized to enable a fair comparison between the m metric spaces. (Without this, $dist^i$ may have a different scale to $dist^j$.) Our framework maintains $\mathcal{P} = \{P^1, ..., P^m\}$, where $P^j = \{\langle o_i^j, mdist_i^j \rangle, ...\}$ and $|P^j| = r$.

Time complexity. Let $\mathcal{T}(n)$ be the time to build a proximity graph. This phase needs $\mathcal{T}(n) \times m$ time to build m proximity graphs and $\mathcal{O}(rn) = \mathcal{O}(n) \times m$ time to obtain $\mathcal{P}$. Therefore, this phase needs $\mathcal{O}(\mathcal{T}(n) + nm)$ time. For example, when a proximity graph is KGraph [18], $\mathcal{T}(n) = \mathcal{O}(n)$ for a fixed degree [5,6]. This offline processing, hence, can be done in time linear to n.

3.4 k-NNS Phase

Given q, k, and W, we first determine the access order of metric spaces and a start node. Algorithm 2 details this procedure. For each metric space and for each representative node o_i, we compute $dist^j(o_i, q)$ to obtain the node (object) closest to q among the representative nodes w.r.t. $dist^j$. We thus have at most m such nodes. We rank them based on $w_j \times mdist^j$ (smaller is better). The rank order corresponds to the access order of metric spaces. In addition, we use the first-ranked representative node as the start node.

After that, we use the greedy algorithm [26] on the proximity graph of each metric space. For the first metric space, we run it from the start node obtained above. When the search converges, we move to the proximity graph of the next metric space based on the determined access order. We set the intermediate nearest node to q as the start node and again run the greedy algorithm. Note that if an object o has been accessed in the former metric space, we do not traverse the node again. (That is, traversed nodes/objects are shared among the

---

[4] Similar results would be obtained by similar greedy algorithms [1,2].
[5] If the proximity graph has a so-called long-range edge, it is straightforward to ignore it.

Algorithm 2: DETERMINE-START-NODE

Input: q, W, and $\mathcal{P}$
Output: o_s (a start node) and I (a list of the access order of metric spaces)
1 $S, I \leftarrow \varnothing$
2 **foreach** $j \in [1, m]$ **do**
3 $\quad\lfloor\ S \leftarrow S \cup \{\langle j, o_i^j, w_j \cdot mdist_i^j\rangle\}$, where $o_i^j = \arg\min_{\mathcal{P}^j} dist^j(o_i, q)$
4 Sort the triplets in S in ascending order of the last element
5 **foreach** $\langle\langle j, o_i^j, w_j \cdot mdist_i^j\rangle \in S$ **do**
6 $\quad\lfloor\ I \leftarrow I \cup \{j\}$ ▷ determine access order of metric spaces
7 $o_s \leftarrow S[0].o_i$ ▷ determine the start node

---

m metric spaces.) This procedure is also done for the subsequent metric spaces. Algorithm 3 details this phase.

Time complexity. Let θ be the average degree of the proximity graphs. Also, let h^i be the traversal hops in the i-th metric space. The time complexity of Algorithm 2 is $\mathcal{O}(m)$ since $r = \mathcal{O}(1)$. We here assume that the distance computation for each metric is done in $\mathcal{O}(1)$ time, so $dist^W$ is done in $\mathcal{O}(m)$ time. Then, the time complexity of Algorithm 3 is $\mathcal{O}(m) + \mathcal{O}(\sum_m \theta \cdot h^i \cdot m) = \mathcal{O}(\sum_m \theta \cdot h^i \cdot m)$. Notice that θ and m are regarded as fixed parameters because they are system- or data-dependent parameters. Since our framework aims at reducing h^i, Algorithm 3 practically yields a small h^i. This is confirmed in the next section.

4 Experiment

This section reports our experimental results. All experiments were conducted on a Ubuntu 22.04 LTS machine with Intel Core i9-9980XE@3.0GHz CPU and 128GB RAM.

4.1 Setting

Dataset. We used two real-world multi-metric datasets: Corel Image Features[6] and NUS-WIDE.[7] Table 1 shows their statistics. Corel Image Features consists of Color Histogram (CH), Layout Histogram (LH), Color Moments (CM), and COOC Texture (CT). We used L_1 norm for CH and LH, whereas we used L_2 norm for the other modals. NUS-WIDE consists of CH, Color Correlogram (CORR), Edge Direction Histogram (EDH), Wavelet Texture (WT), and blockwise Color Moments extracted over 5 x 5 fixed grid partitions (CM55). We used L_1 norm for CH, CORR and EDH, whereas we used L_2 norm for the other modals. For each dataset, we picked random 1,000 objects and used them as queries. For each query, each $w_i \in W$ was randomly determined.

---

[6] https://kdd.ics.uci.edu/databases/CorelFeatures/CorelFeatures.html.
[7] https://lms.comp.nus.edu.sg/wp-content/uploads/2019/research/nuswide/NUSWIDE.html.

Algorithm 3: Our Framework

Input: $O, \mathcal{G}, q, k, W$, and ϵ

1    $(o_s, I) \leftarrow$ DETERMINE-START-NODE
2    $F \leftarrow \{o_s\}$
3    $C \leftarrow \emptyset$
4   **foreach** $i \in I$ such that $w_i > 0$ **do**
5      $C' \leftarrow \emptyset$
6   **if** $C = \emptyset$ **then**
7        $C, C' \leftarrow \{\langle o_s, dist^W(o_s, q)\rangle\}$
8   **else**
9        $o_s \leftarrow \arg\min_{o \in C} dist^i(o^i, q^i)$
10       $C' \leftarrow C \cup \{\langle o_s, dist^W(o_s, q)\rangle\}$
11   **while** $C' \neq \emptyset$ **do**
12        $o_x \leftarrow$ pop C'
13   **foreach** $o_y \in E^i(o_x)$ s.t. $o_y \notin F$ **do**
14          $F \leftarrow F \cup \{o_y\}$
15          $\tau \leftarrow$ the $\epsilon \cdot k$-th distance in C
16   **if** $dist^W(o_y, q) < \tau$ **then**
17            $C \leftarrow C \cup \{\langle o_y, dist^W(o_y, q)\rangle\}$, $C' \leftarrow C' \cup \{\langle o_y, dist^W(o_y, q)\rangle\}$
18   Sort the tuples in C and C' in ascending order of the second element
19   **if** $|C| > \epsilon \cdot k$ **then**
20   Erase the last tuple in C

21 **return** the first k objects in C

Table 1. Statistics of dataset

Dataset	#objects	Dimensionality	#modals	Distance function
Corel image features	66,616	9–32	4	L_1 norm and L_2 norm
NUS-WIDE	269,648	64–228	5	L_1 norm and L_2 norm

Evaluated algorithms. We evaluated our framework, denoted by Ours,[8] Algorithm 1, denoted by Baseline, HJG[9] [34], and Linear-scan, which simply scans O and computes $dist^W(o, q)$ for each $o \in O$ and returns the exact k-NNS result. These algorithms were single-threaded, implemented in C++, and compiled by g++ 11.4.0 with -O3 optimization. We enabled SIMD instruction for distance computations. As for Ours, we set $r = 1,000$.

Proximity graph. For each metric space, we built an undirected approximate K-NN graph [7], where $K = 20$, because this proximity graph satisfies the

---
[8] https://github.com/leonuemura/Multimetric-ANNS-Framework.
[9] https://github.com/ZJU-DAILY/HJG.

requirements of our framework.[10] We used NNDESCENT [18] to build a directed approximate K-NN graph and then made it undirected.

Evaluation criteria. We measured the average running time vs. average recall when issuing 1,000 queries. Let R' be a set of k objects returned by our framework or Algorithm 1. Its recall is obtained by

$$Recall(R') = \frac{|R \cap R'|}{k}.$$

4.2 Result

As for Ours and Baseline, we obtained time-recall curves by varying ϵ (e.g., see Algorithm 3), which are shown in Figs. 4 and 5. Recall that Linear-scan returns the exact result, and its search time is shown in Table 2. Note that HJG is slower than Linear-scan on Corel Image Features although HJG cannot provide the exact result, which is also shown in Table 2. In addition, HJG did not run on NUS-WIDE because of the segmentation fault error (so Table 2 shows "-" in the corresponding cells.) Therefore, we hereinafter omit the results of HJG.

Table 2. Search time and recall of Linear-scan and HJG ($k = 10$)

	Running time [msec]		Recall	
Dataset	Linear-scan	HJG	Linear-scan	HJG
Corel image features	32.28	42.83	1	0.97
NUS-WIDE	227.69	–	1	–

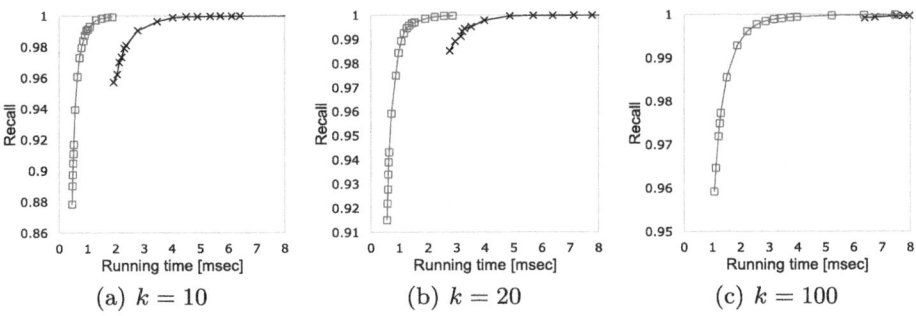

(a) $k = 10$ (b) $k = 20$ (c) $k = 100$

Fig. 4. Result on corel image features. "×" shows Baseline and "□" shows Ours.

---

[10] Famous proximity graphs HNSW [25], NSG [20], and Vamana [22] fix the start node, so they cannot be employed.

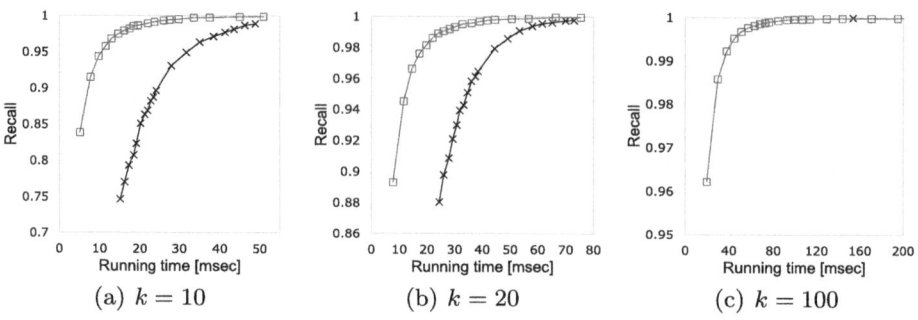

Fig. 5. Result on NUS-WIDE. "×" shows Baseline and "□" shows Ours.

Result on Corel Image Features. Fig. 4 shows the time-recall curves on Corel Image Features with $k = 10$, $k = 20$, and $k = 100$. We see that Ours yields a better time-recall trade-off, i.e., Ours outperforms Baseline. To better understand this result, we investigated the decomposed times of Ours and Baseline under the same recall (0.97). Table 3 shows the decomposed times. As Baseline runs a k-NNS for each metric space independently, it took a similar running time for each metric space. On the other hand, Ours exploits the intermediate k-NN result to determine a start node, so it took *only a few dozen microseoconds* for each $\mathcal{M}_i$ ($i \in [2, m]$), where $\mathcal{M}_i$ represents the i-th accessed metric space. (Note that Ours and Baseline *do not* share the same access order of metric spaces, i.e., $\mathcal{M}_i$ of Ours and $\mathcal{M}_i$ of Baseline can be different.) Ours thus is faster than Baseline under the same recall on Corel Image Features. This result confirms the efficacy of our first main idea described in Sect. 3.1.

In addition, comparing the results in Fig. 4 and Table 2, Ours returns almost perfect k-NN result much faster than Linear-scan. For example, when $k = 10$, Ours returns 0.96 recall in 0.64 [msec], which is 50 times faster than Linear-scan.

Result on NUS-WIDE. Figure 4 shows the time-recall curves on NUS-WIDE with $k = 10$, $k = 20$, and $k = 100$. The results are essentially similar to those on Corel Image Features. Ours always outperforms Baseline. Furthermore, Ours obtains a reasonably high recall in 10 microseconds, which is much faster than Linear-scan.

Table 3. Average running time [msec] under the same recall (Corel Image Features and $k = 10$). $\mathcal{M}_i$ represents the i-the accessed metric space.

	Algorithm 2	$\mathcal{M}_1$	$\mathcal{M}_2$	$\mathcal{M}_3$	$\mathcal{M}_4$	Total	Recall
Baseline	–	0.54	0.62	0.44	0.28	2.10	0.97
Ours	0.22	0.43	0.02	0.01	0.01	0.70	0.97
Ours by a random start node	–	0.96	0.02	0.01	0.01	1.01	0.97

Table 4. Average running time [msec] under the same recall (NUS-WIDE and $k = 10$). $\mathcal{M}_i$ represents the i-the accessed metric space.

	Algorithm 2	$\mathcal{M}_1$	$\mathcal{M}_2$	$\mathcal{M}_3$	$\mathcal{M}_4$	$\mathcal{M}_5$	Total	Recall
Baseline	–	3.16	4.44	4.74	3.17	7.54	24.33	0.90
Ours	0.55	6.03	0.19	0.12	0.08	0.08	7.09	0.90
Ours by a random start node	–	9.85	0.21	0.16	0.09	0.08	10.38	0.90

Efficacy of our start node selection. Last, we investigate the efficacy of representative nodes. To show this, we compare Ours with its variant that used a random node as the start node. Tables 3 and 4 show the results. We need to focus on the running time on $\mathcal{M}_1$, and there is an obvious improvement, which yields about 30% total time improvement under the same recall. This clarifies that our second idea described in Sect. 3.1 also functions well in practice.

5 Conclusion

Due to the proliferation of multi-modal data and embedding techniques, objects often contain multiple high-dimensional data and the importance of multi-metric search has been increasing. This paper hence addressed the multi-metric k nearest neighbor search problem in high dimensions. The state-of-the-art algorithm for this problem does not consider the correlation between multi-metric spaces, so it cannot exploit the search results of the other metric spaces. We overcame this inefficiency issue by proposing a new framework for this problem. Our framework can use any proximity graph that can employ an arbitrary start node and be used in any metric space. Our analysis of the time complexity of our framework suggests that our framework functions better than the baseline algorithm (under correlated multi-metric spaces). We conducted experiments by using real-world datasets, and the results demonstrate that our framework outperforms the baseline algorithm.

Acknowledgements.. This work was partially supported by AIP Acceleration Research JPMJCR23U2, JST, and JSPS KAKENHI Grant Number 24K14961.

References

1. Amagata, D.: Diversity maximization in the presence of outliers. In: AAAI, pp. 12338–12345 (2023)
2. Amagata, D.: Fair k-center clustering with outliers. In: AISTATS, pp. 10–18 (2024)
3. Amagata, D., Arai, Y., Fujita, S., Hara, T.: Learned k-nn distance estimation. In: SIGSPATIAL, pp. 1–4 (2022)
4. Amagata, D., Hara, T., Xiao, C.: Dynamic set knn self-join. In: ICDE, pp. 818–829 (2019)

5. Amagata, D., Onizuka, M., Hara, T.: Fast and exact outlier detection in metric spaces: a proximity graph-based approach. In: SIGMOD, pp. 36–48 (2021)
6. Amagata, D., Onizuka, M., Hara, T.: Fast, exact, and parallel-friendly outlier detection algorithms with proximity graph in metric spaces. VLDB J. **31**(4), 797–821 (2022)
7. Arai, Y., Amagata, D., Fujita, S., Hara, T.: Lgtm: a fast and accurate knn search algorithm in high-dimensional spaces. In: DEXA, pp. 220–231 (2021)
8. Arthur, D., Vassilvitskii, S.: K-means++ the advantages of careful seeding. In: SODA, pp. 1027–1035 (2007)
9. Aumüller, M., Bernhardsson, E., Faithfull, A.: Ann-benchmarks: a benchmarking tool for approximate nearest neighbor algorithms. In: SISAP, pp. 34–49 (2017)
10. Bustos, B., Keim, D., Schreck, T.: A pivot-based index structure for combination of feature vectors. In: SAC, pp. 1180–1184 (2005)
11. Bustos, B., Kreft, S., Skopal, T.: Adapting metric indexes for searching in multi-metric spaces. Multimedia Tools and Applications **58**(3), 467–496 (2012)
12. Bustos, B., Skopal, T.: Dynamic similarity search in multi-metric spaces. In: MIR, pp. 137–146 (2006)
13. Chávez, E., Navarro, G., Baeza-Yates, R., Marroquín, J.L.: Searching in metric spaces. ACM Comput. Surv. **33**(3), 273–321 (2001)
14. Chen, L., Gao, Y., Song, X., Li, Z., Zhu, Y., Miao, X., Jensen, C.S.: Indexing metric spaces for exact similarity search. ACM Comput. Surv. **55**(6), 1–39 (2022)
15. Chen, L., Gao, Y., Zheng, B., Jensen, C.S., Yang, H., Yang, K.: Pivot-based metric indexing. Proceedings of the VLDB Endowment **10**(10), 1058–1069 (2017)
16. Ciaccia, P., Patella, M.: The m2-tree: processing complex multi-feature queries with just one index. In: DELOS (2000)
17. Ciaccia, P., Patella, M., Zezula, P., et al.: M-tree: an efficient access method for similarity search in metric spaces. In: VLDB, pp. 426–435 (1997)
18. Dong, W., Moses, C., Li, K.: Efficient k-nearest neighbor graph construction for generic similarity measures. In: WWW, pp. 577–586 (2011)
19. Franzke, M., Emrich, T., Züfle, A., Renz, M.: Indexing multi-metric data. In: ICDE, pp. 1122–1133 (2016)
20. Fu, C., Xiang, C., Wang, C., Cai, D.: Fast approximate nearest neighbor search with the navigating spreading-out graph. Proceedings of the VLDB Endowment **12**(5), 461–474 (2019)
21. Harwood, B., Drummond, T.: Fanng: Fast approximate nearest neighbour graphs. In: CVPR, pp. 5713–5722 (2016)
22. Jayaram Subramanya, S., Devvrit, F., Simhadri, H.V., Krishnawamy, R., Kadekodi, R.: Diskann: fast accurate billion-point nearest neighbor search on a single node. NeurIPS **32** (2019)
23. Jo, S., Trummer, I.: Demonstration of thalamusdb: answering complex sql queries with natural language predicates on multi-modal data. In: SIGMOD, pp. 179–182 (2023)
24. Li, W., Zhang, Y., Sun, Y., Wang, W., Li, M., Zhang, W., Lin, X.: Approximate nearest neighbor search on high dimensional data-experiments, analyses, and improvement. IEEE Trans. Knowl. Data Eng. **32**(8), 1475–1488 (2020)
25. Malkov, Y.A., Yashunin, D.A.: Efficient and robust approximate nearest neighbor search using hierarchical navigable small world graphs. IEEE Trans. Pattern Anal. Mach. Intell. **42**(4), 824–836 (2018)
26. Malkov, Y., Ponomarenko, A., Logvinov, A., Krylov, V.: Approximate nearest neighbor algorithm based on navigable small world graphs. Inf. Syst. **45**, 61–68 (2014)

27. Patroumpas, K., Zeakis, A., Skoutas, D., Santoro, R.: Multi-attribute similarity search for interactive data exploration. In: EDBT/ICDE Workshop (2021)
28. Tian, Y., Zhao, X., Zhou, X.: Db-lsh: Locality-sensitive hashing with query-based dynamic bucketing. In: ICDE, pp. 2250–2262 (2022)
29. Wang, J., Yi, X., Guo, R., Jin, H., Xu, P., Li, S., Wang, X., Guo, X., Li, C., Xu, X., et al.: Milvus: a purpose-built vector data management system. In: SIGMOD, pp. 2614–2627 (2021)
30. Wang, M., Xu, X., Yue, Q., Wang, Y.: A comprehensive survey and experimental comparison of graph-based approximate nearest neighbor search. Proceedings of the VLDB Endowment **14**(11), 1964–1978 (2021)
31. Wang, R., Deng, D.: Deltapq: lossless product quantization code compression for high dimensional similarity search. Proceedings of the VLDB Endowment **13**(13), 3603–3616 (2020)
32. Yu, J., Wang, Z., Vasudevan, V., Yeung, L., Seyedhosseini, M., Wu, Y.: Coca: contrastive captioners are image-text foundation models. arXiv preprint arXiv:2205.01917 (2022)
33. Zhu, Y., Chen, L., Gao, Y., Jensen, C.S.: Pivot selection algorithms in metric spaces: a survey and experimental study. VLDB J. **31**(1), 23–47 (2022)
34. Zhu, Y., Chen, L., Gao, Y., Ma, R., Zheng, B., Zhao, J.: Hjg: an effective hierarchical joint graph for approximate nearest neighbour search in multi-metric spaces. In: ICDE, pp. 4275–4287 (2024)
35. Zhu, Y., Chen, L., Gao, Y., Zheng, B., Wang, P.: Desire: An efficient dynamic cluster-based forest indexing for similarity search in multi-metric spaces. Proceedings of the VLDB Endowment **15**(10), 2121–2133 (2022)

REHAB24-6: Physical Therapy Dataset for Analyzing Pose Estimation Methods

Andrej Černek[1]({{envelope}}), Jan Sedmidubsky[1], and Petra Budikova[2]

[1] Faculty of Informatics, Masaryk University, Brno, Czech Republic
cernek@mail.muni.cz
[2] VisionCraft, Brno, Czech Republic

Abstract. One of the prospective domains in remote healthcare is monitoring home physical rehabilitation using mobile phones and providing patients with real-time feedback on their exercise performance. Assessing such performance involves analyzing the similarity of spatio-temporal features extracted from human motion data. State-of-the-art research provides multiple tools for estimating human motion from mobile camera video streams. However, their applicability to physical therapy monitoring is not sufficiently explored. To address this problem, we introduce a new rehabilitation dataset (REHAB24-6), which provides untrimmed RGB videos, 2D and 3D skeletal ground truth of human motion, and temporal segmentation for six rehabilitation exercises. We also propose a novel pose transformation technique to evaluate existing 2D and 3D pose estimation methods trained on different datasets with distinct body models. Our experiments explore the current limitations of the state-of-the-art, particularly the depth estimation, and offer recommendations for selecting appropriate models. Finally, we propose similarity-based techniques to assess the ability of estimated pose sequences to discern exercise performance and report promising results of current pose detectors for rehabilitation assistance.

Keywords: pose estimation · motion capture · rehabilitation exercise · skeleton body model · kNN retrieval

1 Introduction

The objective of *human pose estimation* (*HPE*) methods is to determine the 2D or 3D positions of important body joints in consecutive frames of a single RGB video stream. The estimated joint positions constitute a simplified spatio-temporal representation of human motion known as a *skeleton sequence*. The analysis of the skeleton representation opens unprecedented application potential in many domains [12]. One of the currently most prospective domains is healthcare, where computer-assisted motion monitoring can be used, for example, to improve the quality of physical therapy treatments.

Research in this paper is motivated by the real-world needs of a rehabilitation-support software that aims at assisting patients during the home-exercising part of a physical therapy treatment. In this software, correct reference exercises will be recorded for individual patients at a physiotherapist clinic and then compared against the motions captured during patients' home exercising. These motions are supposed to be represented by skeleton sequences obtained by applying a suitable HPE method on an RGB video stream obtained from an ordinary smartphone. To allow high-reliability monitoring of exercise correctness, it is vital to select an HPE method that is both efficient and sufficiently precise. However, how do we choose such a method from the wide offering of state-of-the-art research and commercial models?

Naturally, we are not the first ones to ask questions about HPE quality. There exist several well-known benchmarks and datasets on which HPE methods are routinely evaluated [4]. However, the results of these evaluations are only partial indicators of the suitability of a backbone deep *neural network* (*NN*) model adopted by HPE. First, the existing testbeds typically feature basic daylife motions (e.g., sitting, walking, picking up a phone), which may have properties different from physical therapy exercises. Second, there exist several different human body models that are adopted by different NNs; existing testbeds typically only provide a comparison between NNs that utilize the *same* body model (i.e., the same skeleton format) as the dataset. Third, the standard testbeds only evaluate the estimated skeleton data in terms of joint-coordinate accuracy but provide no information about the usefulness of such data in the context of application semantics that require the similarity evaluation of skeleton data.

Considering all the above, we realize that a new testbed for rehabilitation exercising is needed, as well as a methodology for comparing HPE quality across different body models and with attention to application usefulness. In particular, this paper contributes the following:

- A new dataset of rehabilitation exercises (REHAB24-6) captured by two synchronized RGB cameras and by a motion-capture system that provides accurate ground truth of 3D joint coordinates; the dataset is accompanied by multiple descriptive attributes to test different aspects of HPE quality.
- Quality estimation methodology for assessing individual HPE methods for a given rehabilitation exercise based on several measures.
- Deep pose-estimation analysis with recommendations contributing to a more stable estimation of human body joints (e.g., different body-model alignment functions or different underlying 2D detectors).
- Application-aware verification of HPE quality based on Silhouette coefficient and k-nearest neighbor queries that distinguish between correctly and incorrectly performed rehabilitation exercises.

2 State-of-the-Art Approaches

The development and evaluation of HPE methods are closely linked to the availability of training/test data. In this section, we first provide a brief survey of state-of-the-art HPE approaches and highlight those 2D and 3D HPE methods that are highly ranked for general-purpose scenarios. Then, we focus on available datasets of human motion data and discuss their limitations in terms of scope and available metadata.

2.1 Human Pose Estimation

HPE entails extracting time series of joints from videos, with the most common representation being a skeleton (a set of connected joints or keypoints in general). We differentiate between 2D estimation, which infers positions of joints in the pixel space of video frames, and 3D, where the positions are set in a virtual coordinate system, often normalized using the position of one joint in the origin.

State-of-the-art multi-person 2D pose estimation approaches comprise detecting the object location (e.g., using the YOLOv3 object detector), followed by a single-person pose estimation on each object, like in the cases of HRNet [13], AlphaPose [5] and BlazePose [2]. As a result, if the object's bounding box is incorrectly selected, the 2D model cannot detect anything meaningful. YOLOv7 [10,17] avoids this by detecting the bounding boxes and keypoints in a single pass.

Joint-based monocular (from a single camera stream) 3D pose estimation can also be done in a similar vein. However, the low diversity of datasets with 3D ground truth often leads to difficulties in generalization. The state-of-the-art instead lies in 2D-to-3D lifting, which focuses on predicting the depth of 2D poses generated by a 2D pose detector.

Since a single view cannot capture all the potential depth information, monocular models like MotionBERT [21], MHFormer [8] or STCFormer [15] try to recover additional depth clues from the temporal domain by using a sliding window over the 2D pose sequences generated by off-the-shelf model, resulting in smoothing. Some commercial solutions provide the entire pipeline, e.g., MediaPipe Pose uses modified BlazePose for 2D estimation and then executes lifting to 3D with the help of the mesh-based GHUM [19].

2.2 Human Motion Datasets and Body Models

High-quality datasets for training are vital to the effectiveness of the pose estimation. Table 1 compares the existing datasets with regard to the attributes relevant to the physical therapy correctness discrimination. Overall, the weakness of existing fitness/rehabilitation datasets is a very limited range of exercises with available RGB videos and missing information about the correctness of each exercise repetition. Both of these characteristics are important for deciding which HPE method is suitable for exercise monitoring. To fill this gap, we design the REHAB24-6 dataset.

We also recognize a need for a different evaluation protocol from the one used in standard benchmarks, as standard protocols rely on the dataset and the detectors using the same body model. However, there is no standardized body model, and a variety of body models can be found in the datasets, differing not only in the number of joints but also in their exact position on the human

Table 1. Comparison of properties of general-purpose (MS-COCO and Human3.6M) and fitness-based (the rest) human motion datasets.

Dataset	Modality			RGB props		Repetitions		Dimensions: # of		
	RGB	2Dj	3Dj	#Cams	Lighting	Segment.	Correct.	Subjs	Exercs	Frames
MS-COCO [9]	✓	✓	✗	1	✗	✗	✗	25 K	N/A	330 K
Human3.6M [7]	✓	✓	✓	4	✗	✗	✗	11	N/A	3.6 M
Fit3D [6]	✓	✓	✓	4	✗	✓	✗	11	47	> 3 M
Lower body [16]	N/A	N/A	N/A	1	✗	N/A	✗	20	31	1.9 M
FLAG3D [14]	✓	N/A	✓	1	✗	N/A	✗	24	60	20 M
mRI [1]	✓	✓	✓	2	✗	✗	✗	20	12	160 K
EC3D [20]	N/A	✓	✓	4	✗	✓	✓	4	3	30 K
KIMORE [3]	N/A	✗	✓	1	✗	✗	✗	78	5	N/A
InfiniteForm [18]	✓	✓	✓	N/A	✓	N/A	✗	N/A	15	60 K
REHAB24-6	✓	✓	✓	2	✓	✓	✓	10	6	370 K

The "N/A" symbol denotes that clear information about a given feature is not provided or the feature is not publicly available

body. For example, the most widely used 2D training and benchmarking dataset is MS COCO [9], which features a 17-joint body model extended to 33 joints by BlazePose. On the other hand, the primary dataset for 3D pose estimation is Human3.6M [7] with a different body model of 17 joints, which differs from COCO in joints on the head and body core.

3 REHAB24-6: Dataset Description

To enable the evaluation of HPE models and the development of exercise feedback systems, we produced a new rehabilitation dataset (REHAB24-6), which is publicly available in the Zenodo repository: https://doi.org/10.5281/zenodo.13305825. The main focus is on a diverse range of exercises, views, body heights, lighting conditions, and exercise mistakes. With the RGB videos, skeleton sequences, repetition segmentation, and exercise correctness labels, this dataset offers the most comprehensive testbed for exercise-correctness-related tasks.

3.1 Recording Conditions

Our laboratory setup included 18 synchronized sensors (2 RGB video cameras, 16 ultra-wide motion capture cameras) spread around an 8.2 × 7 m room, as shown in Fig. 1. The RGB cameras were located in the corners of the room, one in a horizontal position (hor.), providing a larger field of view (FoV), and one in a vertical (ver.), resulting in a narrower FoV. Both types of cameras were synchronized with a sampling frequency of 30 frames per second (*FPS*).

The subjects wore motion capture body suits with 41 markers attached to them, which were detected by optical cameras. The OptiTrack Motive 2.3.0

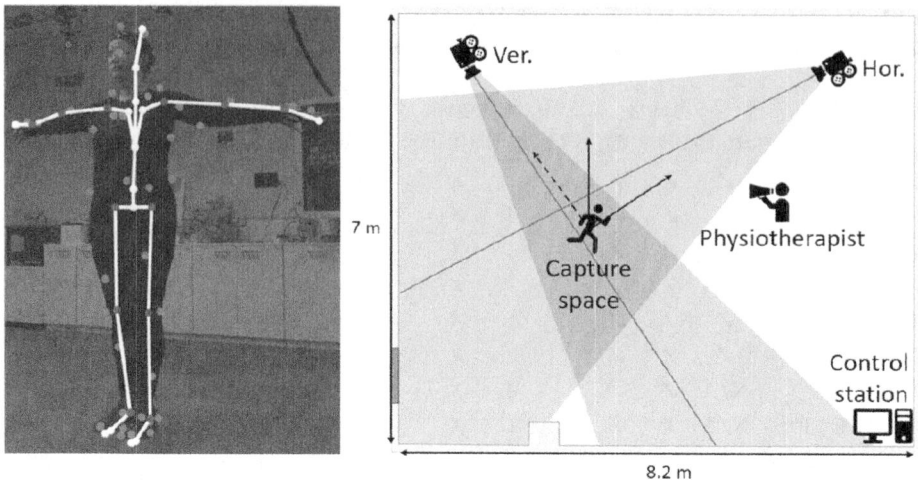

Fig. 1. The floor plan shows the placement of the RGB cameras, the capture space they form, the direction of exercising and an example of a (cropped) vertical camera frame with the body model (white skeleton) and marker (grey points) projection.

software inferred the 3D positions of the markers in *virtual centimeters* and converted them into a skeleton with 26 joints, forming our human pose 3D *ground truth* (*GT*).

To acquire a 2D version of the ground truth in *pixel coordinates*, we applied a projection of the virtual coordinates into the camera using the simplified pinhole model. We estimated the parameters for this projection as follows. First, the virtual position of the cameras was estimated using measuring tape and knowledge of the virtual origin. Then, the orientation of the cameras was optimized by matching the virtual marker positions with their position in the videos.

We also simulated changes in lighting conditions: a few videos were shot in the natural evening light, which resulted in worse visibility, while the rest were under artificial lighting.

3.2 Description of Exercises

Ten subjects participated in our recording and consented to release the data publicly: six males and four females of different ages (from 25 to 50) and fitness levels. The following six types of exercises, which constitute a representative sample for rehabilitating various body parts, were recorded:

Ex1. Arm abduction: sideway raising of the straightened right arm;
Ex2. Arm VW: fluent transition of arms between V (arms straight up) and W (elbows down, hands up) shape;
Ex3. Push-ups: push-ups with hands on a table;
Ex4. Leg abduction: sideway raising of the straightened leg;

Ex5. Leg lunge: pushing a knee of the back leg down while keeping a right angle on the front knee;
Ex6. Squats.

A physiotherapist instructed the subjects on how to perform the exercises so that at least five repetitions were done in what he deemed the correct way and five more incorrectly. The participants had a certain degree of freedom, e.g., in which leg they used in Ex4 and Ex5. Similarly, the physiotherapist suggested different exercise mistakes for each subject.

Every exercise was also executed in two directions, resulting in different *views* of the subject depending on the camera. Facing the horizontal camera resulted in a *front* view for that camera and a *profile* from the other. Facing the wall between the cameras shows the subject from *half-profile* in both cameras. In Fig. 1, these directions can be seen as filled arrows. A rare direction, only used for push-ups due to the use of the table, was facing the vertical camera, shown as a dashed arrow in Fig. 1, with the views being reversed compared to the first orientation.

In summary, the recorded dataset contains the following data:

- 65 recordings (184,825 frames, 30 FPS):
    - RGB videos from two cameras (hor./ver.);
    - 3D and 2D projected positions of 41 motion capture marker;
    - 3D and 2D projected positions of 26 skeleton joints;
- Annotation of 1072 exercise repetitions:
    - Temporal segmentation (start/end frame, most between 2–5 s);
    - Binary correctness label (around 90 from each category in each exercise, except Ex3 with around 50);
    - Exercise direction (around 90 from each direction in each exercise);
    - Lighting conditions label.

4 Evaluation Methodology: Principles and Measures

Standard evaluation protocols measure the per-frame similarity of joint positions between the ground truth and the predictions, assuming that their body models match. Since the models we test are not trained on our dataset to best capture the in-the-wild performance, this assumption does not hold. Furthermore, we would like to explore whether the similarity of joint positions translates well to application semantics, i.e., exercise correctness discrimination.

4.1 Body Models and Joint Mapping

Existing body models differ in the number of joints and their relative positions. To reconcile the former difference, we identify the greatest common subset of joints among all the body models, which includes three joints for each limb, as shown in red in Fig. 1. Only these 12 joints are used in our evaluations.

By matching the joints with similar semantics, we might still keep the systematic error that stems from different specifications of the same joint, e.g., a hip. We analyze these discrepancies by looking at each joint error's mean and standard deviation in 2D (the distance between the predicted and GT joint position, see Fig. 2). If the deviation is high, it suggests that the error is random (due to incorrect pose estimation). On the other hand, if the standard deviation is low and the error high, this might point in the direction of a systematic shift in the location compared to the GT body model. These systematic errors were inconsistent across the views and body orientations, so it is hardly possible to automatically account for them.

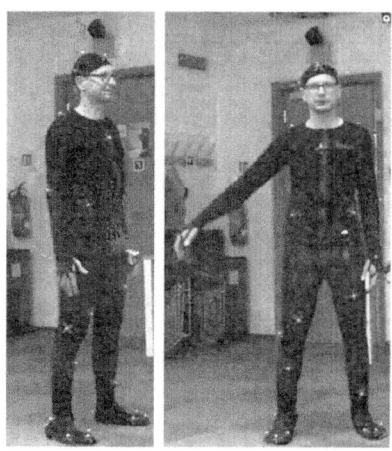

Fig. 2. Visualization of GT (white points) and predicted (red points) positions of a mostly static left-hip joint from the whole video: a high standard deviation indicates incorrect detections (left), while a low standard deviation along with high error indicates the difference between body models (right).

4.2 2D Error and Bounding Box Measures

The 2D pose detectors return the joint positions in pixels, which are directly comparable to the ground truth projected to 2D. Hence, we can utilize the standard Mean Per Joint Position Error (MPJPE) [7], which calculates the mean Euclidean distance between the matched joints across all poses and videos.

Matching the skeletons, of course, only makes sense if the skeletons represent the same person. As the object detection parts of the pose detectors might detect another object in the video (e.g., the physiotherapist or a hallucination), we propose to measure these errors in object detection and joint position separately. To determine whether the same person is represented in GT and HPE output, we can apply the standard Intersect over Union with a low threshold on the bounding boxes encompassing the GT and predicted skeletons in 2D. We only use the leg and shoulder joints to minimize the influence of differing body models and not to make the bounding box too sparse, like in Fig. 3-left. Finally, we filter these incorrectly detected frames from the joint position error calculations.

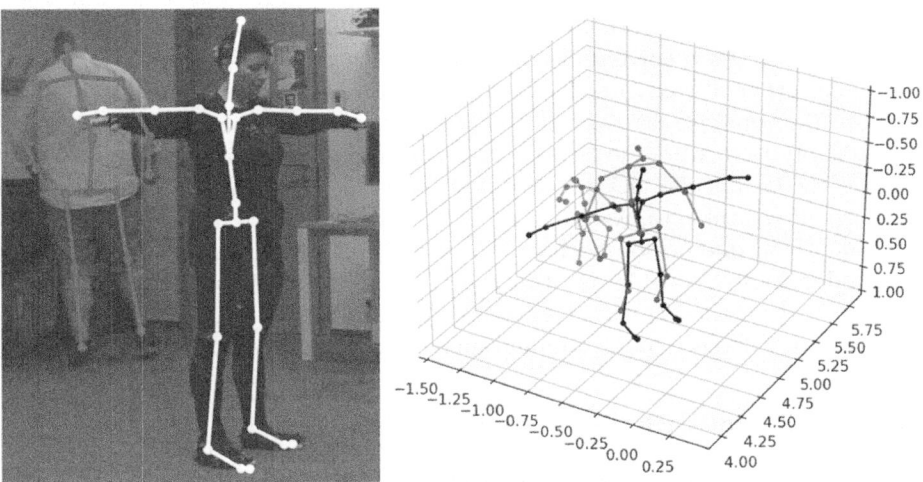

Fig. 3. Scenario in which there would be an overlap of GT bounding box (white) with the prediction (red) if arm joints were used (left). Comparison of the *BestFit* (blue) and *BBox* (red) transformations against GT (black) in such a scenario (right).

4.3 3D Error and Similarity Transformations

With the 3D ground truth in virtual centimeters and predictions in arbitrary units, we must first transform the predictions to be able to compare them. Traditional implementations of the 3D MPJPE use poses normalized with one joint in the origin (usually the center of the pelvis). Since this joint is inconsistent across different body models, we analyze the traditionally used *Best Fit* transformation and identify its weaknesses. To remedy them, we propose an alternative *Bounding Box* transformation.

Best Fit (BestFit) Transformation. To fit the predicted joints onto the ground truth joints, we can perform the so-called Procrustes transformation, which scales, translates and rotates the coordinates to produce the minimal MPJPE. Scaling provides the same units as ground truth, while the translation and rotation move the matched joints close to each other. Such transformation can mask specific issues, particularly in cases where the poses do not match the direction or are not even overlapping, like in Fig. 3-right. It also might not correctly show which joints contribute most to the error.

Bounding Box (BBox) Transformation. Since we have approximate camera parameters, we can rotate the ground truth to match what NNs try to predict and forgo the rotation step from the previous transformation. Similarly, we can calculate specific translations in the image axes based on the positions of the 2D and 3D bounding boxes, i.e., the distance between the 2D bounding box centers relative to their sizes should match the distance between the 3D bounding boxes. Therefore, the only steps that are taken from the BestFit manner are scaling (to have the correct units) and shifting in depth (there is no universal reference to set the predicted depth correctly).

Among other advantages, the proposed BBox transformation allows us to calculate the separate error in 2D (height and width of the image) and depth more reliably than with the BestFit transformation, providing insight into how effective the monocular 3D pose estimation is compared to 2D. However, the transformation still retains minor issues: it cannot detect distance differences in the depth (if one person occludes the other and the NN tracks the incorrect one), and it doesn't consider the fact that 2D and 3D predictions may not match perfectly (e.g., if the 3D model smooths the frames). Nevertheless, these issues are rare and, therefore, have minimal impact on the mean errors.

4.4 Discriminating Exercise Correctness

Physiotherapy assistance tools should mainly distinguish between correct and incorrect exercise execution, so we also look for metrics that would evaluate the discriminating power of individual HPE results. The correct exercise instructions, as provided by the physiotherapist, use angle thresholds rather than absolute joint positions. Accordingly, we calculate a set of main body angles as a baseline representation of each pose. In particular, we use 8 angles defined by limb roots (roughly armpits and groin) and middle points (knees, elbows). The angle size was computed as a Cosine distance between the two vectors defined by three joints that determine each angle. Thus each exercise repetition (motion) is represented as an 8-dimensional time series. To determine the similarity of two motions (i.e., 8D time series) of variable lengths, we apply the Dynamic Time Warping with the Euclidean distance as the internal metric.

Our first approach is to model the correctness assessment as a classification problem, assuming that correct exercise repetitions will be similar. We apply the 1-nearest neighbor (1-NN) algorithm for the classification and evaluate its accuracy for each subject.

From another perspective, we want the similarity between correct exercises to be smaller than the similarity with incorrect repetitions. For this purpose, we propose to employ the Silhouette coefficient [11] and calculate the mean of these coefficients for each subject using the correct repetitions.

5 Experimental Evaluation

The collected REHAB24-6 dataset gives us a unique opportunity to gain insight into the effectiveness and efficiency of 2D and 3D pose detectors on in-the-wild rehabilitation videos and the exercise discrimination task.

5.1 Models Overview

The experiments included four 3D pose estimators with various configurations: MotionBERT (two network sizes, both with a smoothing window of 243 frames), MHFormer (with a window of 351), STCFormer (with a window of 81) and MediaPipe Pose (three network sizes). We selected the models trained with the widest available smoothing windows.

Aside from MediaPipe Pose, these models demand 2D joints as input, for which we tested AlphaPose (Fast Pose trained on the Halpe dataset, as recommended by MotionBERT), HRNet (w48, recommended by MHFormer), YOLOv7 and even the MediaPipe Pose (heavy).

We mostly used existing implementations, including object detection and pose tracking, or implemented a naive approach if unavailable. However, we decided not to analyze the influence of these parts of the pose estimation pipeline.

5.2 Invalid Frames and 2D Error

As discussed in Sect. 4.2, the analysis of the effects of object detection and pose-tracking methods was not the aim of this work, so we filtered out the frames that were not detected and those where the predictions had low bounding box overlap with ground truth. Table 2 shows that the amount of invalid frames is insignificant outside of exercises Ex3 and Ex4. In particular, the drop in AlphaPose coverage was caused by the pose tracking method that provided fragmented pose sequences from which we used only the longest one.

Table 2. The mean percentage of valid frames (in %). **Best** and second best (not rounded) result in each column is highlighted.

2D model	Ex1	Ex2	Ex3	Ex4	Ex5	Ex6
AlphaPose	99.9	97.9	80.9	92.1	96.9	97.7
HRNet	99.9	**100.0**	**94.9**	**97.0**	**100.0**	**100.0**
YOLOv7	98.6	99.6	91.4	95.6	97.9	99.6
MPP-heavy	**100.0**	99.9	**97.7**	**99.1**	99.9	99.9

Table 3. The 2D MPJPE (in pixels).

2D model	Hor. camera						Ver. camera					
	Ex1	Ex2	Ex3	Ex4	Ex5	Ex6	Ex1	Ex2	Ex3	Ex4	Ex5	Ex6
AlphaPose	19	21	26	19	19	21	36	40	29	35	40	35
HRNet	19	20	24	18	18	19	32	36	31	33	38	30
YOLOv7	18	21	27	20	21	21	34	40	33	33	46	32
MPP-heavy	22	22	27	23	20	22	43	38	36	35	36	32

With the invalid frames filtered out, we can look at the errors of 2D pose detectors. Since the 3D detectors use their outputs, we can expect that their choice might also affect the overall 3D performance. Table 3 shows HRNet achieving the best scores and MediaPipe the worst.

It is important to realize that the pixel error is relative to the image and person size: In the videos from the Horizontal camera, the standing person had a height of approximately 700–900 pixels, depending on their actual height. On the other hand, in the Vertical videos, the heights ranged around 1000–1300, roughly 1.5 times more, and as a result, the errors are more significant as well.

5.3 3D and Depth Error

Moving from the 2D to 3D detectors, we are particularly interested in how the added depth affects the error. The BBox transformation allows us to see the error broken down into the error in depth and the error in width and height (this should correlate with 2D error). This separation cannot be correctly achieved with the traditional BestFit transformation due to the rotations, but the BBox transformation also causes more significant reported errors and, unlike BestFit, is not directly comparable to existing literature. Looking at Table 4 with the errors aggregated across all videos (i.e., all exercises and cameras), the MediaPipe Pose overcame the other models thanks to its lower depth error. Still, the higher depth error than the 2D error points to the main bottleneck of monocular (single camera view) pose estimation methods.

We can also observe that changing the input 2D model changes the 3D error very little, by less than a centimeter, so the percentage of valid frames and the efficiency matter more for the correct choice. However, we can see more significant differences in the best views for each exercise, like in Table 5. Here, MediaPipe Pose tops the other models only in a single exercise, and the best 2D models also vary. It is still worth noting that even at this level of detail, the choice of 2D detectors influences the errors less than that of the 3D model.

Table 4. The 3D MPJPE (in centimeters).

3D model	2D model	BestFit 3D	BBox 3D	BBox 2D	BBox Depth
MotionBERT-full	AlphaPose	9.2	14.0	6.5	11.2
MotionBERT-lite	AlphaPose	9.4	13.5	6.3	10.7
	HRNet	9.2	13.6	6.0	11.1
	YOLOv7	9.2	13.6	6.2	10.9
STCFormer-81	HRNet	10.9	15.8	7.7	12.3
MHFormer-351	HRNet	9.0	14.6	6.4	12.1
	YOLOv7	9.4	15.0	6.7	12.4
	MPP-Heavy	9.4	15.3	7.2	12.3
MPP-heavy		8.7	_12.0_	6.6	8.7
MPP-full		**8.4**	**11.7**	6.6	8.4
MPP-lite		_8.6_	12.3	6.8	8.9

Table 5. The best model on each exercise

Exercise	3D model	2D model	Camera	View	3D MPJPE BestFit	3D MPJPE BBox
Ex1	MHFormer-351	YOLOv7	hor.	half-profile	5.7	8.4
Ex2	MHFormer-351	HRNet	hor.	front	7.2	9.2
Ex3	MPP-heavy		hor.	profile	8.1	11.4
Ex4	MotionBERT-lite	AlphaPose	hor.	front	5.5	8.1
Ex5	MotionBERT-lite	HRNet	hor.	half-profile	8.8	9.7
Ex6	MotionBERT-lite	AlphaPose	hor.	half-profile	7.3	9.0

5.4 Discriminating Exercise Correctness

While joint position errors give us comparisons between the models, they do not show how well the models generally behave. Conversely, the 1-NN classification and Silhouette coefficients can also be calculated on ground truth, offering additional insight. Table 6 shows such a comparison on three exercises (Ex1, Ex5, Ex6) and views that were selected for covering the basic movements. Note that the results on ground truth do not achieve 100 % 1-NN accuracy as well as the Silhouette coefficient of 1.0 since the simple sequence of 8-dimensional angle features might not be the most suitable representation for the discrimination task. Sometimes, the results of HPE models are slightly better compared to GT – this is mainly caused by improper estimation of joint coordinates on incorrectly performed repetitions, which better contributes to the discrimination.

The 1-NN classification shows minimal variation between the models and GT, which is promising for exercise correctness detection but provides no idea

Table 6. Comparison of angle quality metrics to joint position metrics (all from hor. camera with front view on Ex1 and half-profile on the rest).

3D model	2D model	3D MPJPE			1-NN acc.			Silhouette coef.		
		Ex1	Ex5	Ex6	Ex1	Ex5	Ex6	Ex1	Ex5	Ex6
Ground truth					97%	**96%**	93%	0.47	**0.47**	**0.51**
MotionBERT-lite	AlphaPose	10.0	10.4	9.0	94%	95%	89%	**0.54**	0.37	0.40
	HRNet	10.2	9.7	9.1	97%	96%	93%	0.53	0.39	0.40
MHFormer-351	HRNet	10.4	10.8	10.0	98%	96%	94%	0.47	0.44	0.47
	YOLOv7	11.3	13.5	10.2	98%	93%	95%	0.47	0.45	0.45
MPP-heavy		9.9	10.6	10.5	95%	95%	92%	0.48	0.39	0.38

Table 7. The runtime speed of HPE NNs in the frame-per-second rate (FPS).

Model	Type	GPU	FPS
AlphaPose	2D	✓	5.7
HRNet	2D	✓	6.2
YOLOv7	2D	✓	3.3
MotionBERT-full	3D	✓	212.6
MotionBERT-lite	3D	✓	308.1
STCFormer-81	3D	✓	20.4
MHFormer-351	3D	✓	13.2
MPP-heavy	2D & 3D	✗	5.0
MPP-full	2D & 3D	✗	14.2
MPP-lite	2D & 3D	✗	18.3

which model to select. Interestingly, the MotionBERT with AlphaPose 2D inputs is the worst, even though it is among the best in 3D MPJPE position error.

Silhouette coefficients ($\in [-1, 1]$), on the other hand, support MHFormer as the best model across several exercises, despite its MPJPE only being the best on exercise Ex1. It is essential to mention that the coefficients varied between the subjects and, in rare cases, dropped the scores under zero (this indicates that correctly and incorrectly performed repetitions are not well discriminated).

Nevertheless, both metrics lack correlation with the 3D errors, which gives credence to the idea that positional error is not necessarily the only sufficient metric to judge the pose detector quality. Note that, unlike 1-NN, the Silhouette coefficient considers all distances and ideally expects a single cluster, so these two metrics also do not correlate.

5.5 Efficiency

All models were executed on a Windows 10 machine (Intel i5-2400 CPU, 8 GB RAM) with Nvidia GeForce GTX 960 GPU with 4 GB VRAM. The only models that were tested on CPU only were the MediaPipe Pose ones, as the TensorFlow Lite 2 focuses primarily on mobile devices, including GPU support.

Nonetheless, MediaPipe Pose finished the computation fastest if we consider that the 3D models require the 2D inputs first. MotionBERT models achieved high speed thanks to the official implementation supporting parallelization, unlike the other networks, and theoretically, it could compete with MPP given a fast 2D model. Overall, the 2D detectors create a bottleneck that limits the 2D-to-3D lifting strategy in efficiency. Generally, the camera orientation did not affect speed, except for YOLOv7, where we used higher input image resolution for vertical videos, as that led to better effectiveness.

6 Conclusions

This paper introduces the new rehabilitation dataset REHAB24-6, which provides a variety of exercises, views, body heights, lighting conditions, and exercise mistakes. The dataset, together with the proposed BBox transformation, also serves as a testbed for evaluating 2D and 3D pose estimation methods trained on other datasets with distinct body models to get more realistic performance on in-the-wild videos. Our experiments on a selection of state-of-the-art 3D HPE models show higher joint position errors than in the standard benchmarks, partially due to the mismatch of body models but also due to the current limitations of HPE methods. We demonstrate that depth estimation is truly the main challenge of HPE – the choice of 2D detectors influences the errors less than that of the 3D model, so we recommend preferring 2D pose estimators over 3D if possible. The MediaPipe Pose method has the lowest 3D position error in general, but different models provide the best results on individual exercises. Further experiments on the specific task of exercising correctness discrimination show no clear winner among the HPE methods either. However, they give us the intuition that 3D skeletons produced by HPE methods could work as well as the GT skeleton sequences in exercise monitoring applications. From the efficiency point of view, the MediaPipe Pose provides the best results, working fast even on CPUs. In the future, it would be interesting to investigate the adaptability of HPE models to different body postures (e.g., a lying person) or to the specific requirements of patients with varying levels of mobility.

Acknowledgments. This work is co-financed from the state budget by the Technology Agency of the Czech Republic under the TREND Programme; project "VisioTherapy: Supporting physiotherapy treatments using computer-based movement analysis" (No. FW09020055).

References

1. An, S., Li, Y., Ogras, U.: mRI: Multi-modal 3d human pose estimation dataset using mmwave, RGB-d, and inertial sensors. In: 36th Conference on Neural Information Processing Systems Datasets and Benchmarks Track (2022)
2. Bazarevsky, V., Grishchenko, I., Raveendran, K., Zhu, T., Zhang, F., Grundmann, M.: Blazepose: On-device real-time body pose tracking (2020)
3. Capecci, M., Ceravolo, M.G., Ferracuti, F., Iarlori, S., Monteriù, A., Romeo, L., Verdini, F.: The kimore dataset: kinematic assessment of movement and clinical scores for remote monitoring of physical rehabilitation. IEEE Trans. Neural Syst. Rehabil. Eng. (TNSRE) **27**(7), 1436–1448 (2019)
4. Dubey, S., Dixit, M.: A comprehensive survey on human pose estimation approaches. Multimedia Syst. 1–29 (2022). 10.1007/s00530-022-00980-0
5. Fang, H.S., Li, J., Tang, H., Xu, C., Zhu, H., Xiu, Y., Li, Y.L., Lu, C.: Alphapose: whole-body regional multi-person pose estimation and tracking in real-time. IEEE Trans. Pattern Anal. Mach. Intell. (2022)
6. Fieraru, M., Zanfir, M., Pirlea, S.C., Olaru, V., Sminchisescu, C.: Aifit: automatic 3d human-interpretable feedback models for fitness training. In: IEEE/CVF Conference on Computer Vision and Pattern Recognition (CVPR) (2021)
7. Ionescu, C., Papava, D., Olaru, V., Sminchisescu, C.: Human3.6m: large scale datasets and predictive methods for 3d human sensing in natural environments. IEEE Trans. Pattern Anal. Mach. Intell. (TPAMI) **36**(7), 1325–1339 (2014)
8. Li, W., Liu, H., Tang, H., Wang, P., Van Gool, L.: Mhformer: multi-hypothesis transformer for 3d human pose estimation. In: IEEE/CVF Conference on Computer Vision and Pattern Recognition (CVPR), pp. 13147–13156 (2022)
9. Lin, T.Y., Maire, M., Belongie, S., Bourdev, L., Girshick, R., Hays, J., Perona, P., Ramanan, D., Zitnick, C.L., Dollár, P.: Microsoft coco: common objects in context (2015)
10. Maji, D., Nagori, S., Mathew, M., Poddar, D.: Yolo-pose: enhancing yolo for multi person pose estimation using object keypoint similarity loss (2022)
11. Rousseeuw, P.J.: Silhouettes: a graphical aid to the interpretation and validation of cluster analysis. J. Comput. Appl. Math. **20**, 53–65 (1987)
12. Sedmidubsky, J., Elias, P., Budikova, P., Zezula, P.: Content-based management of human motion data: survey and challenges. IEEE Access **9**, 64241–64255 (2021)
13. Sun, K., Xiao, B., Liu, D., Wang, J.: Deep high-resolution representation learning for human pose estimation. In: CVPR (2019)
14. Tang, Y., Liu, J., Liu, A., Yang, B., Dai, W., Rao, Y., Lu, J., Zhou, J., Li, X.: Flag3d: a 3d fitness activity dataset with language instruction (2023)
15. Tang, Z., Qiu, Z., Hao, Y., Hong, R., Yao, T.: 3d human pose estimation with spatio-temporal criss-cross attention. In: IEEE/CVF Conference on Computer Vision and Pattern Recognition (CVPR), pp. 4790–4799 (2023)
16. Wang, C., Li, Y., Xiong, Z., Luo, Y., Cao, Y.: Lower body rehabilitation dataset and model optimization. In: 2021 IEEE International Conference on Multimedia and Expo (ICME), pp. 1–6 (2021)
17. Wang, C.Y., Bochkovskiy, A., Liao, H.Y.M.: Yolov7: trainable bag-of-freebies sets new state-of-the-art for real-time object detectors (2022)
18. Weitz, A., Colucci, L., Primas, S., Bent, B.: Infiniteform: a synthetic, minimal bias dataset for fitness applications (2021)
19. Xu, H., Bazavan, E.G., Zanfir, A., Freeman, W.T., Sukthankar, R., Sminchisescu, C.: GHUM & GHUML: generative 3d human shape and articulated pose models.

In: IEEE/CVF Conference on Computer Vision and Pattern Recognition (CVPR) (2020)
20. Zhao, Z., Kiciroglu, S., Vinzant, H., Cheng, Y., Katircioglu, I., Salzmann, M., Fua, P.: 3d pose based feedback for physical exercises (2022)
21. Zhu, W., Ma, X., Liu, Z., Liu, L., Wu, W., Wang, Y.: Motionbert: a unified perspective on learning human motion representations. In: IEEE/CVF International Conference on Computer Vision (ICCV), pp. 15085–15099 (2023)

ETDD70: Eye-Tracking Dataset for Classification of Dyslexia Using AI-Based Methods

Jan Sedmidubsky[1]( ), Nicol Dostalova[2], Roman Svaricek[2], and Wolf Culemann[3]

[1] Faculty of Informatics, Masaryk University, Brno, Czech Republic
xsedmid@fi.muni.cz
[2] Faculty of Arts, Masaryk University, Brno, Czech Republic
[3] University of Duisburg-Essen, Essen, Germany

Abstract. Dyslexia, a specific learning disorder, poses challenges in reading and language processing. Traditional diagnostic methods often rely on subjective assessments, leading to inaccuracies and delays in intervention. This work proposes classifying dyslexia using AI-based methods applied to eye-tracking data captured during text reading tasks. To facilitate future research in this domain, we collect a novel dataset (ETDD70) comprising eye-tracking recordings of 70 individuals for three reading tasks. In particular, the dataset contains high-frequency and accurate time series of 2D positions of eye movements and many derived characteristics extracted from eye movement patterns. By leveraging similarity-search approaches and deep learning models, we demonstrate the utility of such data in training several classification models, the best of which can distinguish between dyslexic and non-dyslexic individuals with an accuracy of around 90 %. Both the dataset and evaluated models provide a valuable resource for researchers to further advance AI-based methods for dyslexia classification.

Keywords: dyslexia · eye tracking · time-series data · classification · k-nearest neighbor query · multilayer perceptron · residual networks

1 Introduction

Dyslexia is a specific learning disorder characterized by severe text-reading difficulties, such as errors in decoding and spelling. These symptoms subsequently lead to a decreased overall reading process and text comprehension, negatively impacting an individual's social and emotional development [5]. For this reason, it is crucial to diagnose dyslexia as soon as possible, facilitating timely interventions and personalized learning strategies for affected individuals.

Dyslexia classification, sometimes also referred to as dyslexia detection, is a complex task for psychologists to decide how prone a given subject (usually an 8–10 year child) is to dyslexia disorder. Since up to 15 % of primary school pupils suffer from dyslexia [6,21], it might be time- and cost-consuming to find psychological specialists. The objective of this paper is to detect dyslexia without any specialist by applying AI-based principles to *eye-tracking data* that are captured during the subject's text-reading activity.

Eye-tracking data provide a rich source of information about the cognitive processes underlying reading, offering insights into how individuals interact with written text. *Raw* eye-tracking data can be represented as a sequence of 2D points that reflect the x/y pixel positions at a monitor screen where a given eye is looking over time. Since such raw data usually have a high sampling frequency (e.g., 250 frames per second), they are further processed to extract high-level gaze *events*; the most common events are *fixations* and *saccades*. Fixations, the periods during which the eyes remain relatively still on a particular point, reflect cognitive processing and attention allocation. Short fixations may indicate efficient processing, while longer fixations could suggest difficulty in decoding or comprehension. From the spatial dimension point of view, the fixations can be perceived as clusters of temporally close 2D raw data points, where each fixation is represented by its center, radius, and duration corresponding to the difference between the oldest and youngest points in the cluster. Saccades, rapid eye movements between fixations, signify the transition between words or phrases and indicate reading fluency and scanning behavior. Both fixations and saccades can be further processed to derive other characteristics, such as fixation duration, saccade length, or regression frequency, that could additionally provide valuable insights into reading behavior and cognitive processing.

In this paper, we introduce and make publicly available a new dataset, called Eye-Tracking Dyslexia Dataset (ETDD70), that contains fixations, saccades, and many derived characteristics from 35 dyslexic and 35 non-dyslexic individuals. We further propose new compact representations extracted from such eye-tracking data characteristics, introduce several methods to classify dyslexia disorder based on such representations, and thus establish initial baselines for the proposed ETDD70 dataset. The main contributions of this paper are:

- Dataset of derived characteristics of eye-tracking data of 70 subjects for three text-reading tasks suitable for analysis of dyslexia disorder;
- Pre-processing methods for extraction of feature-based and visualization-based representations from eye-tracking data for the particular tasks;
- Proposal of several similarity-search and deep-learning approaches for classification of the extracted representations.

2 Related Work

In the area of dyslexia research, there is a relatively substantial amount of research papers that target the classification of dyslexia from eye-tracking data. We briefly survey existing classification methods and eye-tracking datasets.

2.1 Dyslexia Datasets

There are several publicly available datasets [1,10,12] that provide recordings of eye movements captured during text reading in dyslexic and non-dyslexic readers (both adults and children). We especially highlight two representative datasets [1,10] that provide either the original raw eye-tracking data, or processed data in the form of event-based characteristics.

A large dataset in [1] contains raw eye-tracking recordings of 185 Swedish subjects aged 9–10 years (3rd grade of elementary school), of which 97 are high-risk readers (dyslexic) and 88 low-risk readers (non-dyslexic). Eye-tracking data are captured using special eye-tracking glasses at a sampling rate of 100 Hz. All participants read silently a text printed on a common paper [15]. However, having the paper at different distances and capturing eye movements using glasses leads to "desynchronized" eye-tracking data that are quite hardly comparable across various tasks and participants.

The CopCo (The Copenhagen Corpus of Eye-Tracking Recordings from Natural Reading) dataset [10] investigates reading behavior in the Danish language in different types of populations. It contains eye movement recordings of adult native Danish speakers, both from the 39 non-dyslexic readers and 19 dyslexic readers. The data are gathered using the EyeLink 1000 Plus eye-tracking device running at a sampling rate of 1000 Hz. This dataset contains raw data files as well as event-based characteristics in the form of fixations, saccades, and word-level features for a text-reading task (participants read different texts several times) [11].

Since both Swedish and Danish languages are quite specific for dyslexia classification, there is an increasing need for broad and homogeneous datasets, including those from different linguistic groups, such as Slavic languages. In this paper, we contribute to the community by introducing the ETDD70 dataset, which offers rich data from the Czech language environment. At the same time, our dataset is the first one in Czechia and also provides a considerably homogeneous sample of eye-tracking data specifically designed for children aged 9–10 years. Additionally, our dataset consists of *three* types of reading tasks, which potentially enables detecting dyslexia in different reading environments, in contrast to other single-task datasets.

2.2 Classification of Dyslexia

Numerous studies have focused on recognizing dyslexia by applying various machine-learning techniques to eye-tracking data. Rello and Ballesteros [17] were among the first attempting to classify dyslexia with an accuracy of ~80% using the Support Vector Machine (SVM) approach on eye-tracking data from 97 participants. The Linear SVM approach is used in [15] by achieving an accuracy of ~95% on the Swedish-language dataset [1]. A similar accuracy is achieved using a CNN [14] and a hybrid SVM [13] applied to specific saccade parameters and fixations. A similar high accuracy is also achieved on a text-reading task in Greek [19] or Serbian [20] languages.

The main observation is that traditional machine learning methods, such as SVM, random forests, or k-nearest neighbor (kNN) approaches, are commonly applied to statistical features that are extracted from eye-tracking data in a handcrafted way. Such statistical features usually ignore the actual positions of fixations or saccades. The surprising fact is that only a limited number of approaches employ deep learning that perfectly suits analyzing complex data, such as eye-tracking data. In this paper, we present two types of eye-tracking data representations, classify them using deep neural networks, and compare them against the standard kNN retrieval approach.

3 ETDD70: Eye-Tracking Dyslexia Dataset

Our collected dataset ETDD70 (Eye-Tracking Dyslexia Dataset) comprises raw eye-tracking data and several types of event-based characteristics to evaluate dyslexic and non-dyslexic readers' performance during the completion of three types of text-reading tasks: Task1: *syllables reading*, Task2: *meaningful-text reading*, and Task3: *pseudo-text reading*. The following sections describe these tasks together with the data recording process and key aspects of the dataset.

3.1 Participants

The eye-tracking data were captured from 70 participants: 35 dyslexic (19 female and 16 male) and 35 non-dyslexic (also 19 female and 16 male) readers. In all cases, all participants are elementary school pupils aged 9–10 years (i.e., 4th grade of elementary school). The recruitment of suitable participants was conducted in cooperation with the psychological counseling center, which facilitated the recruitment of pupils diagnosed with dyslexia. The non-dyslexic readers, who showed no symptoms of dyslexia, were recruited in cooperation with the counseling facilities of selected elementary schools. The dataset was collected between October 2022 and August 2023. The legal representatives of all participants were properly informed about the research procedure and agreed to participate in the study, for which they subsequently received compensation.

3.2 Description of Text-Reading Tasks

We designed three verbal tasks based on standardized paper-based dyslexia diagnostics used in Czechia. These source texts were transferred to the digital version in a controlled form (e.g., amount of text, font size, line spacing, background color, etc.) for the requirements of eye-tracking measurements. These tasks are described in more detail below.

Task1 (syllables) contains a total of 90 syllables (i.e., parts of words) arranged in the 9×10 matrix (9 rows and 10 columns). The following syllables are frequently encountered in the Czech language. The individual rows of syllables were categorized according to syllable composition (based on linguistic aspects) as follows:

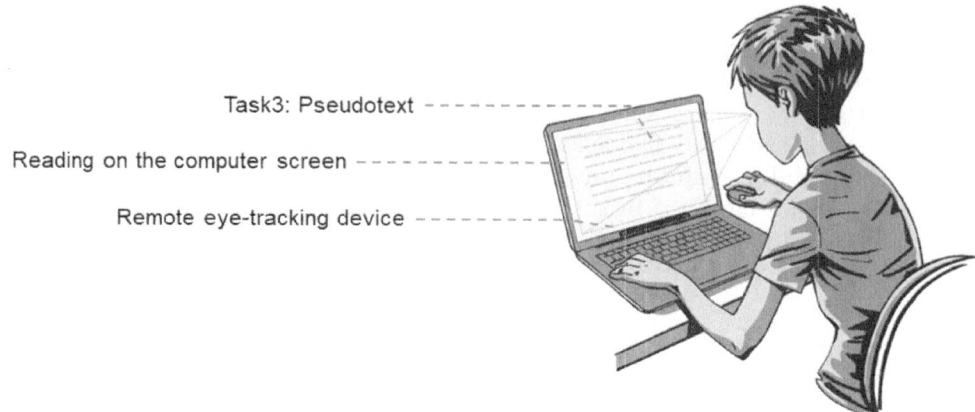

Fig. 1. An illustration of the data recording setup using a remote eye-tracking device attached to a laptop screen. The device captures the child's eye movements when reading, and the screen then displays the particular stimulus (Task3 in this example).

open syllables with no meaning, i.e., consonant + vowel (e.g., "ta", "na"), closed syllables with a central vowel bearing a meaning, i.e., consonant + vowel + consonant (e.g., "suk", "mák"), meaningless syllables consisting of two consonants (e.g., "vl", "bz"), a meaningless syllable formed by a cluster of two consonants ending in a vowel (e.g., "tle", "mra"), and finally a meaningful syllable formed by a cluster of 3 consonants with one vowel in the 3rd position (e.g., "mrak", "vlak"). All syllables were presented in black font, with Times New Roman on a gray background. The objective of the task is to read aloud all syllables from left to right and from top to bottom. A fixation cross was placed in the lower right corner for gaze-contingent task closure – when the participant looks at this cross, the recording is automatically terminated.

Task2 (meaningful text) consists of a text themed around the story of a young boy who watches a squirrel from his window. This text is intended for elementary school readers in grades 3 and 4. The stimulus text contains a total of 7 text lines with 6 logical sentences. The text is again written in black-colored font with double line spacing on a grey background and the fixation cross in the lower right corner. The aim of the task is to read the entire text aloud.

Task3 (pseudo text) comprises a text that is composed of fictional, meaningless words. This text has a total of 7 lines with 15 artificial sentences. The text formatting, as well as the ending fixation cross, are the same as in Task2. The objective of the task is to read the entire text aloud as smoothly as possible.

3.3 Capturing Procedure

The underlying functionality and visualization of each task was implemented in Python, mainly using the PsychoPy [16] library, in which the entire experiment was configured. All tasks were presented on a monitor screen with a refresh rate set to 60 Hz and a constant screen resolution of 1680×1050 pixels. All participants had their heads positioned throughout the measurements to maintain a consistent eye-to-screen distance of 60–70 cm. This contributes to better synchronized eye-tracking data that can be compared across different tasks, in contrast to inconsistently moving glasses in [1]. The eye-tracking data were recorded by the SMI RED 250 remote eye-tracking device with a sampling rate of 250 Hz. A schematic illustration of the capturing process is visualized in Fig. 1.

All data were collected in a specially designated room at Masaryk University, which met all conditions for eye-tracking laboratory measurements. In the case of some non-dyslexic participants, measurements were also performed in specially reserved rooms in selected elementary schools. The participants completed the tasks in the following order: Task1, Task2, Task3. Before data recording, each participant was given verbal instructions about the entire capturing procedure. If the participant understood the instructions, the procedure continued with the calibration and subsequent validation of the eye-tracking device. Before the start of each task, the participant underwent a testing attempt on "random" data to verify the participant's understanding of the task. This helped to avoid common mistakes and misunderstandings when recording the final data.

3.4 Provided Data: Fixations, Saccades, and Derived Characteristics

The raw eye-tracking data recorded for each task were further processed to extract event-based characteristics – fixations, saccades, and dozens of derived statistical characteristics (e.g., total number of fixations, average saccadic amplitude, etc.). All the data as well as extracted feature-based and visual-based representations (further introduced in Sect. 4) are made publicly available on the dataset web page [4]: https://doi.org/10.5281/zenodo.13332134.

The fixations were detected using the i2mc algorithm [9], as it was specifically designed to be noise-robust for measurements in children. The minimum fixation duration was set to 40 ms. To account for possible inaccuracies of the eye tracker, the fixations were further corrected for vertical errors using a line assignment algorithm, as commonly applied in eye-tracking reading studies. In particular, we adopted the warping algorithm from [2], which is very robust even in the presence of strong drift. The saccades were then defined as the transitions between all pairs of consecutive fixations.

The derived characteristics provide additional information about how participants interact with text. These characteristics are divided into *whole-task* and *region-of-interest* (ROI) characteristics. While the whole-task characteristics describe the semantics on the global level of the whole screen, the ROI ones characterize the semantics on the local level of a small rectangular area. In this paper, we define ROIs on the level of *words*. Whole-task characteristics include, for example, the number of fixations, saccades, regressions (within or between lines), mean and sum of fixation durations, or saccade amplitude. ROI characteristics include dwell time (sum of fixation and saccade durations), the number and duration of all corresponding fixations, fixations during the first ROI visits, or the number of revisits to a ROI.

4 Extraction of Characteristic Representations

Even though we observed some small deviations between the characteristics of dyslexic and non-dyslexic subjects (e.g., in total reading time), they do not provide statistically significant differences usable for classification. Therefore, we introduce two types of compact representations – *feature-based* and *visualization-based* – that can be more effectively processed by AI-based classifiers than very complex *raw* eye-tracking data.

4.1 Feature-Based Representations

We define feature-based representations as *high-dimensional vectors* of real numbers. For each task, the vector of a specific length is extracted. Specifically, we extract 450-, 406-, and 414-dimensional vectors for Task1, Task2, and Task3, respectively. Each *feature* (i.e., the value of a given vector dimension) has its own specific semantics. We extracted *whole-task* features as well as *region-of-interest* features on the level of words. The ROI features correspond to 90, 100, and 102 areas encapsulating individual words displayed on the screen for Task1, Task2, and Task3, respectively. The specific task-level and ROI-level features for individual tasks are defined in Table 1. Finally, these features are concatenated to form a single high-dimensional vector defined for particular tasks as follows:

- Task1 (450-dim. vector) – 5 features for each out of 90 defined ROIs;
- Task2 (406-dim. vector) – 6 whole-task features + 4 features for each out of 100 defined ROIs;
- Task3 (414-dim. vector) – 6 whole-task features + 4 features for each out of 102 defined ROIs.

In summary, these features were selected by psychological experts and primarily reflect high-level statistics about changes in eye movements. The ROI features focus on how much subjects move with their eyes when reading a specific word and how many times they repeatedly return their eye to the same word during text reading. Such statistics could be potentially used to analyze which words are the most difficult to read.

4.2 Visualization-Based Representations

Transforming spatio-temporal data into 2D image representations that are further classified by deep-learning methods is a successful approach in different domains, e.g., in human motion data classification [18]. Inspired by the recent transformation of eye-tracking fixations into a 2D image [3], we generate a visualization of eye movement patterns for each individual and each task. The output is a fixed-size image that specifically visualizes information about fixations. As illustrated in Fig. 2, the image consists of ellipses where each ellipse corresponds to a single fixation. The center of each ellipse denotes the fixation's position, while its x/y dimensions reflect the dispersion along the respective axes. In other words, the number of visualized ellipses corresponds to the number of fixations. The color of each ellipse represents the fixation duration which is normalized for each task independently. The color spectrum ranges from black (very short duration), through red and orange, to yellow color (very long duration). The output image is finally cropped to encompass only the area containing the words pertinent to the specific task. The visualizations are also publicly available [4].

Table 1. Description of the extracted feature-based representations for individual tasks. While Task1 is represented by 5 features for each ROI, Task2 and Task3 have 4 features for each ROI and additional 6 features defined on the level of the whole task.

Task	Level	Feature description
Task1	ROI	First fixation duration
		Average fixation duration
		Number of fixations
		Number of fixations and saccades, w/o the incoming/outgoing saccade
		Number of revisits – incoming saccades hitting this ROI from outside
Task2 & Task3	whole task	Number of regressions
		Ratio of progressive/regressive saccades
		Average saccadic amplitude
		Total reading duration
		Average fixation duration
		Number of fixations
	ROI	Average fixation duration
		Number of fixations
		Number of revisits – incoming saccades hitting this ROI from outside
		Landing position of the first fixation

These features are concatenated to form a single high-dimensional vector for each task

5 Classification of Extracted Representations

We adopt three principally different types of classifiers to assess the suitability of the proposed representations for distinguishing between dyslexic and non-dyslexic subjects. In particular, (1) to classify *feature-based* representations, we design the kNN similarity-search approach and several variants of Multi-layer-perceptron architectures, and (2) to classify *visualization-based* representations, we fine-tune existing residual neural networks.

5.1 kNN Classifier for Feature-Based Representations

As a baseline, we adopt the k-Nearest Neighbor (kNN) classification which requires defining the concept of similarity between input feature-based representations. We apply the commonly employed Euclidean distance metric to quantify the proximity between a pair of feature-based representations, i.e., 450-, 406-, and 414-dimensional vectors for Task1, Task2, and Task3, respectively. The class prediction for a given input vector is determined by a majority vote among the retrieved k nearest neighbors, where k is a predefined hyperparameter.

Before applying the kNN classifier, we pre-process the input feature vectors by scaling individual dimensions to prevent features with larger scales from dominating the Euclidean distance computations. We apply the standard scaling mechanism by subtracting the mean from each feature and then scaling it by dividing it by the standard deviation. This results in a more balanced influence of each feature on the distance calculation.

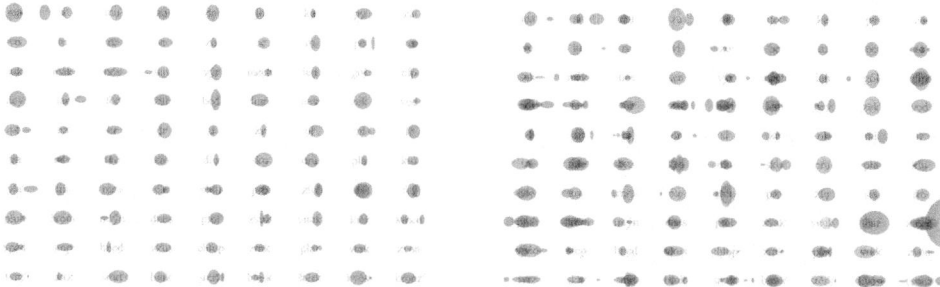

Fig. 2. Generated visualization-based representations for Task1 – reading 10 × 9 standalone syllables – for a non-dyslexic subject (left) and dyslexic subject (right).

5.2 MLP Classifiers for Feature-Based Representations

Multi-Layer Perceptrons (MLPs) serve as fundamental architectures for classification tasks, especially when dealing with fixed-size vector data. We construct a baseline MLP architecture consisting of an input layer, one or more hidden layers, and an output layer with softmax activation for two-class classification.

We mainly explore the effect of altering the number of hidden layers to evaluate the network's capacity to capture more complex relationships within the input data. We experiment with architectures ranging from shallow networks with a single hidden layer to deeper architectures with two or three hidden layers. The size of a hidden layer is set as half of the previous layer; the size of the first hidden layer corresponds to half of the input layer.

The size of the input MLP layer corresponds to the size of the input vector data. The MLP architecture with three hidden layers applied to the 450-dimensional input vectors of Task1 is depicted in Fig. 3. We initialize the weights and biases of the network using appropriate initialization techniques to prevent vanishing or exploding gradients. We also analyze the effect of learning rate on MLP convergence and generalization. The learning rate governs gradient descent step size, affecting convergence speed and stability. We systematically adjust the learning rate across a predefined range. We also incorporate dropout regularization to prevent overfitting and improve generalization. Dropout randomly deactivates a fraction of neurons during training, forcing the network to learn redundant representations and reducing the reliance on specific features.

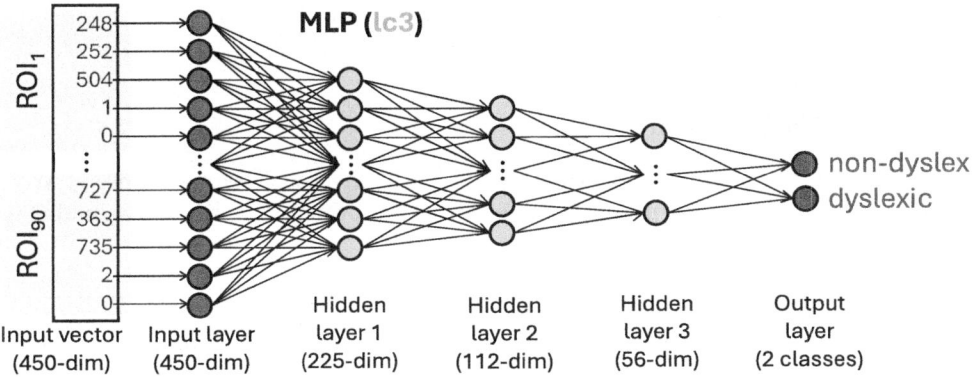

Fig. 3. Illustration of the used MLP classifier with three hidden layers.

5.3 ResNet Classifiers for Visualization-Based Representations

There is a possibility of applying Vision Transformers or convolutional neural networks to classify the visualization-based representation [7]. We select two variants of residual networks (ResNet18 and ResNet50) [8] that provide a suitable trade-off between classification accuracy and network complexity. Since our dataset contains a limited number of training data, we employ the pre-trained network weights of both the ResNet$X$ models trained on the general domain of images and fine-tune them on the generated visualization-based representations. To do this, we need to change the original classification layer (corresponding to

1 k classes of general-purpose images) to recognize only two classes – dyslexic and non-dyslexic individuals.

6 Experimental Evaluation

We define the evaluation protocol for the introduced ETDD70 dataset and evaluate qualitative and quantitative metrics for each of the proposed classifiers, trained on corresponding feature-based or visualization-based representations.

6.1 Evaluation Methodology

To be fair, we apply the 5-fold cross-validation procedure for each task, which results in five independently evaluated experiments whose results are finally averaged. The dataset of 70 subjects is split so that each fold contains the same number of dyslexic and non-dyslexic subjects. In this way, each experiment uses four folds (i.e., 56 subjects) for training and one fold (i.e., 14 subjects) for validating classifiers. For each experiment, we train all introduced classifiers and quantify their quality by *classification accuracy* as the ratio of correctly classified subjects from the validation fold. We also quantify the performance of each trained classification model by measuring its training time and space complexity.

6.2 Qualitative Results

We evaluate the quality of classifiers by varying selected hyperparameters. We start with the proposed most promising MLP approach, for which we mainly

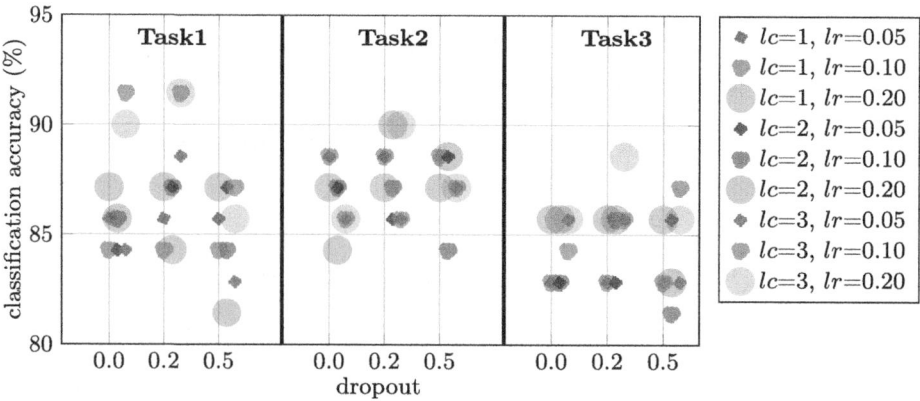

Fig. 4. Classification accuracy of MLP classifiers – with a different number of lc layers (different color in the legend) and varying hyperparameters of learning rate lr (different shape in the legend) and dropout (axis x) – individually depicted for each of the three tasks. Note that the same accuracy can be achieved by different configurations, so there can be multiple overlapping points in the same position of the figure.

test a different number of hidden layers (lc) and different values of dropout and *learning rate* (lr). The classification accuracy for each task and different combinations of hyperparameters is graphically depicted in Fig. 4. Since dropout is only applicable in case of >1 hidden layers, the results for $lc = 1$ and fixed lr are the same for a changing dropout value. The dropout value of 0.2 is especially useful for MLP with three hidden layers and higher values of lr. The globally clear winner over all the tasks is the configuration with $lc = 3$ layers, dropout of 0.2, and $lr = 0.20$ learning rate (i.e., "big blue circle"), achieving the classification accuracy around 90 %.

We compare MLP with the baseline similarity-search kNN classifier on the same feature-based representations (as specified in Sect. 4.1). For the kNN classifier, we test $k \in \{1, 3, 5\}$, which are sufficient numbers for 28 positive (i.e., either dyslexic, or non-dyslexic) subjects available in the training data of each fold. The results in Table 2 show that the best accuracy of ~69 % is achieved over the tasks for $k = 1$, which demonstrates that the proposed MLP approach can achieve significantly better results (~90 %) than the similarity-search baseline on the same feature representations. In the same table, we also compare the results of ResNet18/50 classifiers fine-tuned by visualization-based representations from a given training fold. On average, the results of more-complex ResNet50 are better than those of ResNet18; but the differences are very tiny. More importantly, the results of the proposed MLP approach are better than any ResNet18/50 over all the tasks. We can observe that the best MLP configuration also achieves quite low standard deviations over individual folds compared to other classifiers.

We have also experimented with different regions of interest: instead of considering ROIs on the level of individual words, we have defined coarse-grained ROIs on the level of whole lines consisting of several words. In particular, a sin-

Table 2. Comparison of the classification accuracy and standard deviations for all the tested classifiers and tasks (the best results are emphasized in bold, the second best results are underlined)

Approach	Hyperparams	Task1	Task2	Task3
kNN	$k = 1$	68.57 ± 7.28	68.57 ± 8.57	71.43 ± 12.78
	$k = 3$	65.71 ± 8.33	65.71 ± 8.33	68.57 ± 8.57
	$k = 5$	68.57 ± 3.50	67.14 ± 9.69	67.14 ± 9.69
MLP	$lr = 0.05$	88.57 ± 7.28	<u>85.71</u> ± 4.52	<u>85.71</u> ± 9.04
	$lr = 0.1$	**91.43** ± 5.35	<u>85.71</u> ± 4.52	<u>85.71</u> ± 9.04
	$lr = 0.2$	**91.43** ± 5.35	**90.00** ± 7.28	**88.57** ± 5.71
RestNet18	$lr = 0.00001$	75.71 ± 11.61	81.43 ± 10.69	82.86 ± 9.69
	$lr = 0.0001$	81.43 ± 7.28	81.43 ± 13.25	**88.57** ± 8.57
	$lr = 0.001$	84.29 ± 5.35	84.29 ± 2.86	<u>85.71</u> ± 7.82
RestNet50	$lr = 0.00001$	78.57 ± 4.52	84.29 ± 8.33	82.86 ± 7.28
	$lr = 0.0001$	82.86 ± 7.28	82.86 ± 10.69	**88.57** ± 7.28
	$lr = 0.001$	82.86 ± 5.71	84.29 ± 5.35	84.29 ± 8.33

Table 3. Quantitative results of evaluated classification models: training time and model complexity (averaged over individual folds and tasks)

Metric	kNN	MLP	ResNet18	ResNet50
Training time (s)	–	0.12–0.16 s	160.00 s	340.00 s
Model size (MB)	0.2 MB	0.8–1.6 MB	90.0 MB	189.0 MB

gle ROI of 10 syllables in Task1 and a single ROI of a variable number of words occurring within one text line in Task2/Task3. We have also tried to consider the fusion of both word-level and line-level ROIs. While the changed number of ROIs leads to different sizes of features-based representations, the change in classification accuracy was negligible.

6.3 Quantitative Results

We report quantitative results regarding the time needed to train a classification model and regarding the size of the trained model. The training process of classifiers supports GPU implementation and is evaluated on the NVIDIA RTX A4000 graphics card. We average the results over 5 folds and individual tasks (Task1–Task3) as the tasks are very similar from the complexity point of view. The results are available in Table 3.

Although the kNN model does not require any training phase and occupies the least space, the actual model size and inference time depend on the dataset size, as kNN stores the entire dataset in memory. Training the MLP model for 20 epochs takes less than 1 second and the model size occupies only around 1 MB, slightly depending on the number of hidden layers (0.12/0.16 s and 0.8/1.6 MB for 1/3 hidden layers, respectively). Training the ResNet18/50 models for 50 epochs takes approximately 13/28 minutes, which reflects the increased complexity and depth of ResNet50 (189 MB) compared to ResNet18 (90 MB).

7 Conclusions

This work currently constitutes the largest study on dyslexia classification of the Czech children population using AI-based methods. The ETDD70 dataset collected in the last two years is publicly available with the hope of inspiring other researchers to analyze novel forms of similarity and develop robust classification models for dyslexia, enabling early identification and tailored interventions for individuals with such conditions. The complexity of dyslexia detection is demonstrated by a limited classification accuracy ($\sim$69 %) of the kNN classifier that determines the similarity of individuals based on carefully chosen feature-based representations. High classification accuracy ($\sim$90 %) is achieved by the specifically designed MLP model for the same feature-based representations, outperforming the visualization-based representations classified by ResNet18/50.

Together with low training time and small model size, the MLP approach constitutes a promising approach for future studies on dyslexia classification. In the future, we would like to focus on designing various eye-tracking data representations (e.g., visualizing saccades), adopting alternative classifiers (e.g., popular SVMs or other kinds of deep neural networks), and investigating the explainability of the trained classification models.

Acknowledgments. This work is co-financed from the state budget by the Technology Agency of the Czech Republic under the ÉTA Programme; project "Diagnostics of dyslexia using eye-tracking and artificial intelligence" No. TL05000177.

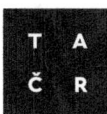

References

1. Benfatto, M.N., Seimyr, G.Ö., Ygge, J., Pansell, T., Rydberg, A., Jacobson, C.: Screening for dyslexia using eye tracking during. Reading (2016). https://doi.org/10.6084/M9.FIGSHARE.C.3521379.V1
2. Carr, J.W., Pescuma, V.N., Furlan, M., Ktori, M., Crepaldi, D.: Algorithms for the automated correction of vertical drift in eye-tracking data. Behav. Res. Methods **54**(1), 287–310 (2022). https://doi.org/10.3758/s13428-021-01554-0
3. Cernek, A.: Recognition of reading disorder based on eye-tracking data. Master's thesis, Masaryk University (2023)
4. Dostalova, N., Svaricek, R., Sedmidubsky, J., Culemann, W., Sasinka, C., Zezula, P., Cenek, J.: ETDD70: Eye-tracking dyslexia dataset. [Data set], Zenodo (2024). https://doi.org/10.5281/zenodo.13332134
5. Emily, M., Livingston, L.S.S., Ribary, U.: Developmental dyslexia: emotional impact and consequences. Aust. J. Learn. Difficulties **23**(2), 107–135 (2018). https://doi.org/10.1080/19404158.2018.1479975
6. Habib, M.: The neurological basis of developmental dyslexia: an overview and working hypothesis. Brain **123**(12), 2373–2399 (2000). https://doi.org/10.1093/brain/123.12.2373
7. Han, K., Wang, Y., Chen, H., Chen, X., Guo, J., Liu, Z., Tang, Y., Xiao, A., Xu, C., Xu, Y., Yang, Z., Zhang, Y., Tao, D.: A survey on vision transformer. IEEE Trans. Pattern Anal. Mach. Intell. **45**(1), 87–110 (2023). https://doi.org/10.1109/TPAMI.2022.3152247
8. He, K., Zhang, X., Ren, S., Sun, J.: Deep residual learning for image recognition. In: IEEE Conference on Computer Vision and Pattern Recognition (CVPR) (2016)
9. Hessels, R.S., Niehorster, D.C., Kemner, C., Hooge, I.T.C.: Noise-robust fixation detection in eye movement data: identification by two-means clustering (I2MC). Behav. Res. Methods **49**(5), 1802–1823 (2017). https://doi.org/10.3758/s13428-016-0822-1
10. Hollenstein, N., Barrett, M., Björnsdóttir, M.: CopCo: The Copenhagen Corpus of Eye-Tracking Recordings from Natural Reading. Open Science Framework (2022). https://doi.org/10.17605/OSF.IO/UD8S5
11. Hollenstein, N., Barrett, M., Björnsdóttir, M.: The copenhagen corpus of eye tracking recordings from natural reading of Danish texts. In: Proceedings of the Thirteenth Language Resources and Evaluation Conference, pp. 1712–1720. European Language Resources Association (2022)

12. Jakovljević, T., Janković, M.M., Savić, A.M., Soldatović, I., Čolić, G., Jakulin, T.J., Papa, G.,Ković, V.: The relation between physiological parameters and colour modifications in text background and overlay during reading in children with and without dyslexia. Brain Sci. **11**(5), 539 (2021). https://doi.org/10.3390/brainsci11050539
13. Jothi Prabha, A., Bhargavi, R.: Predictive model for dyslexia from fixations and saccadic eye movement events. Comput. Methods Programs Biomed. **195**, 105538 (2020). https://doi.org/10.1016/j.cmpb.2020.105538
14. Nerušil, B., Polec, J., Škunda, J., Kačur, J.: Eye tracking based dyslexia detection using a holistic approach. Sci. Rep. **11**(1), 15687 (2021). https://doi.org/10.1038/s41598-021-95275-1
15. Nilsson Benfatto, M., Öqvist Seimyr, G., Ygge, J., Pansell, T., Rydberg, A., Jacobson, C.: Screening for dyslexia using eye tracking during reading. PLoS ONE **11**(12), e0165508 (2016). https://doi.org/10.1371/journal.pone.0165508
16. Peirce, J., Gray, J.R., Simpson, S., MacAskill, M., Höchenberger, R., Sogo, H., Kastman, E., Lindeløv, J.K.: Psychopy2: experiments in behavior made easy. Behav. Res. Methods **51**, 195–203 (2019)
17. Rello, L., Ballesteros, M.: Detecting readers with dyslexia using machine learning with eye tracking measures. In: Proceedings of the 12th International Web for All Conference, pp. 1–8. ACM, Florence Italy (2015). https://doi.org/10.1145/2745555.2746644
18. Sedmidubsky, J., Elias, P., Zezula, P.: Effective and efficient similarity searching in motion capture data. Multimedia Tools Appl. **77**(10), 12073–12094 (2018). https://doi.org/10.1007/s11042-017-4859-7
19. Smyrnakis, I., Andreadakis, V., Selimis, V., Kalaitzakis, M., Bachourou, T., Kaloutsakis, G., Kymionis, G.D., Smirnakis, S., Aslanides, I.M.: RADAR: a novel fast-screening method for reading difficulties with special focus on dyslexia. PLoS ONE **12**(8), e0182597 (2017). https://doi.org/10.1371/journal.pone.0182597
20. Vajs, I.,Ković, V., Papić, T., Savić, A.M., Janković, M.M.: Spatiotemporal eye-tracking feature set for improved recognition of dyslexic reading patterns in children. Sensors **22**(13), 4900 (2022). https://doi.org/10.3390/s22134900
21. Yang, L., Li, C., Li, X., Zhai, M., An, Q., Zhang, Y., Zhao, J., Weng, X.: Prevalence of developmental dyslexia in primary school children: a systematic review and meta-analysis. Brain Sci. **12**(2), 240 (2022). https://doi.org/10.3390/brainsci12020240

Demonstrating the Efficacy of Polyadic Queries

Ben Claydon[1](✉)[iD], Richard Connor[1](✉)[iD], Alan Dearle[1][iD], and Lucia Vadicamo[2][iD]

[1] University of St Andrews, St Andrews, Scotland, UK
rchc@st-andrews.ac.uk, al@st-andrews.ac.uk, lucia.vadicamo@isti.cnr.it
[2] Institute of Information Science and Technologies, CNR, Pisa, Italy

Abstract. Similarity search is normally defined to be the task of identifying those objects, from a large collection, that are most similar to a further single object presented as a query. Using *polyadic queries*, a small set of objects are presented to the system, with the intent of finding those objects most similar to all elements of the query set. A few scenarios have previously demonstrated the usefulness of this notion. For example, we may be searching for images similar to a red balloon over a lake. With a single query, it is impossible to tell if the intent is to search for other images of balloons over lakes, or for other red balloons in any background. If instead we could present a system with a few different images of balloons, all of which are either all red, or all over lakes, the similarity search engine may be able to respond more appropriately. In this paper we demonstrate software which permits the user to provide explicit feedback by selecting the best few results from an intermediate set which are best suited to their original information need. A polyadic query can be formed from this set, which should give better results with a minimum of user interaction.

Keywords: Similarity Search · Scalable Search · Polyadic Query · HNSW · SED · MSED · Divergence Function · f-Divergence

1 Introduction

This paper is a demonstration of a recently reported approach [7], which extends the notion of similarity search to perform similarity queries with multiple independent query objects.

Similarity search is normally described as the process of finding, from a large database of objects, those that are the most similar to another presented as a query. Traditional nearest neighbour search algorithms have proven effective in retrieving similar objects based on queries with a single query object, such as finding images similar to a given query image.

In this paper, we expand on the notion of polyadic query, and show they can be used in the context of *relevance feedback*, to interactively refine queries in order to satisfy an information need. The hypothesis is that by repeatedly performing polyadic queries we can refine results more quickly than by using single-subject queries. This should allow a user to find better solutions than could be found using a single object queries, and to find them with less user interaction. The demonstrated software allows a user to refine a search and interactively explore a data collection.

2 Example

We illustrate the process with the example shown in Fig. 1. In this example the user is interested in finding pictures of Border Collie dogs.

The result of an initial single-subject query is shown in Fig. 1. This image shows the 60 nearest-neighbours of the query (the user interface gives 100 scrollable images), ordered left-to-right and top-to-bottom, with the query itself in the query panel at the top of the page. We start with a query image of a dog, but not a Border Collie dog. Figures 1a–c, show the results of the repeated application of polyadic queries using the images we (subjectively) judged to be the closest to Border Collies. In each case, the number of Border Collies increases until 95 Border Collies have been found after the third application. The demonstration software described in this paper permits the user to perform this process interactively in order to understand the efficacy of polyadic queries.

We have compared this process with a similar set of iterations using single-image search, the results of which are not shown for brevity. In that experiment we repeatedly searched using the single (subjective) best results of the previous attempt. After three iterations, we did not get close to the same quality of result.

3 Technological Underpinnings of the Demonstration

3.1 Polyadic Queries Using Divergence Functions

Similarity search is normally defined over a dissimilarity space (U, d), where U is some universal domain and $d : U \times U \to \mathbb{R}^+$ is a dissimilarity function. Search is performed with respect to a large finite data set $S \subset U$. The general requirement is to efficiently find members of S which are most similar to an arbitrary member of U given as a query, where the function d gives the only way by which any two objects may be compared. Zezula et al. [16] and Chavez et al. [2] summarise a large volume of research in this domain.

In a polyadic query, instead of searching using a single element of U given as a query, we search using a set of elements $Q = \{q_1, ..., q_n\}$ where $q_i \in U$. For a single query q, the result is usually defined by $\{s \in S \mid d(q, s) \leq t\}$, for some *threshold* value t which gives a useful size of result set

- The definitions of spaces (U, d) and (S, d) as above are maintained
- Instead of a query $q \in U$, search is defined in terms of a query set $Q = \{q_1, ..., q_n\}$ where $q_i \in U$

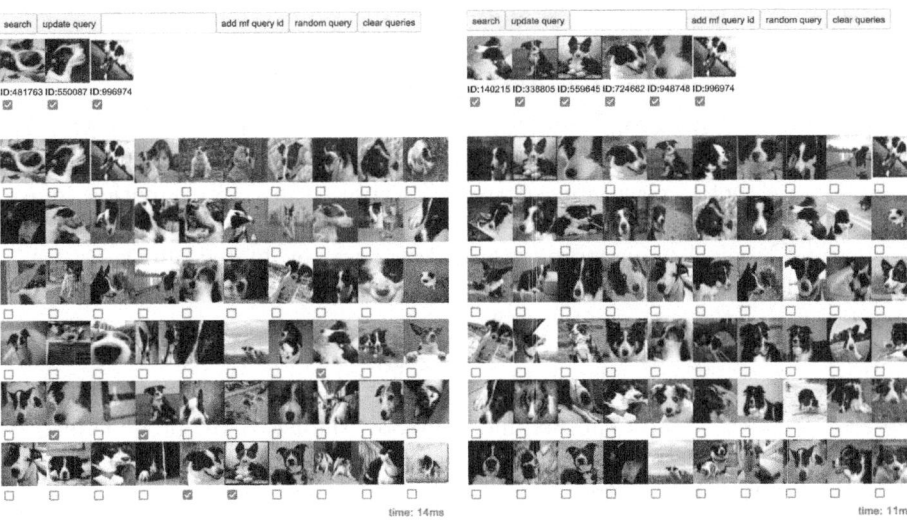

(a) Initial Query Results (using the image shown in the query panel) - None of the results is a Border Collie, but some results (ticked) are black dogs with pointed muzzles; we performed a polyadic query on these two images.

(b) Second Iteration of Results (using the images in the query panel as a polyadic query) - The results now contain several dogs which we deem more like Border Collies. We select the ticked dogs to form the next query.

(c) Results (using the images in the query panel as a polyadic query). We find several Border Collies (ticked), which we use to form the next polyadic query.

(d) Final polyadic query result. We believe that all of the query panel images are Border Collies, as are the majority of the results returned by this polyadic query.

Fig. 1. Example refinement queries searching for Border Collie dogs

– A more general numeric dissimilarity function $\theta : \mathscr{P}(U) \times U \to \mathbb{R}^+$ is required where $\mathscr{P}(U)$ is the powerset of U. θ then defines an ordering on S according to the query Q and each element $s \in S$, so that the result of a query will be $\{s \in S \,|\, \theta(Q,s) \leq t\}$ for some suitable threshold t

In the literature, some θ measures have been proposed for searching with multiple query objects, typically combining distances from individual queries using methods like linear combinations or fairness measures [3,9]. In this paper, however, we focus on divergence functions since they were found to work best in our earlier experiments [7]. We define a *divergence* function δ as a (positive, numeric) dissimilarity measure over a finite subsetof a domain U, i.e. $\delta : \mathscr{P}(U) \to \mathbb{R}^+$. The notion is that the function returns some indicator of a general dissimilarity over all the elements of its argument; for example, a collection of identical objects should return 0, while a collection of objects that have little in common with each other should return a high value.

A normal binary dissimilarity function would thus be a special case of this more general divergence: given a divergence function δ, a dissimilarity function d can be defined as $d(x,y) = \delta(\{x,y\})$. In the context of polyadic query, a divergence function δ can be used by ordering the search space in terms of the divergence of the query Q, with each element of S added in turn: the nearest neighbour to a query Q is thus the database object which gives the smallest divergence when that object is added to the query set. Formally, we use $\theta(Q,s) = \delta(Q \cup \{s\})$, and thus the solution to a polyadic query Q over a finite set S is $\{s \in S \,|\, \delta(Q \cup \{s\}) \leq t\}$ for some appropriate value of t.

3.2 Polyadic Search Using *MSED*

One example of a divergence function is *MSED* [8], which is derived from the information-theoretic metric Structural Entropic Distance (*SED*). While initially proposed as a metric over labelled tree structures, the core evaluation is over *probability* vectors, that is any domain $\mathbb{R}^m$ where for each element of $v_i \in v$, $v_i \geq 0$ and $\sum_i v_i = 1$. This more general metric was evaluated in [13].

SED is defined as:

$$SED(v,w) = \frac{C(\frac{v+w}{2})}{\sqrt{C(v)C(w)}} - 1 \qquad (1)$$

where $C(x) = e^{H(x)}$, and the Shannon entropy H of two vector is defined by $H(v) = -\sum_i v_i \ln v_i$ [15]. *SED* gives an outcome in the range $[0,1]$ with 0 implying that the two input vectors are identical and 1 implying that no individual dimension has a non-zero value in both input vectors, i.e. their dot product is zero and they are therefore orthogonal. Connor et al. observed [8] that the *SED* metric [4] generalises to a variadic input, rather than just a pair of values. A normalised form of this notion may derived, for a set of m vectors:

$$MSED(\boldsymbol{v}_1, \ldots, \boldsymbol{v}_m) = \frac{1}{m-1} \left(\frac{C\left(\sum_i \frac{v_i}{m}\right)}{\sqrt[m]{\prod_i C(\boldsymbol{v}_i)}} - 1 \right) \qquad (2)$$

As before, an outcome of 0 implies all elements of V are identical, and an outcome of 1 implies that all elements of V are mutually orthogonal.

3.3 Optimising *MSED* for Efficient Indexing and Querying

Most indexing systems require a single distance function to be provided which is used for both index construction and search. We have recently shown that it is possible to construct a navigational search index using *SED* which is subsequently searched using *MSED* [5].

It is noteworthy that, for calculations of the term $MSED(Q \cup \{s\})$ where Q is fixed for many different values of s, the majority of the cost can be amortised given prior knowledge of Q. Specifically, $dMSED(Q, s) = MSED(Q \cup \{s\})$ can be computed as

$$dMSED(Q, s) = \frac{1}{m} \left(\frac{C\left(\sum_j \frac{\sum_{q \in Q} q_j}{m+1} + \frac{s_j}{m+1}\right)}{\sqrt[m+1]{\left(\prod_{q \in Q} C(q)\right) C(s)}} - 1 \right) \quad (3)$$

where $m = |Q|$, q_j and s_j are the jth components of q and s, respectively. The point here is that the terms $M = \frac{\sum_{q \in |Q|} q_j}{m+1}$ and $P = \prod_{q \in Q} C(q)$ can be calculated once per query, therefore minimising the cost of the calculation for each individual s.

From here, as shown in [5], we can derive a generic form of the function *gMSED* which works over any pair of argument sets Q_i, Q_j such that $gMSED(Q_i, Q_j) = MSED(Q_i \cup Q_j)$. To achieve this, each subset Q_i is represented as a triple $V_{Q_i} = \{m_{Q_i}, P_{Q_i}, M_{Q_i}\}$ for the values m, P, M above specific to the subset in question, and *gMSED* is defined:

$$gMSED(Q_1, Q_2) = \frac{1}{m_{Q_1} + m_{Q_2} - 1} \left(\frac{C\left(\frac{(m_{Q_1}+1)M_{Q_1} + (m_{Q_2}+1)M_{Q_2}}{m_{Q_1} + m_{Q_2}}\right)}{(P_{Q_1} P_{Q_2})^{1/(m_{Q_1}+m_{Q_2})}} - 1 \right) \quad (4)$$

Since HNSW [12] builds an index by constructing a graph over pairs of points, we require a dyadic form of *MSED* over pairs of points. In order to construct an index, each point may be expressed as the triple $V_s = \{1, C(s), s/2\}$ and the *gMSED* function used calculate the divergences between the pairs. In a similar manner, *gMSED* may also be used to query the index by calculating *gMSED(Q, s)*, where Q is the query set and s is any point in the index.

4 The Demonstrator System

4.1 The Index

To construct the index used in the demonstration system we used Kuperus' Java implementation [11] of HNSW [12]. This creates an approximate index over the space using the dyadic form of the *gMSED* divergence function over pairs of points with the parameters $M = 15$ and $EfConstruction = 125$. The index is implemented as a Jelmerk `com.github.jelmerk.knn.hnsw.HnswIndex` which is precomputed and stored in a Java serialised form.

4.2 Data Preparation

The images used in the demonstration are drawn from the MirFlickr 1M image set [10]; they encompass a wide range of subjects. The size of this database is sufficient that one might reasonably expect to find multiple images in many popular subject domains.

The images have been encoded using the DINOv2 network [14] which generates a 384-dimensional feature vector for each image. Although DINOv2 was trained to minimise Cosine distance over the embedding vectors of similar images, as our metric requires probability distributions, a conversion from Euclidean space to information space is required. We exploit recent results on the conversion of Euclidean spaces to information spaces [6] to convert the original vectors to probability vectors as required by *SED/MSED*. For the DinoV2 vectors we apply l_2-normalisation, followed by a softmax function using a temperature of 10 which results in a probability vector. This produces a near perfect correlation between the *SED* function over the resulting space and Cosine distance over the original [5].

4.3 The User Interface

The user interface of the demonstrator is shown in Fig. 1. It permits (repeated) polyadic queries to be performed over the 1M MirFlickr dataset [10]. The functionality of the interface components is as follows:

1. The `search` button performs a polyadic search with whichever images are selected in the *query panel* (below the buttons). This causes the 100 nearest neighbours of the polyadic query formed using the selected images to be displayed in the *results panel* (the main part of the page). The time taken in ms to perform the query is shown at the foot of the page.
2. The `update query` button adds any images selected from the *results panel* to the *query panel*.
3. The `add mf query id` adds the image corresponding to the integer entered in the the *query_id box*. For example if the *query_id box* contains the image id "181360" a picture of a Bugatti would be added.
4. The `random query` replaces the contents of the *query panel* with five new images and selects them all.
5. The `clear queries` button removes all the entries in the *query panel*.
6. The `copy result IDs` button copies the MirFlickr IDs of the results currently on screen to the clipboard.
7. The `information toggle checkboxes` (not shown in Fig. 1) permit the rank ordering, MirFlickr IDs, and distances to be shown below the results.

8. The `examples panel` (also not shown in Fig. 1) contains example queries. Any image may be clicked on to add it to the *query panel*.
9. Lastly, any of the images in the result set or *query panel* may be clicked on to enlarge them. The enlarged image may be clicked on to close it.

The process of iteratively searching the database consists of refining the set of images used in a query by selecting the best query results from the result set and repeating a search.

Implementing The User Interface The user interface is built using the Google Web Toolkit (GWT) [1] and comprises client and server code written in Java. The server provides two entry points to initialise the above mentioned index and provide the query functionality. The query interface takes a set of image ids as a parameter and returns the set of image ids corresponding to the results. On the client side the images for the thumb nails are computed and requested from a server using static paths.

4.4 Running and Downloading the Demonstration

The demonstrator software may run from the Web from the following URL: https://similarity.cs.st-andrews.ac.uk/polydemo. The code for the demonstrator system may be found in GitHub at the following address: https://github.com/MetricSearch/SISAP2024Demo.

5 Conclusion

The concept of a polyadic query, which extends the notion of a query from a single parameter to a set of more than one query object, was previously identified. Previously scenarios were demonstrated where this approach delivers more suitable results than using a single query object. These include: query specialisation in which a user may be using a query image in order to find images of some specific type of item; query generalisation where the user may be seeking images of some general class of objects; and subject combination query in which the user wishes to find objects that have commonality to all of the individual query items.

In this paper we present a demonstration system, in the domain of query specialisation. We have demonstrated how polyadic query techniques may be applied by interactively refining a set of results according to user preference. By the application of polyadic query specialisation the user can refine query images in order to obtain results that are both more suitable than the initial query results or those that could be found using the repeated application of single image queries.

Acknowledgements. This work is partly supported by ESRC grant ES/W010321/1 "2022-2026 ADR UK Programme" and Adobe Systems, and MUCES - a MUltimedia platform for Content Enrichment and Search in audiovisual archive project (CUP B53D23026090001)

References

1. The google web tookkit. https://www.gwtproject.org. Accessed 24-06-2024
2. Chávez, E., Navarro, G., Baeza-Yates, R., Marroquín, J.L.: Searching in metric spaces. ACM Comput. Surv. **33**(3), 273–321 (2001). https://doi.org/10.1145/502807.502808
3. Ciaccia, P., Patella, M., Zezula, P.: Processing complex similarity queries with distance-based access methods. In: International Conference on Extending Database Technology, pp. 9–23. Springer (1998)
4. Connor, R.: A tale of four metrics. In: 9th International Conference on Similarity Search and Applications, SISAP 2016, pp. 210–217. Springer (2016)
5. Connor, R., Dearle, A., Claydon, B.: Scaling polyadic queries. In: Proceedings of the 17th International Conference on Similarity Search and Applications. Springer (2024), to-appear
6. Connor, R., Dearle, A., Claydon, B., Vadicamo, L.: Correlations of cross-entropy loss in machine learning. Entropy **26**(6) (2024)
7. Connor, R., Dearle, A., Morrison, D., Chávez, E.: Similarity search with multiple-object queries. In: Pedreira, O., Estivill-Castro, V. (eds.) Similarity Search and Applications, pp. 223–237. Springer (2023)
8. Connor, R., Simeoni, F., Iakovos, M., Moss, R.: A bounded distance metric for comparing tree structure. Inf. Syst. **36**(4), 748–764 (2011)
9. Hetland, M.L., Hummel, H.: Fairest neighbors: Tradeoffs between metric queries. In: Similarity Search and Applications: 14th International Conference, SISAP 2021, Dortmund, Germany, September 29–October 1, 2021, Proceedings 14, pp. 148–156. Springer (2021)
10. Huiskes, M.J., Lew, M.S.: The mir flickr retrieval evaluation. In: Proceedings of the 1st ACM International Conference on Multimedia Information Retrieval, p. 39-43. MIR '08, Association for Computing Machinery (2008)
11. Kuperus, J.: Java library for approximate nearest neighbors search using hierarchical navigable small world graphs. https://github.com/jelmerk/hnswlib/tree/master (2024). Accessed 31-05-2024
12. Malkov, Y., Ponomarenko, A., Krylov, V.: Scalable distributed algorithm for approximate nearest neighbor search problem in high dimensional general metric spaces. In: Similarity Search and Applications, vol. 7404, pp. 132–147 (2012)
13. Moss, R., Connor, R.: A multi-way divergence metric for vector spaces. In: Proceedings of the 6th International Conference on Similarity Search and Applications, vol. 8199, p. 169–174. SISAP 2013, Springer-Verlag, Berlin, Heidelberg (2013)
14. Oquab, M., Darcet, T., Moutakanni, T., Vo, H., Szafraniec, M., Khalidov, V., Fernandez, P., Haziza, D., Massa, F., El-Nouby, A., et al.: Dinov2: learning robust visual features without supervision. arXiv preprint arXiv:2304.07193 (2023)
15. Shannon, C.E.: A mathematical theory of communication. The Bell System Technical Journal **27**, 379–423 (1948)
16. Zezula, P., Amato, G., Dohnal, V., Batko, M.: Similarity Search: The Metric Space Approach, vol. 32. Springer Science & Business Media (2006)

Scalable Polyadic Queries

Richard Connor[✉][iD], Alan Dearle[✉][iD], and Ben Claydon[iD]

University of St. Andrews, St. Andrews, Scotland, UK
{rchc,al,bc89}@st-andrews.ac.uk

Abstract. In previous work, the notion of polyadic similarity query was introduced. Normally, similarity queries take a single argument and attempt to find those objects within a large collection which are most similar to that argument. The idea of polyadic queries is to generalise this notion, by taking a number of query arguments, and giving results based on some combination of their characteristics. It was previously shown how polyadic queries could be of use in various contexts.

The initial work on polyadic queries provided a proof of concept but left many unanswered questions. In particular, it did not show a proper semantic basis for the polyadic query function used or how to achieve sub-linear query times for polyadic searches over large data.

Here, we address these issues. This work demonstrates that the polyadic query mechanism can scale to large data, and gives results which are better than those obtained by executing simple queries over each of the arguments individually.

Keywords: Similarity Search · Polyadic Query · SED · MSED

1 Introduction

Previous work introduced the notion of a *polyadic* or *conjunctive* similarity query [2]. The notion of a polyadic query is very simple: rather than querying a collection of objects using a single argument from the same domain, a (typically small) set of arguments may be used instead. It was demonstrated that the polyadic dissimilarity function Multi-SED (MSED) seemed to perform reasonably well as the basis for a polyadic query function over collections of images.

Three scenarios were identified where such a mechanism could be useful:

query specialisation: multiple query objects give a more specific representation of the search subject than possible using a single object,

query generalisation: to avoid query results being too specialised with respect to a single query, multiple objects from a broader domain may give outcomes drawn from that broad domain, and

subject combination query: objects representing different domains may return results which exhibit features of all the query domains.

The quantified examples in this article are all in the category of query specialisation.

© The Author(s), under exclusive license to Springer Nature Switzerland AG 2024
E. Chávez et al. (Eds.): SISAP 2024, LNCS 15268, pp. 57–64, 2024.
https://doi.org/10.1007/978-3-031-75823-2_5

1.1 Contribution of This Article

Previous work introduced the dissimilarity functions Structural Entropic Distance (SED) and its polyadic variant MSED which was used to perform polyadic similarity queries. The previous work demonstrated significant promise, but two issues remained unresolved:

Semantics of MSED: Although evidence was presented that MSED gave desirable query outcomes in some specific contexts there was no attempt to quantify this. MSED by definition is a generalised form of SED with multiple arguments. The intention is that MSED captures aspects of dissimilarity amongst all of its arguments giving a coherent polyadic query model. Section 4 presents experiments which quantify these desired effects.

Scalability: Secondly, there was no known way of using the polyadic form MSED to construct a scalable search, and all examples previously shown used an exhaustive strategy. Without scalability, the usefulness of polyadic query would be severely limited. Section 3 shows properties of SED and MSED that can be used to perform pre-processing over a SED-governed space, which can be subsequently queried using MSED in sublinear time. In Sect. 4 we show measurements demonstrating this.

The rest of this article is structured as follows. In Sect. 2 we recap the definitions of the dyadic function SED and its polyadic generalisation MSED, and show an optimised form dMSED that we use for polyadic query. In Sect. 3 we show the relationship between SED and MSED which allows a navigational data structure to be built using SED, and queried in sublinear time using MSED. Section 4 gives details of a study showing both scalability and useful semantic outcomes of polyadic query against single aggregated queries, and in Sect. 5 we conclude.

2 Definitions of SED, MSED and dMSED

Structured Entropic Distance (SED, [3]) is a dissimilarity function which operates over discrete sets of probabilities, i.e. it is an f-divergence. Data suitable for an f-divergence function can be achieved by applying the softmax function to arbitrary vectors as described in [1].

SED is defined:

$$SED(\boldsymbol{v}, \boldsymbol{w}) = \frac{C\left(\frac{\boldsymbol{v}+\boldsymbol{w}}{2}\right)}{\sqrt{C(\boldsymbol{v}) \cdot C(\boldsymbol{w})}} - 1 \qquad (1)$$

where $C(\boldsymbol{v}) = x^{H_x(v)}$, H being the Shannon entropy of $\boldsymbol{v}$. Note that SED is essentially a comparison of the entropy of the arithmetic mean of its arguments with the geometric mean of their individual entropies, giving a dissimilarity outcome in the range $[0, 1]$.

Our work on polyadic queries uses a polyadic divergence function in place of a dyadic one. That is, we consider a polyadic dissimilarity space (U, d) where

$d : \mathscr{P}(\mathcal{U}) \to \mathbb{R}_{\geq 0}$, where $\mathscr{P}(\mathcal{U})$ denotes the powerset of $\mathcal{U}$ in place of the usual $d : U \times U \to \mathbb{R}_{\geq 0}$. When comparing a polyadic query $Q \in \mathscr{P}(\mathcal{U})$ against a large finite set $S \subset U$, we rank the results according to the application of the function $d(Q \cup \{s\})$ for all $s \in S$.

MSED [8] is a polyadic generalisation of SED. MSED over m arguments is defined:

$$MSED(\boldsymbol{v}_1, \ldots, \boldsymbol{v}_m) = \frac{1}{m-1}\left(\frac{C\left(\sum_i \frac{\boldsymbol{v}_i}{m}\right)}{\sqrt[m]{\prod_i C(\boldsymbol{v}_i)}} - 1\right) \quad (2)$$

which can be seen to have the same intent as the dyadic version, but now generalised over multiple arguments. Again this form is scaled into the interval $[0, 1]$, giving an outcome of 0 for a set of identical inputs and 1 for a set of orthogonal inputs.

This allows the evaluation of polyadic queries. However, for a large finite set S, the naive application of MSED to each element would be infeasibly expensive. We address this issue by first identifying an optimisable form of MSED, denoted dMSED. This function takes two arguments: a query set Q and a singleton value $\boldsymbol{u}$, and has the property that $\text{dMSED}(Q, \boldsymbol{u}) = \text{MSED}(Q \cup \{\boldsymbol{u}\})$

$$dMSED(Q, \boldsymbol{u}) = \frac{1}{m}\left(\frac{C\left(\left(\sum_{q \in Q} \frac{q}{m+1}\right) + \frac{\boldsymbol{u}}{m+1}\right)}{\sqrt[m+1]{\left(\prod_{q \in Q} C(q)\right) C(\boldsymbol{u})}} - 1\right) \quad (3)$$

where $m = |Q|$.

Given this form, the majority of the calculation can be performed only once, as soon as the query Q becomes available, rather than for each iteration of $s \in S$. Therefore when evaluating $\text{dMSED}(Q, s)$, the terms $M = \sum_{q \in Q} \frac{q}{m+1}$ and $P = \prod_{q \in Q} C(q)$ can be calculated once per query, and then for each element $s \in S$ only the calculation

$$dMSED(Q, \boldsymbol{s}) = \frac{1}{m}\left(\frac{C\left(M + \frac{\boldsymbol{s}}{m+1}\right)}{\sqrt[m+1]{P \cdot C(\boldsymbol{s})}} - 1\right) \quad (4)$$

requires to be performed. Note also that the term $C(\boldsymbol{s})$ can be calculated for each $\boldsymbol{s}$ during pre-processing of the finite data, reducing the cost of this residual calculation to a single complexity evaluation over the calculated vector, followed by some simple arithmetic.

This optimisation reduces the cost of an MSED calculation to that of the simple SED function independent of the size of the set argument. However this is insufficient to allow scalability for use over very large data. We address this issue in Sect. 3 where we show a relationship between SED and dMSED that allows sublinear query cost.

3 Scalable Search with MSED

Pre-processing a set S with the intent of optimising polyadic queries is clearly a different prospect to that of pre-processing with respect to a binary dissimilarity function. Pragmatically, the cost of any pre-processing must be at most $O(N \log N)$ where N is the size of the finite data set. In itself, this requirement clearly rules out any analysis involving multiple finite subsets of S, as might be naturally considered.

The next consideration is whether there may be some relationship between SED and dMSED which would allow the reuse of previous work in scalable dyadic search. Previous work by Dong et al. [4] uses the near-neighbour overlap property to construct an index over pairs of points using a near-neighbour graph. Clearly, a polyadic index cannot be created in this manner due to the variable number of parameters and their different permutations. However, if there exists commonality between the near-neighbours governed by the dyadic SED function and dMSED, we may use an index constructed using the former and query using the latter. We now demonstrate that such a relationship does exist, based on the near neighbour overlap property, and that it is possible to construct such an index.

Near-Neighbour Overlap We express the k nearest-neighbours of a value q within a finite space S as

$$kNN(q) = \{s \in S \mid d(s,q) \leq t_q^k\}$$

noting the threshold distance t_q^k implied by this definition. Importantly, this is the value of the threshold distance that implies the near-neighbour relation, i.e. $d(q,s) \leq t_q^k$ implies that $s \in kNN(q)$.

The near-neighbour overlap property implied in [4] can be informally stated as follows:

> If, for some appropriate values of k, m, where s_1 is in the kNN of q, and s_2 is in the mNN of s_1, then there is a significant probability p that s_2 is also within the kNN of q.

The actual values required for k, m and p depend on various factors, including the dimensionality of the data, and the accuracy required for the results. We can denote this property in terms of dyadic distance functions as follows:

$$\begin{aligned} \text{if} \quad & d(q, s_1) \leq t_q^k \\ \text{and} \quad & d(s_1, s_2) \leq t_{s_1}^m \\ \text{then} \quad & \mathcal{P}(d(q, s_2) \leq t_q^k) \geq p \end{aligned}$$

where $\mathcal{P}$ denotes probability, for some appropriate value of p. Now, this property is directly reusable for polyadic query as follows:

$$\text{if} \quad d(Q, s_1) \leq t_Q^k$$
$$\text{and} \quad d(s_1, s_2) \leq t_{s_1}^m$$
$$\text{then} \quad \mathcal{P}(d(Q, s_2) \leq t_Q^k) \geq p$$

for the same meanings of k, m and p.

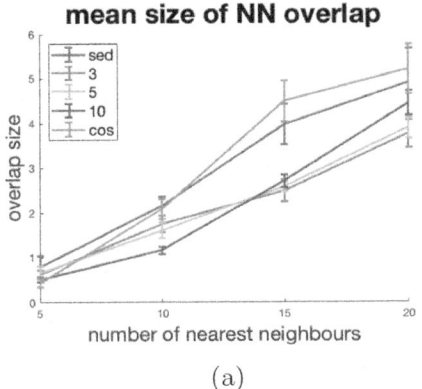 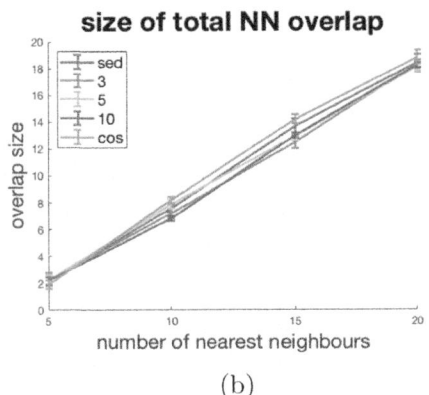

Fig. 1. Near-neighbour overlap for polyadic and single query. Each data series represents polyadic queries of size 3, 5 and 10, also compared with single-subject queries using SED and Cosine distance.

Figure 1 shows the experimental quantification of this value p for DinoV2s [9] encodings of the MirFlickr

As a control, the measurements are taken for the Cosine metric over the original space. Other data series plotted correspond to SED and dMSED for query sizes of 3, 5 and 10. In each case a large number of random queries were selected and the mean results reported.

For each set of queries $\{Q\}$ we calculate $K = kNN(Q)$ and for each element $k_i \in K$ we measure the size of the intersection of $kNN(k_i)$ and K.

Figure 1a shows the mean size of the intersection of K and each $kNN(k_i)$. Figure 1b shows the intersection of $kNN(Q)$ and the union of $kNN(k_i)$. Both diagrams shown a significant near neighbour overlap between MSED and SED.

As previously mentioned, the value of p required depends on many factors related to details of the search mechanism, some of which are not well understood. However, this experiment convincingly demonstrates that any mechanism relying on the NN-overlap property which supports scalable search for Cosine distance over the original dyadic space will also perform well for polyquery using dMSED. Scalability results supporting this are given in the next section.

4 Experiments

In the experiments described in this paper, we use Wikipedia GloVe 100D feature vectors ($N = 4 \times 10^5$) [10] which should be searched using Euclidean or Cosine distance. Since SED/MSED is f-divergence over a probability space, we apply the softmax function to the original vectors with $t = 10$ and use the MSED function over the resulting space. This gives near-perfect pairwise distance correlations with the original metric space [1].

Approximate nearest-neighbour tables were built using SED according to the algorithm described in Dong et al. [4]. *PyNNDescent*, a Python implementation [6] of the approach described by Dong et al. [4] was used.

Approximate queries over this table were performed using the search algorithm used in HSNW, as described by Malkov and Yashunin [5]. This algorithm is supplied with the dMSED function calculated per query, and is implemented in our own Java code. The search algorithm uses a randomly chosen fixed entry point. This code was used to determine the speed of the indexing mechanism, the number of distance calculations performed and the recall.

All experiments were run on an AMD Ryzen 7 3700X 8-Core Processor and 64GB RAM. All the code used in this paper may be found in the GitHub repository https://github.com/MetricSearch/SISAP2024PolyadicQuery.

Quality of Polyadic Search We used word embeddings to perform a fully objective test of polyadic query, using the idea that the synonyms of words should be included the near-neighbours of those words. Using an externally sourced definition of synonyms, we were able to compare the results of polyadic queries formed from sets of synonyms with the aggregation of results from simple queries using each element of the set separately.

A database containing a number of synonyms of common English words was drawn from The Oxford 5000 corpus.[1] For each word in the corpus a list of synonyms was extracted from the WordNet [7] thesaurus[2] which are also in the Oxford 5000 thesaurus. Words with at least three synonyms were used. This gives a ground truth of synonyms which may be searched individually or in combination.

Embeddings are obtained for each word in the corpus from the Wikipedia GloVe 100D dataset [10].

For each word w_i in the corpus we use its synonyms to construct all possible single and polyadic queries, and calculate the position at which w_i is found in the result set. To measure quality, we employ a *rank score* defined as 100 - the position of w_i in the 100 near neighbours of the query (with the query terms removed), or 0 if w_i is not present. In Fig. 2a, we report the mean values for rank score for over all w_i, for all words which have exactly 5 synonyms; the pattern reported is consistent amongst other numbers of synonyms.

---

[1] https://github.com/tyypgzl/Oxford-5000-words.
[2] Found at https://github.com/zaibacu/thesaurus.

In Fig. 2b, we contrast the recall of the algorithm against a brute force query using different numbers of synonyms. At the time of writing, we do not understand why the recall drops in this manner.

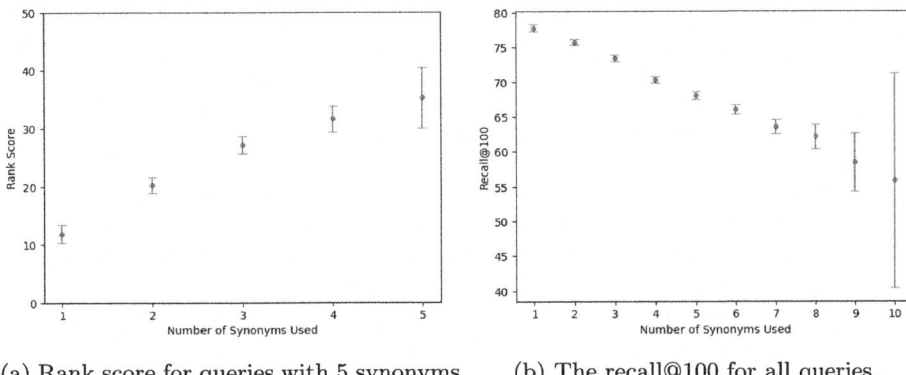

(a) Rank score for queries with 5 synonyms (b) The recall@100 for all queries.

Fig. 2. Rank score and recall for Oxford 5000 polyadic queries.

Efficiency of Scalable Search To establish the efficiency of metric indexing we perform polyadic queries with different numbers of parameters (synonyms) as described above. The performance of the indexed search in comparison to the non-indexed version is shown in Fig. 3.

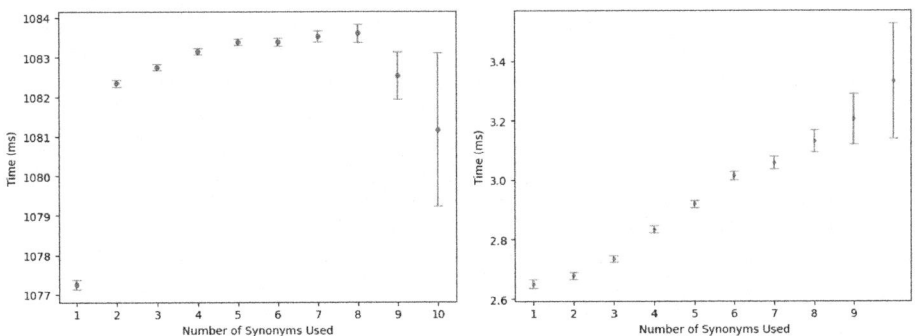

(a) Timings for exhaustive polyadic queries (b) Timings for indexed polyadic queries

Fig. 3. Comparison of the performance of indexed and non-indexed polyadic queries. *Note* the y-axes scale differences in the two diagrams.

Summary of Synonym Results The value of these semantic results is that the experiment is purely objective. It thus convincingly demonstrates that polyadic queries using small subsets of synonyms outperform those using single-subject queries, thus proving that the polyadic query using MSED does indeed capture aspects of all the query elements. In terms of scalability, the indexed version is 2–3 orders of magnitude faster than exhaustive search using dMSED, and the cost does not vary greatly with larger query sets.

5 Conclusions and Further Work

Here we have significantly developed the idea of polyadic queries as originally outlined in [2]. We have shown that the MSED abstraction produces results based on common properties of its multiple arguments and that polyadic queries can be made scalable.

Section 4 uses the WordNet [7] database, with a ground-truth of word synonyms, to rigorously semantically quantify the effectiveness of the mechanism.

Finally, we prove through the property of near-neighbour overlap discussed in Sect. 3 that a class of search algorithm may used to perform polyadic queries in sublinear time. This allows polyadic queries to scale to large datasets which was not previously possible. Polyadic query based on MSED has now shifted from what was an interesting idea in outline, to a methodology of proven value.

References

1. Connor, R., Dearle, A., Claydon, B., Vadicamo, L.: Correlations of cross-entropy loss in machine learning. Entropy **26**(6) (2024)
2. Connor, R., Dearle, A., Morrison, D., Chávez, E.: Similarity search with multiple-object queries. In: Similarity Search and Applications, pp. 223–237 (2023)
3. Connor, R., Simeoni, F., Iakovos, M., Moss, R.: A bounded distance metric for comparing tree structure. Inf. Syst. **36**(4), 748–764 (2011)
4. Dong, W., Moses, C., Li, K.: Efficient k-nearest neighbor graph construction for generic similarity measures. In: Proceedings of 20th International WWW Conference, pp. 577–586 (2011)
5. Malkov, Y.A., Yashunin, D.A.: Efficient and robust approximate nearest neighbor search using hierarchical navigable small world graphs. IEEE Trans. Pattern Anal. Mach. Intell. **42**(4), 824–836 (2020)
6. McInnis, L.: A python nearest neighbor descent for approximate nearest neighbors. https://github.com/lmcinnes/pynndescent. Accessed 31 May 2024
7. Miller, G.A.: Wordnet: a lexical database for English. Commun. ACM **38**(11), 39–41 (1995)
8. Moss, R., Connor, R.: A multi-way divergence metric for vector spaces. In: Similarity Search and Applications, pp. 169–174 (2013)
9. Oquab, M., Darcet, T., Moutakanni, T., Vo, H., Szafraniec, M., Khalidov, V., et al.: Dinov2: Learning Robust Visual Features Without Supervision (2023). https://arxiv.org/abs/2304.07193
10. Pennington, J., Socher, R., Manning, C.: Glove: global vectors for word representation. In: Proceedings of EMNLP 2014, pp. 1532–1543 (2014)

A Dynamic Evaluation Metric for Feature Selection

Muhammad Rajabinasab[✉], Anton D. Lautrup, Tobias Hyrup, and Arthur Zimek

University of Southern Denmark, Odense, Denmark
rajabinasab@imada.sdu.dk

Abstract. Expressive evaluation metrics are indispensable for informative experiments in all areas, and while several metrics are established in some areas, in others, such as feature selection, only indirect or otherwise limited evaluation metrics are found. In this paper, we propose a novel evaluation metric to address several problems of its predecessors and allow for flexible and reliable evaluation of feature selection algorithms. The proposed metric is a dynamic metric with two properties that can be used to evaluate both the performance and the stability of a feature selection algorithm. We conduct several empirical experiments to illustrate the use of the proposed metric in the successful evaluation of feature selection algorithms. We also provide a comparison and analysis to show the different aspects involved in the evaluation of the feature selection algorithms. The results indicate that the proposed metric is successful in carrying out the evaluation task for feature selection algorithms.

Keywords: Feature Selection · Evaluation Metric · Performance Analysis · Stability Analysis

1 Introduction

Feature selection is the task of selecting the most informative features for the target machine learning or data mining task and removing redundant features. There are different aspects of feature selection algorithms that can be investigated, such as performance, amount of dimensionality reduction, stability, and complexity [2]. A combination of these aspects can also be considered as the target of the evaluations. Feature selection algorithms are usually evaluated based on their performance in a downstream machine learning or data mining task. The downstream task can be supervised (e.g., classification) [4,5,11] or unsupervised (e.g., clustering) [6,13]. In this form of evaluation, a classification or clustering

---

The original version of the chapter has been revised. A correction to this chapter can be found at https://doi.org/10.1007/978-3-031-75823-2_26.

algorithm is employed after the feature selection process and its performance is evaluated based on measures such as accuracy, F1-score, clustering accuracy, and Normalized Mutual Information (NMI). These evaluations and their respective experimental results still yield valuable insights, but are highly dependent on the machine learning model and its characteristics. They also do not take into account the number of selected features.

To the best of our knowledge, there are only a few related works proposing metrics for the evaluation of different characteristics of feature selection algorithms. Nogueira et al. [10] propose a metric to calculate the stability of a feature selection algorithm in the presence of noise. This metric is based on Pearson's correlation coefficient and investigates whether the same features are selected in different scenarios or not. This metric yields values in the range $[-1, 1]$. Values less than 0.40 are considered to indicate poor stability, values between 0.40 and 0.75 imply an intermediate to good level of stability, and values greater than 0.75 suggest the algorithm to be nearly perfectly stable. Another stability measure [8] puts the focus on Sequential Forward Selection (SFS) considering the intersection between two sets of selected features to measure stability. The Baseline Fitness Improvement (BFI) [9] is a normalized metric for the evaluation and comparison of feature selection algorithms which shows the potential gain by applying feature selection. BFI takes into account the performance of the downstream task and the amount of dimensionality reduction.

BFI is the more consistent approach for evaluating feature selection algorithms, but still, it does not address the advantage gained by adding or removing a feature. It also does not provide insights into how stable the performance of the feature selection algorithm is with different numbers of features to be selected. On the other hand, there are usually confounding effects involved with real data. In other words, different features might exist in the dataset which represent the same information. The stability measure [10] is not able to capture their impact on the feature selection procedure as they are considered to be different features, though they code for the same information. Such challenges and shortcomings motivate this study to propose a new feature selection evaluation metric which can better reflect the blind spots of a feature selection method.

In this paper, we propose the *Feature Selection Dynamic Evaluation Metric* (FSDEM).[1] FSDEM is focused on assessing the performance and stability of feature selection methods. FSDEM offers a dynamic framework for the assessment and evaluation of the feature selection algorithm which can be instantiated with any performance measure. Hence, it is able to give insights into different performance aspects of the algorithm. It also yields insights into the stability of the feature selection algorithm. Unlike previous methods, FSDEM defines stability based on the changes in the performance measure as the number of selected features varies. Therefore, it avoids the challenges posed by the previous methods.

---

[1] An extended version of this paper providing more details is available on arXiv [12].

2 Proposed Method

In this paper, FSDEM is proposed to address some of the challenges and shortcomings described above. FSDEM is dynamic, and it can be integrated with any performance measure to provide insights to the feature selection algorithm. FSDEM has two properties that can be effective for the analysis of different aspects of a feature selection method.

2.1 FSDEM Score

Let F be the total number of features. Let $M(f)$ be the value of an arbitrary performance measure with f out of F features selected. Let $g(x)$ be the function approximated from $M(f)$ with different observations of $f \in \{1, ..., F\}$. The FSDEM score will be the area under the curve of $g(x)$:

$$\text{FSDEM} = \frac{\int_a^b g(x)\,dx}{(b-a)} \tag{1}$$

Clearly, the range of possible values for the FSDEM score is bounded and it inherits its bounds from the measure M. Assuming that the values of M are in the range $[0, 1]$ (bounds of most ordinary measures, such as accuracy), we desire to have values closer to 1 as it indicates a better feature selection performance. Linear approximation is the simplest yet most efficient method for approximating a continuous function from the observations of different performance measures at discrete steps. Linear approximation draws a line between two consecutive discrete observations.

There are three main advantages to using the FSDEM score: i) If the number of features is large, we can run fewer experiments and have the curve with a certain degree of precision; ii) Some feature selection algorithms might perform better in selecting a specific number of features for the target downstream task. Using FSDEM, we can set a and b according to our preferred range, and hence select a feature selection method that performs best with respect to our preferences; iii) We can use the first-order derivative as a measure of stability.

2.2 Stability Score

The stability score is calculated based on the first-order derivative of g:

$$S = \frac{\sum_{i=a}^{b} g'(x)}{(b-a)+1} \tag{2}$$

The derivative is calculated using the finite difference method [3] and indicates the amount of change in the performance measure w.r.t. different number of features. Thus, the stability score can be interpreted as a value in the range $[-1, 1]$ (for measures with values in the range $[0, 1]$). Positive and negative values indicate the direction of change in the FSDEM score by selecting more features.

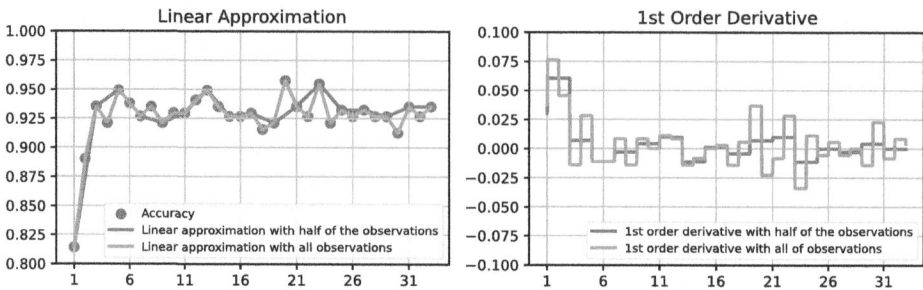

Fig. 1. Example of the function approximation (left) and first order derivative (right) with all and half of the observation, based on the accuracy as the performance measure.

Ideally, we desire to have larger positive values for the stability score since it indicates a better selection of features. It is noteworthy that in real scenarios, the values are often close to 0 as the performance does not change dramatically with additional features. Hence, positive values are considered to be good in general. The stability score can be used effectively in comparisons alongside FSDEM to select the best feature selection algorithm for a given task and dataset.

The stability metric proposed in this paper accounts for the informative value of the selected features and not whether the same features are selected or not. Hence, it does not suffer from the assumption of the previous methods. Consider a scenario where a feature selection algorithm selects two feature subsets $F = \{f_1, f_2, ..., f_m\}$ and $F' = \{f'_1, f'_2, ..., f'_m\}$ on two different noisy versions of dataset D. If $F \cap F' = \emptyset$, the algorithm is considered to be unstable based on the previous metrics [8,10]. Whereas F_1 and F_2 might represent the same information or include the same amount of informative value. FSDEM stability score is calculated based on the effect of the selected feature subset on the final downstream task and hence, it can successfully identify stable or unstable feature selection algorithms based on the informative value of the selected features.

FSDEM can be integrated with any arbitrary performance measure. Figure 1 illustrates an example of the function approximation and its first-order derivative as the backbone of the FSDEM and the stability score based on the accuracy as the underlying measure. The two cases shown are for the approximation with half and all of the observations. As shown in Fig. 1, even when approximating the function with half of the observations, both results show adequate consistency.

3 Analysis and Empirical Results

In this section, we present the analysis and empirical results to provide insights into the proposed feature selection evaluation metric. First, we present the capabilities of FSDEM in identifying the best feature selection algorithm w.r.t. the desired range for the number of selected features. Then, we investigate the stability score of FSDEM by analyzing a case in which the stable feature selection algorithm is identified by FSDEM and not by the previous

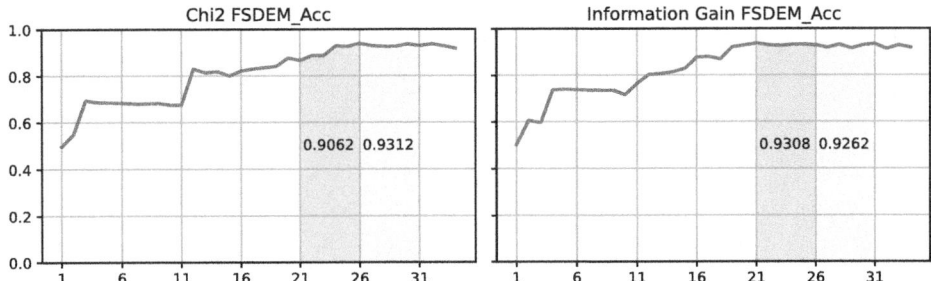

Fig. 2. FSDEM score based on accuracy for two different feature selection methods and target feature number ranges.

metric [10]. We provide empirical results for the FSDEM and its stability score based on 20 different datasets with different numbers of instances and features from the UCI repository [7] (for details see extended version [12]). We use random feature selection, Information gain, chi2, and wrapper-based feature selection based on random forest as feature selection algorithms.

Evaluation Based on the Target Number of Features : Figure 2 shows the accuracy curve for chi2 and information gain and the FSDEM score for different ranges of target feature numbers. For a target feature number in the range [21, 26], information gain is a better choice as it yields a higher overall accuracy. But for a target feature number in the range [26, 31], chi2 performs better. FSDEM successfully provides insights based on the target number of features and can contrast different feature selection methods in this respect. This enables selecting the best method based on a specific number of selected features.

Stability Score Based on the Informative Value: Previous methods proposed for investigating the stability of feature selection algorithms measure stability as the ability to select the same features in different scenarios [10]. However, in some cases, the same information might be included in different features. An algorithm that selects different features in different scenarios is considered unstable by these methods, regardless of the information included in the features.

As an example, consider a dummy dataset with 500 samples and 6 features. The features include: {*age, salary in EUR, salary in USD, size of the residence, distance to the city center in kilometers, distance to the city center in miles*}. The target variable is *wealth*, which is a binary variable that indicates whether a person is wealthy or not. Consider algorithm A which is used to select half of the features in the dataset. The features ranking by A in two different scenarios are {*salary in EUR, size of the residence, distance to the city center in kilometers, salary in USD, distance to the city center in miles, age*} and {*salary in*

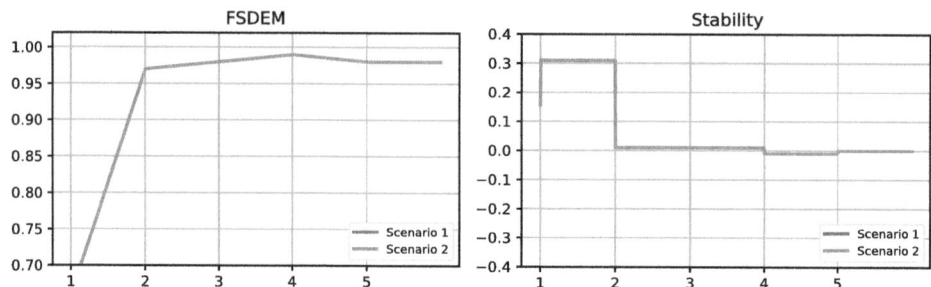

Fig. 3. FSDEM and stability score for the two scenarios.

USD, size of the residence, distance to the city center in miles, salary in EUR, distance to the city center in kilometers, age}. The stability measure proposed by Noguiera et al. [10] yields a value of -0.3333 for A. This value is indicative of poor stability. However, the information represented in both of the feature subsets is the same. From an information-theoretic point of view, the algorithm must be considered stable as it is able to select the same amount of information in both cases. In Fig. 3, an illustration of the FSDEM and its stability score for the two scenarios is shown.

As FSDEM takes performance into account, it shows the same values in both scenarios. The average value for the stability score of the two scenarios is 0.0466 which is positive and indicates good stability w.r.t. accuracy. FSDEM relies on the integrated performance measure as a factor for stability. In this case, despite different selected features, the feature subsets represent the same information. Hence, the underlying feature selection algorithm must not be considered unstable. FSDEM successfully captures the informative aspect of stability instead of strictly requiring the same features to be selected, allowing for more robust measures in data with confounding effects.

Empirical Results: We provide empirical results to illustrate the effectiveness of FSDEM and its stability score in the successful evaluation of feature selection algorithms. In Fig. 4, we display the experimental results in terms of FSDEM score and its associated stability value for two different performance measures, namely (classification) accuracy and clustering accuracy (CLACC) [1].

The experimental results indicate that, as expected, the wrapper-based feature selection algorithm based on random forest has the best overall performance compared to the other algorithms. The stability results show that the most stable algorithm in feature selection is the random feature selection procedure. This is expected as in random feature selection, a feature is selected without considering any specific criteria. Features are selected randomly at every step. Hence, by selecting more features, it is more likely to include the informative features. As a result, the final performance is improved by increasing the number of features. On the other hand, when it comes to the other methods, it might be more beneficial to select fewer features as it is more likely that the most informative

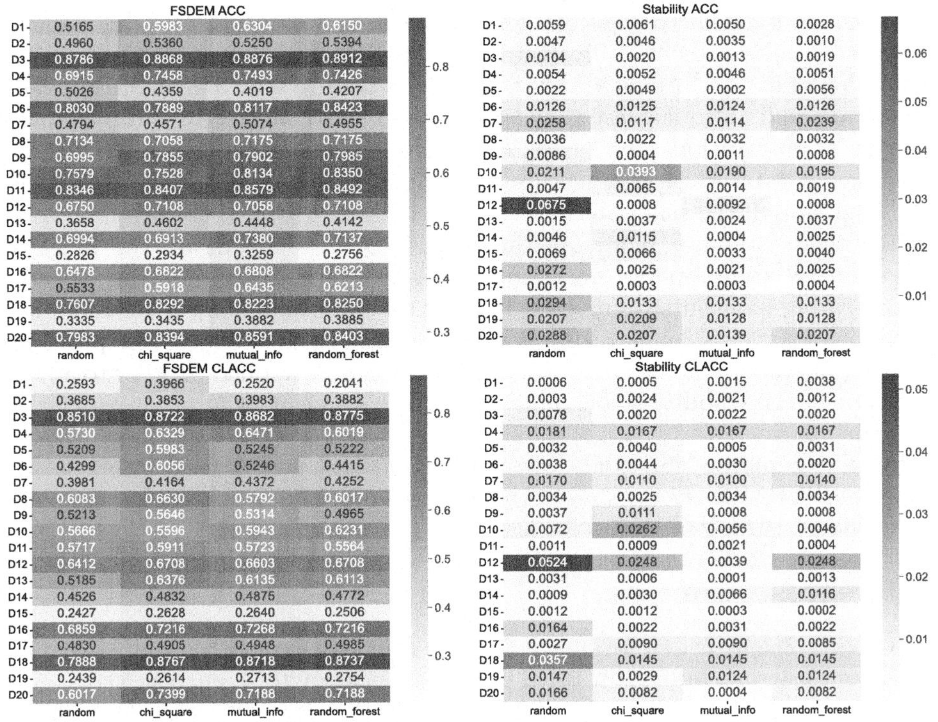

Fig. 4. FSDEM score and its associated stability for different algorithms and datasets based on accuracy and clustering accuracy. Note that the ranges are different.

features are selected in the first steps. Investigating the values of the FSDEM and stability score together allows one to select the most efficient and stable feature selection algorithm for an arbitrary task. We also repeated the experiments with half of the observations. The results (for details see extended version [12]) showed an insignificant change in the FSDEM and stability score. Hence, it is feasible to conduct the experiments effectively with fewer observations.

4 Conclusion

In this paper, we proposed FSDEM, a novel dynamic metric to evaluate the performance and stability of feature selection algorithms. FSDEM can integrate any performance measure to provide insights into the effectiveness and stability of a feature selection algorithm. FSDEM is able to provide a thorough analysis of a feature selection algorithm w.r.t. its performance with different numbers of features selected. Additionally, FSDEM can help to select the best feature selection algorithm for a desirable target range for the number of features. The stability score provided by FSDEM can assess the stability of a feature selection algorithm w.r.t. the informative value of the selected feature subset. It can also reflect the overall effectiveness of a feature selection algorithm. The analysis and empirical results presented illustrate that FSDEM is an effective measure for evaluation of the performance and stability of a feature selection algorithm. The empirical results also suggest that FSDEM can work effectively with a limited number of observations and hence, is efficient for experiments involving high-dimensional datasets or slow feature selection algorithms.

Despite of all the strengths and capabilities of FSDEM, there is a limitation related to its stability score. As the stability score depends on the function approximated from a performance measure which in most cases ranges between 0 and 1, it often outputs values close to 0. As the most important use-case for the evaluation metrics is comparison, this might not be considered a severe problem. However, methods such as correction for chance can be applied in order to adjust the value of the stability score. This can be a main focus of future research.

Acknowledgement. This study was funded by Innovation Fund Denmark in the project "PREPARE: PERSONALIZED RISK ESTIMATION AND PREVENTION OF CARDIOVASCULAR DISEASE".

References

1. Cao, X., Zhang, C., Fu, H., Liu, S., Zhang, H.: Diversity-induced Multi-view subspace clustering. In: CVPR. IEEE Computer Society (2015), pp. 586–594
2. Chandrashekar, G., Sahin, F.: A survey on feature selection methods. Comput. Electr. Eng. **40**(1), 16–28 (2014)
3. Tomas B. Co.: Methods of Applied Mathematics for Engineers and Scientists. Cambridge University Press (2013)
4. Fan, Y., Liu, J., Tang, J., Liu, P., Lin, Y., Du, Y.: Learning correlation information for multi-label feature selection. Pattern Recogn. **145**, 109899 (2024)
5. Guha, R., Khan, H.A., Singh, P.K., Sarkar, R., Bhattacharjee, D.: CGA: a new feature selection model for visual human action recognition. Neural Comput. Appl. **33**(10), 5267–5286 (2021)
6. Guo, Y., Sun, Y., Wang, Z., Nie, F., Wang, F.: Double-structured sparsity guided flexible embedding learning for unsupervised feature selection. IEEE Trans. Neural Netw. Learn. Syst. 1–14 (2023)
7. Kelly, M., Longjohn, R., Nottingham, K.: The UCI Machine Learning Repository. https://archive.ics.uci.edu (2024)
8. Kuncheva, L.I.: A stability index for feature selection. In: Artificial Intelligence and Applications. IASTED/ACTA Press (2007), pp. 421–427
9. Mostert, W., Malan, K.M., Engelbrecht, A.P.: A feature selection algorithm performance metric for comparative analysis. Algorithms **14**(3), 100 (2021)
10. Nogueira, S., Sechidis, K., Brown, G.: On the stability of feature selection algorithms. J. Mach. Learn. Res. **18**, 174:1–174:54 (2017)
11. Pan, H., Chen, S., Xiong, H.: A high-dimensional feature selection method based on modified Gray Wolf optimization. Appl. Soft Comput. **135**, 110031 (2023)
12. Rajabinasab, M., Lautrup, A.D., Hyrup, T., Zimek, A.: FSDEM: Feature Selection Dynamic Evaluation Metric (2024). arXiv: 2408.14234 [cs.LG]. https://arxiv.org/abs/2408.14234
13. Wang, C., Wang, J., Gu, Z., Wei, J.-M., Liu, J.: Unsupervised feature selection by learning exponential weights. Pattern Recogn. **148**, 110183 (2024)

Personalized Similarity Models for Evaluating Rehabilitation Exercises from Monocular Videos

Miriama Jánošová[1], Petra Budikova[2], and Jan Sedmidubsky[1]

[1] Faculty of Informatics, Masaryk University, Brno, Czech Republic
424615@mail.muni.cz
[2] VisionCraft, Brno, Czech Republic

Abstract. Automatic monitoring of exercise correctness during home physical rehabilitation could significantly increase the impact of rehabilitation treatments. To evaluate exercise quality effectively, it is necessary to extract relevant spatio-temporal motion features and compare them to an ideal exercise pattern. We argue that the features should be personalized to the patient's needs, as the movement abilities of each patient are specifically limited and also change over time. Towards this end, we utilize the MediaPipe Pose tool to estimate 2D and 3D coordinates of skeleton joints from a monocular video stream. The joint coordinates are then processed to extract specific spatio-temporal features that are automatically weighted for each patient. This allows for personalized similarity based on the individual's exercise patterns while requiring minimal training data and possibly offering explainable evaluations. The proposed approach is tested on the REHAB24-6 rehabilitation dataset, reaching superior effectiveness and being about 2–3 orders of magnitude more efficient than state-of-the-art solutions.

Keywords: pose estimation · skeleton sequence · rehabilitation exercise · human body keypoint · exercise quality assessment · exercise similarity · personalized similarity · kNN retrieval

1 Introduction

According to the World Health Organization (WHO) [5], roughly one in three people on Earth has some health problem that could be alleviated by rehabilitation, and the number of people with movement difficulties is expected to further grow due to lifestyle changes and population development. In most countries, the healthcare system is already struggling with the physical therapist shortage. At the same time, there is a rapidly developing field of AI that deals with human motion recognition and analysis through the structured *skeleton* modality [13]. In particular, research in Human Pose Estimation (HPE) [3] aims to extract skeleton sequences from ordinary video recordings. Recently, HPE methods have become sufficiently precise and efficient to be run on widely available devices such as smartphones. This opens the possibility of software-assisted rehabilitation.

In a standard rehabilitation scenario, a patient visits a therapist, who diagnoses the patient, performs some hands-on therapy, and then selects suitable exercises for the patient. A successful treatment then requires a long-term cooperation of the patient, who must perform the recommended exercises at home without the direct supervision of healthcare personnel. However, exercising without guidance and feedback is difficult, which leads to frequent negligence of home exercising and inefficient recovery. There is wide agreement in the healthcare community that a software tool able to provide monitoring and real-time feedback could significantly improve rehabilitation efficiency, when respecting the specific requirements of rehabilitation exercises. In particular, home exercises are typically personalized to the specific patient's needs, and the optimal way of performing the exercises is limited by the patient's current abilities – e.g., a post-surgery patient won't be able to perform a full extent of some movements at the beginning of the rehabilitation treatment.

In the last years, several software tools have emerged that exploit HPE and motion analysis to offer exercise monitoring [12]. These tools focus on the domains of fitness and preventive exercising and evaluate the quality of users' performance by comparing their movements to a universal model exercise or set of rules that define a correct "textbook" exercise. However, this assumes that all users are able to achieve the quality of textbook exercise, which is not true for physical therapy patients. Therefore, the existing exercise-control tools are hardly applicable for rehabilitation monitoring.

In this paper, we present a novel technology that provides *personalized* control of exercise quality. We call this technology the *True Motion Twin*, as it enables evaluating a user's rehabilitation performance with respect to his/her personal abilities and limitations. The technology exploits similarity-based data analysis, which compares a patient's exercising to his/her personal model based on personalized motion similarity. We design our solution to be applicable in real-world rehabilitation scenarios where only a very limited number of training examples are available. The main contributions of this paper are:

– Compact exercise features that are specifically designed for each type of rehabilitation exercise.
– Personalized feature thresholds and weights that are auto-tuned for each individual to increase similarity between correctly performed exercises while decreasing similarity between correctly and poorly performed ones.
– Qualitative comparison of general and personalized similarity models.

2 Similarity-Based Motion Analysis

There are several possible approaches to assessing the quality of any movement: machine learning [7], rule-based evaluation [17], and similarity-based motion matching [16]. Machine learning models perform well when sufficient training

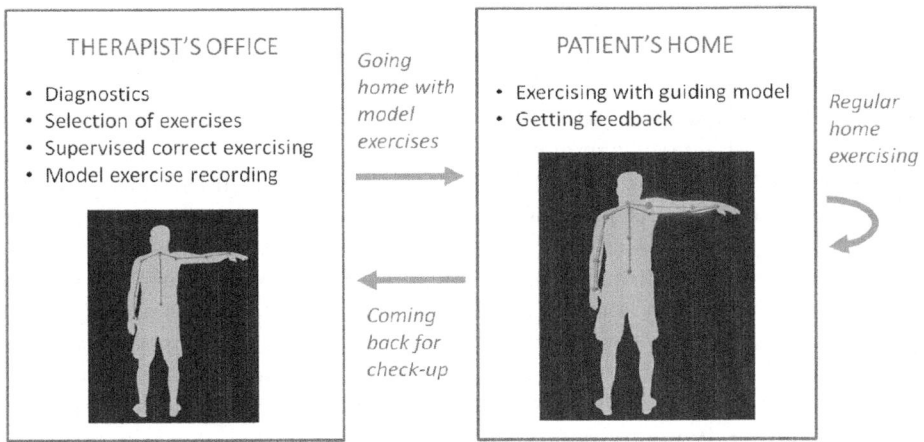

Fig. 1. Use-case overview: software-assisted rehabilitation

data is available. However, this is not the case of personalized exercise monitoring, which requires an adaptation of similarity models to each new individual. Similarly, the rule-based quality evaluation either works with universal good/bad motion rules, or requires new rules for each patient, which is not feasible to define in a real environment. On the other hand, similarity-based motion analysis requires only a small amount of learning data (even only a single example) and provides an explainable evaluation of motion quality [13]. Therefore, we base our approach on this similarity paradigm.

In the following subsections, we provide a high-level overview of our approach, discuss related work, and explain the evaluation methodology used to analyze individual components of our solution. In Sect. 3, we provide a detailed explanation and evaluation of the individual techniques we developed.

2.1 Our Approach

Our objective is to enrich the standard rehabilitation treatment with motion monitoring technologies without forcing the therapists and patients to significantly change their routines. As illustrated in Fig. 1, a rehabilitation treatment involves two steps repeated until the patient is cured: the patient visits the therapist's office, and the patient performs exercises at home. In the first step, the therapist works with the patients and teaches them exercises for home rehabilitation. Even now, therapists often record the exercises for patients using smartphones to provide the patients with video guidance at home. Our approach enriches this by applying an HPE method to extract skeleton sequences from the recorded videos. Since patients often make mistakes even under the therapist's supervision, we can assume that *reference* examples of both correctly and incorrectly performed exercises are recorded. The therapists must manually annotate the examples as correct/incorrect, which is considered acceptable by consulted therapists. In the second step, the recorded reference movements can be used to both guide the patients and evaluate the correctness of their home exercises.

To obtain the skeleton data from video streams, we utilize the MediaPipe Pose (MPP)[1] tool, which is an open-source, cross-platform framework developed by Google. MPP employs BlazePose [1], a lightweight machine learning architecture, to extract 33 2D positions of *joints* (keypoints) on the human body along with the depth of each joint. The main advantage of MPP is its ability to perform real-time HPE from video recordings on both smartphones and PCs.

To determine motion similarity, we must realize that *speed* is an important property of any motion – rehabilitation exercises can be performed at varying speeds without diminishing the exercise quality. Consequently, we need to compensate for possible speed differences during the comparison of the reference and home exercises. A standard algorithm for speed-invariant motion comparison is the Dynamic Time Warping (DTW) [14], which seeks to find the optimal alignment between two time-dependent sequences of varying lengths. Specifically, to align sequences A and B, we evaluate the similarity of each frame in A with each frame in B, resulting in a *cost matrix*. The DTW algorithm then finds an alignment between A and B that minimizes the total cost.

All techniques discussed in this paper utilize DTW for evaluation of motion similarity, but differ in the inner distance function that determines pose similarity for the cost matrix. Our objective is to optimize this pose-similarity distance from both quality and speed perspectives. In particular, we focus on representing each pose by a compact feature, and finding a pose-similarity measure that allows personalized evaluation of motion quality.

2.2 State-of-the-Art

With the development of Microsoft Kinect and HPE methods, many studies on monitoring the rehabilitation process have emerged, as recently surveyed in [2, 12]. However, most of the studies rely on skeleton data estimated by Microsoft Kinect, which does not constitute an affordable and easy-to-use technology for the general public. Among the HPE-based approaches, we especially highlight two recent representative methods [7,10] that focus on a comparative evaluation of human motion.

In [7], the quality of rehabilitation exercises is assessed by a single number based on the extraction of effective spatial features. These features are extracted by a deep-learning architecture that combines a graph convolutional network and a transformer. However, this requires a non-trivial amount of training data, which contradicts our hypothesis that the motion features need to support personalization. Moreover, only a single number is returned as the assessment of each exercise, providing insufficient feedback/guidance for the user on how to improve their performance.

---

[1] https://ai.google.dev/edge/mediapipe/solutions/vision/pose_landmarker.

The recent approach in [10] is principally very similar to ours. The main idea is to analytically compare the similarity between the skeleton movements of a patient and a reference person based on DTW. However, the reference person is a professional physiotherapist, which again contradicts the personalization principle (the reference person should have the same abilities as the patient). Moreover, the cost function (inner distance) inside DTW only sums the Euclidean distances between the positions of corresponding joints, which is costly and doesn't reveal semantic differences in complex exercises.

There also exist several commercial solutions that deal with motion monitoring using plain RGB stream and HPE and have patents for exercise quality control, e.g., Kaia [8], Exer [4], or Google [6]. However, all these approaches are based on the universal model of similarity and do not support personalization, therefore their use is limited to the domains of fitness and preventive exercising.

2.3 Dataset and Evaluation Methodology

Our personalized solution consists of several principles. To assess the usefulness of each principle, we utilize a collection of RGB videos of rehabilitation exercises with both well- and poorly-performed motions. This section outlines the dataset and the methodology we use to evaluate the quality of motion monitoring.

Dataset In this study, we utilize the REHAB24-6 dataset[2], which contains expert-annotated multi-modal recordings of rehabilitation exercises. The dataset consists of 65 recordings of six types of exercises captured from several views, out of which we selected three distinct exercises to demonstrate the viability of our approach. Each exercise was chosen to target different muscle groups and movement patterns:

Exercise1 (arm abduction): lifting the arm sideways from the resting position next to the body at the side to a position above the head.
Exercise2 (leg lunge): step forward with one leg, lowering your hips until both knees are bent at about a 90° angle, then return to the starting position.
Exercise3 (basic squat).

Each actor performed at least 5 *correct* and 5 *incorrect* repetitions of each exercise under the supervision of the therapist, who also suggested various exercise mistakes to the actors. To capture comprehensive footage, two cameras were positioned at 90° angles to each other. Each exercise was recorded with a half-profile view, ensuring coverage from both camera angles. The video recordings were manually segmented into individual exercise repetitions. Altogether, 273 repetitions (31,318 frames at 30 FPS) were used from each camera.

For the purpose of this study, the physiotherapist enriched the REHAB24-6 dataset with detailed correctness assessments for each exercise repetition, assigning grades from 1 to 5, where 1 indicated an exceptionally executed movement

---
[2] https://doi.org/10.5281/zenodo.13305826.

and 5 indicated a subpar performance. In the following, repetitions graded 1, 2, and 3 are considered well-performed (referred to as "correct"), while those graded 4 and 5 are poorly performed ("incorrect").

Evaluation Methodology The main objective of exercise monitoring is distinguishing between correct and incorrect repetitions. To decide the correctness, we apply the straightforward approach of the 1-nearest neighbor (1NN) classifier, which compares the test query to a set of training repetitions with known quality. To ensure the evaluation scenario closely mirrors real-world conditions, each exercise is assessed individually.

Let us denote our set of actors as $patients = \{p_1, p_2, ..., p_m\}$ and introduce a database DB of exercise repetitions, which is internally structured according to the exercise type and the person:

$$DB[id_{exercise}][id_{person}] = \{r_1, r_2, ..., r_k\}.$$

By omitting the person identifier, we get all repetitions of a given exercise:

$$DB[id_{exercise}] = \bigcup_{id_{person} \in patients} DB[id_{exercise}][id_{person}].$$

Given the sparsity of our database and the limited number of correct and incorrect repetitions we expect in a real-life scenario, we evaluate our experiments using the leave-one-out principle. Specifically, each repetition is once used as *unknown/test* instance A, while the remaining $n - 1$ repetitions of the same exercise serve as *known/training* data. Test query A is compared against the training set to find the most similar repetition B by evaluating the 1NN query using DTW. Since we assume the repetitions follow similar patterns, we limit the largest temporal shift from the diagonal by extending DTW with a Sakoe-Chiba band [11] with a radius of 15 (considering the average length of repetition is 90 frames). We compare the correctness labels of both A and B and check whether they match. *Classification accuracy* is finally determined as a ratio of correctly classified repetitions.

In addition to classification accuracy, we also report the execution time of a bottleneck part of the similarity evaluation. Given that the lengths of sequences vary and all the assessed methods use DTW to align them, we consider the alignment time to be constant. The bottleneck is the computation of the *cost matrix* used within the DTW function, for which it is necessary to evaluate the pair-wise distances of all frame pairs of the two sequences being aligned. We, therefore, report the average time needed to calculate the cost matrix for a given pair of repetitions. Let us emphasize that the actual query processing time is at least one order of magnitude higher since the 1NN classifier needs to evaluate the distances between the query and all training repetitions. However, the number of training repetitions is one of the parameters of individual approaches, therefore we rather report the costs of a single comparison.

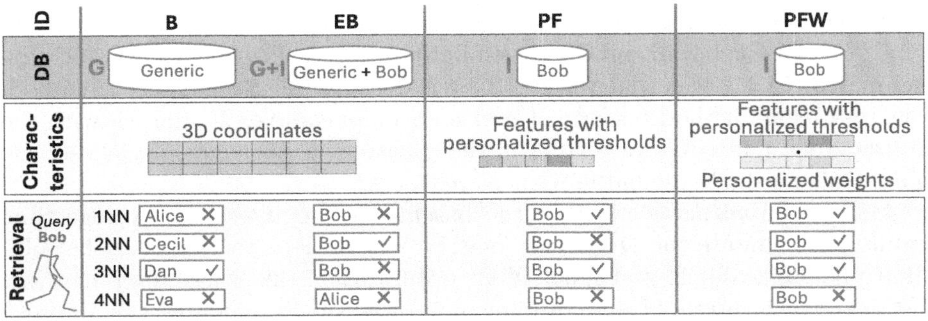

ID	B	EB	PF	PFW
DB	G: Generic	G+I: Generic + Bob	I: Bob	I: Bob
Characteristics	3D coordinates	3D coordinates	Features with personalized thresholds	Features with personalized thresholds / Personalized weights
Retrieval (Query Bob) 1NN	Alice ✗	Bob ✗	Bob ✓	Bob ✓
2NN	Cecil ✗	Bob ✓	Bob ✗	Bob ✓
3NN	Dan ✓	Bob ✗	Bob ✓	Bob ✓
4NN	Eva ✗	Alice ✗	Bob ✗	Bob ✗

Fig. 2. Similarity models for rehabilitation exercises. The first column corresponds to *B* and *EB* techniques that utilize raw 3D coordinates of joints, whereas methods in the second (*PF*) and third column (*PFW*) extract key characteristics inherent for a given exercise. The difference between *B* and *EB* lies merely in the content of the database. The visualization depicts the 4-nearest neighbors to a query repetition performed by Bob. The generic DB ("G") contains repetitions recorded by other patients (Alice, Cecil, Dan, and Eva); individual DB "I" contains only Bob's exercises. Symbols ✓ and ✗ stand for *correctly-* and *incorrectly*-performed exercises, respectively.

3 Towards the True Motion Twin

In this section, we gradually build the True Motion Twin technology for exercise correctness monitoring and experimentally evaluate the contribution of individual concepts we propose. The individual steps are illustrated in Fig. 2. First, we demonstrate that comparing an individual's motion against a generic database is ineffective and not personalized. Next, we explore the advantages of extracting exercise-specific features from raw 2D/3D coordinate skeleton data. A crucial aspect of our research involves developing a computationally efficient solution suitable for real-time execution on mobile devices.

3.1 Baseline

Let G be a generic database containing prerecorded correct and incorrect exercises performed by random individuals $\{p_1, p_2, ..., p_m\}$, who are different from the current exercising person p_{m+1}. The movements of person p_{m+1} are compared with the DB, which corresponds to standard exercise monitoring proposed e.g. in [10]. No data normalizations are applied apart from standard MediaPipe size normalization, which fits each skeleton within a cube of a uniform size.

Following the similarity model suggested in [10], we utilize 3D joint coordinates for skeleton representation. Given the fact that the coordinates of 33 joints are captured, each frame contains 99 float values that need to be compared to compute one cell of the DTW cost matrix. Similar to [10], we consider both Cosine and Euclidean distances for comparison of the 3D coordinates.

Evaluation In Table 1, we report results of the *baseline* approach (denoted by Type "B") for Euclidean and Cosine similarity metrics. The highest classification accuracy of 74.4 % was achieved for Exercise 1 using the $B+Euclidean$ method. The $B+Cosine$ method performed best on a more complex Exercise 3, with the accuracy of 71.1 %. We believe that the incorrect assessments are caused by individual-specific body and movement patterns.

Another important concern of the baseline techniques is the average time required to compute the DTW cost matrix. This time is too high for the solution to be applicable in real time. While optimizations like early stopping when costs exceed a threshold and pruning non-promising candidates can improve speed [15], these methods may hide important semantic differences in movement. Omitting certain joints would result in a smaller cost matrix, but with the risk of losing essential data needed to distinguish between correct and incorrect exercises.

3.2 Importance of Individualized Exercise Models

As discussed earlier, a key requirement for rehabilitation exercise monitoring is the personalization of exercise quality evaluation. This can be achieved in two ways: using individualized motion models, and using personalized motion similarity metrics. In this section, we analyze the importance of using individualized models, i.e., examples of correct and incorrect exercising performed by a particular patient. From an application standpoint, we can assume that correct exercise repetitions, as well as several examples of the particular patient's mistakes, are recorded during the initial consultation with the physiotherapist.

The expected advantages of this approach are threefold. First, by introducing training examples that share the same body type as the query motion, we allow the DTW algorithm to better identify the differences in movements. Second, the individual's correct references allow for different levels of correctness for each patient. Third, each patient is likely to make unique mistakes during exercises; the individualized database can identify these errors more effectively, providing more precise guidance on areas for improvement.

The individual patient's motion recordings can be either used alone as reference points for home exercise quality evaluation, or merged with a generic database of motion models introduced in the previous section. The first possibility is clearly more efficient, since fewer examples need to be compared to each home exercise. However, the generic dataset may provide some additional information that could increase the precision of the 1NN quality evaluation. Therefore, we experiment with both options – an individualized database I^{p_m}, which contains only the motions performed by patient p_m, and a combined database $G + I$ that contains all available examples apart from the query one.

Evaluation Similar to the previous experiment, we evaluate the similarity of motions using either Euclidean or Cosine distance on raw 3D joint coordinates. In Table 1, we report the results for the *extended baseline (EB)* approach, which was evaluated over the combined $G+I$ dataset. The EB approach achieved the highest accuracy of 98.9 % for Exercise1 – using $I+G$ instead of G increased the ratio of correctly classified repetitions by ca. 25 percentage points (37 for Cosine distance) for Exercise1. Similar trends were observed for Exercise2 and Exercise3, although the improvement was less pronounced.

The 1NN query in the EB approach rarely returned a repetition performed by a different individual, indicating a high specificity of each person's movements. Consequently, the results of the same experiment evaluated only on the individualized dataset I (not reported in Table 1 due to space limitations) were almost identical to the EB results. Following these findings, we limit the database to I in all following experiments, which leads to a significant speed-up of the whole query evaluation. However, there still remains the bottleneck of the cost matrix computation, which is prohibitively high when the motions are represented by the raw 3D coordinates.

3.3 Feature-Based Exercise Personalization

The 3D coordinate representation of skeleton poses not only requires costly computations but is also susceptible to noise. Even a small change of overall posture can lead to large differences in the DTW distance, even though exercising is still correct. Also, there can be some movement characteristics that are influenced by the surroundings (e.g., available space) and do not relate to exercise correctness; however, the coordinate-based evaluation will require them to match. Finally, the joint coordinates estimated by MediaPipe introduce a certain level of inaccuracy, especially the depth dimension, further complicating the assessment.

Exercise-Specific Characteristics To develop a more robust solution, we collaborated with a physiotherapist to define a set of distinctive *features* of each exercise. The concept of features is inspired by the work of [9], which utilized a set of 39 generic movement characteristics, often focusing on velocity. In our research, features capture the physiotherapist's criteria for a correct exercise and include measurements such as angles between joints, joint positions relative to a plane, and the relative distances between joints. For instance, one feature could measure the distance of the right knee from a plane defined by a line between the right foot and the right shoulder, indicating whether the patient maintains a straight knee during the exercise.

The exercise features can be based on either 2D or 3D coordinates produced by an HPE method (MediaPipe in our case). In our solution, we utilize a combination of 2D-based and 3D-based features. The reason is that 2D coordinates are typically much more precise, enhancing the system's overall accuracy, but some characteristics need to be considered in the full 3D body model.

Since our objective is to support online motion evaluation, the features are computed per frame, thus enabling frame-by-frame processing. Formally, let $f_k(r_j)$ denote the feature f_k extracted from repetition r_j. Let us emphasize that $f_k(r_j)$ is a time series of the feature values for individual frames of r_j, with $f_k(r_j)[i]$ being the feature's value of the i-th frame.

Binary Exercise Features The first step of our procedure involves extracting the absolute values of features, such as the absolute value of a given angle. In the next step, these absolute features are converted into binary ones using feature thresholds to determine whether a value falls within a specific range. Using the example of the position of the knee with respect to a plane, the binary feature better reflects whether the knee remains close enough to the plane or has shifted too much in the left/right direction. The thresholds are again *personalized* and derived from the data in the personal database.

To establish the tightest possible thresholds, we derive them using only the repetitions with the highest available grades. Specifically, let $best_e^p \subseteq DB[e][p]$ be the set of best repetitions of exercise e for person p. Each repetition $r_j \in best_e^p$ is processed separately. For each feature f_k, we extract the absolute value from each frame of r_j and create an interval $(min(f_k(r_j)), max(f_k(r_j)))$. These intervals are averaged across *all* repetitions r_j in $best_e^p$, yielding the personalized interval for each feature, denoted as:

$$\left(min_{avg}^k, max_{avg}^k\right) = \left(\underset{\forall r_j \in best_e^p}{avg}(min(f_k(r_j))), \underset{\forall r_j \in best_e^p}{avg}(max(f_k(r_j)))\right).$$

To further enhance adaptability to movement variability, we slightly expand these intervals by introducing a *grey zone* (g_1, g_2), typically a small percentage of the interval span. The absolute value of the feature f_k is then compared with $(min_{avg}^k - g_1, max_{avg}^k + g_2)$; if its value falls within this interval, it is flagged as true, otherwise as false. Notice that we specifically distinguish between g_1 and g_2, hinting that the grey zone added to both sides of intervals might differ.

When dealing with complex movements with thresholds spanning from 10° to 180°, we also found that subdividing features into finer intervals proves advantageous. This approach ensures the highest level of personalization but also increases computational time, as it involves extracting more features.

The distance between two frames represented by the binary features is computed by the Hamming distance across all the features. Intuitively, the distance corresponds to the number of flipped true/false values in fixed-length boolean feature vectors for the given two frames, which is very efficient to compute.

Evaluation The effectiveness of the *personalized feature (PF)* approach is evident in the classification accuracy summarized in Table 1. For Exercise2 and Exercise3, the accuracy was comparable to the *EB* methods. For Exercise1, the *PF* solution was slightly worse but still achieved a respectable 92.2 % accuracy. The probable reason for the discrepancy in accuracy across exercises is the difference in the number of extracted features. Exercise1 utilizes only 17 key characteristics, whereas Exercise2 and Exercise3, being more complex, employ a larger set of interval features. This results in a higher number of extracted features and, consequently, higher computational requirements. However, from the accuracy standpoint, these fine-grained features effectively distinguished between correct and incorrect repetitions.

A key advantage of features over previous solutions is the reduction in computational time. The features are represented as binary values, and we use far fewer of them compared to the number of 3D coordinates required in the baseline solution. Combined with a small individualized database I, this enables real-time evaluation on mobile devices. Moreover, the aforementioned optimizations of DTW [15] can also be applied to binary features, further increasing the efficiency gap between the baseline and feature-based personalized approach.

3.4 Automatic Weighting of Features

Extracting features from MediaPipe-based skeleton data proved feasible, though the evaluation indicated room for improvement in classification performance. The PF solution utilized a non-negligible number of features, treating them with equal importance in determining the distance between sequences. However, this approach may reduce the impact of more significant features, especially if an individual tends to make specific mistakes captured by only a few features, while the rest reflect the correct posture.

To address this, we introduce the concept of personalized *feature weight*. In practice, the weight is a decimal number that is derived for each individual once, utilizing both good and bad repetitions of a given exercise. Once determined, they remain constant until the therapist records a new set of repetitions.

The feature weights are used to compute individual entries in the DTW cost matrix. To compute the distance between the x-th frame from repetition r_j and y-th frame from repetition r_o (both are of the same person due to the personalized scenario), we follow a specific process. Each binary value of feature f_k, assuming $false = 0$ and $true = 1$, is multiplied by its corresponding weight. The difference between these weighted features is then calculated using the Manhattan distance. Specifically, the distances are summed for all the features f_k ($k \in [1, n]$):

$$\sum_{k=1}^{n} \left| w_{f_k} \cdot f_k(r_j)[x] - w_{f_k} \cdot f_k(r_o)[y] \right|.$$

The value of the weight w_{f_k} is derived from a modified version of the Kendall Tau correlation coefficient, as detailed below. The weight reflects the ability of feature f_k to distinguish between well- and poorly-performed exercises. The weighted features are compared using the Manhattan distance instead of Euclidean so that the derived weights are not further manipulated by the distance function.

Kendall Tau The objective of the Kendall-Tau coefficient is to decide how well two rankings of the same set of objects match. Our idea is to use a correct repetition of a given person as a query and search the database of repetitions of the same person to rank these repetitions with respect to that query. Then, we compute Kendall Tau between the retrieved ranking and the ranking of the physiotherapist (i.e., repetitions sorted by their grade from 1 to 5). The higher

Algorithm 1: Computation of Feature Weights

Data: $train_e^p$ repetitions, $best_e^p$ repetitions, binary $features$ of $\forall r_j \in train_e^p$

1 $kendalls \leftarrow [\ ][\ ]$
2 **foreach** $r_j \in best_e^p$ **do**
3   **foreach** $f_k \in features$ **do**
4          $dists \leftarrow [\ ]$
5   **foreach** $r_o \in train_e^p \setminus \{r_j\}$ **do**
6              $dists[o] \leftarrow d(f_k(r_j), f_k(r_o))$
7          $ordering \leftarrow$ order repetitions $r_o \in (train_e^p \setminus \{r_j\})$ based on $dists$
8          $pred \leftarrow$ grades of repetitions in order of $ordering$
9          $gt \leftarrow$ sorted grades of repetitions in ascending order
10         $kendalls[k][j] \leftarrow \tau(pred, gt)$
11 for each feature, average the particular $kendalls$ and **return** as $weights_e^p$

Kendall Tau, the more similar ranking to the physiotherapist's ground truth. Note that we needed to slightly modify the standard Kendall Tau computation to consider consecutive repetitions of the same grade as equally ranked.

Computation of Feature Weights The process of determining weights for our features begins with the precondition that *binary features* are already available for each training repetition $train_i$. The complete procedure is formalized in Algorithm 1; in the following paragraphs, we comment on the individual steps.

The computation of weights is performed separately for each feature f_k. For exercise e and person p, we use the best-graded repetitions $best_e^p$ to derive the weights. Let us have a repetition r_j selected from $best_e^p$. We compute similarity between r_j and $r_o \in train_e^p \setminus \{r_j\}$ for the feature f_k (Algorithm 1, lines 5–6). The repetitions are then sorted based on their similarity to r_j. The weight is determined using our *Kendall Tau* definition, which compares the predicted grading and ground truth grading provided by the physiotherapist (lines 8–10).

The algorithm produces a set of $weights_e^p$, represented as averaged Kendall Tau values. Since these values fall into interval $[-1, 1]$, we shift the interval by $+1$ to $[0, 2]$. To emphasize the discrepancy between well-separating and bad-separating features, we square each of $weights_e^p$. This results in personalized feature weights that better capture the movement specifics of each individual.

Evaluation As shown in Table 1, the *personalized features with weights (PFW)* improve classification accuracy as compared to the *PF* method. Across all exercises, we improved classification accuracy by 6.8 % on average, achieving over 95 % classification accuracy for each exercise. This substantial reduction in misclassification rates demonstrates the effectiveness of feature weighting, emphasizing each movement's most critical aspects. In comparison to the baseline method, our final solution increases the classification accuracy by 24–28 percentage points. Additionally, the *PFW* approach maintains fast frame similarity evaluations, confirming its practicality for real-time feedback and analysis.

Table 1. Effectiveness and efficiency results for all three exercises. The best results are highlighted in bold font. Column "DB" denotes whether the patient database is generic ("G") or individualized ("I"). Columns "Thr." and "Wgh." indicate whether the personalized feature thresholds and weights are applied (✓) or not (–), respectively.

Ex.	Extracted characteristics						Distance	Classif. accuracy	Process. time
	ID	Type	DB	Thr.	Wgh.	Size			
Exercise1	B	3D coords [10]	G	–	–	99xF	Euclidean	74.4 %	3.650 s
	B	3D coords [10]	G	–	–	99xF	Cosine	61.1 %	4.163 s
	EB	3D coords	G+I	–	–	99xF	Euclidean	**98.9 %**	3.650 s
	EB	3D coords	G+I	–	–	99xF	Cosine	**98.9 %**	4.163 s
	PF	Features	I	✓	–	17xB	Hamming	92.2 %	**0.037 s**
	PFW	Features	I	✓	✓	17xB	Manhattan	**98.9 %**	0.041 s
Exercise2	B	3D coords [10]	G	–	–	99xF	Euclidean	70.5 %	1.832 s
	B	3D coords [10]	G	–	–	99xF	Cosine	63.9 %	2.097 s
	EB	3D coords	G+I	–	–	99xF	Euclidean	95.3 %	1.832 s
	EB	3D coords	G+I	–	–	99xF	Cosine	94.2 %	2.097 s
	PF	Features	I	✓	–	44xB	Hamming	91.9 %	**0.019 s**
	PFW	Features	I	✓	✓	44xB	Manhattan	**98.8 %**	0.021 s
Exercise3	B	3D coords [10]	G	–	–	99xF	Euclidean	70.1 %	1.299 s
	B	3D coords [10]	G	–	–	99xF	Cosine	71.1 %	1.397 s
	EB	3D coords	G+I	–	–	99xF	Euclidean	91.8 %	1.299 s
	EB	3D coords	G+I	–	–	99xF	Cosine	92.8 %	1.397 s
	PF	Features	I	✓	–	22xB	Hamming	89.7 %	**0.013 s**
	PFW	Features	I	✓	✓	22xB	Manhattan	**94.9 %**	0.014 s

4 Conclusions

In this paper, we explored motion analysis in rehabilitation and outlined the requirements for real-time processing. We evaluated four different approaches to processing rehabilitation exercises. The first baseline approach simulates state-of-the-art methods for motion processing, lacking support for personalized motion monitoring. We demonstrated that introducing individualized motion models significantly improves the accuracy of detecting exercise correctness. However, the simple solution based on 3D coordinates proved to be computationally impractical for real-time applications due to its high processing demands.

In contrast, the proposed personalized features enhanced accuracy and reduced computational complexity. By focusing on key characteristics and converting them into binary features, we effectively addressed the variability in individual movement patterns. We also introduced an auto-tuning method to weight each feature, giving greater importance to specific parts of the movement for each individual. This enhances our ability to distinguish between correctly and incorrectly performed exercises compared to state-of-the-art methods.

In summary, our work highlights the importance of personalized and efficient solutions in the field of rehabilitation. By focusing on individual movement patterns and leveraging simplified data models, we developed an approach that not only meets the technical requirements for real-time processing but also provides significant practical benefits for both patients and healthcare providers. In future work, we plan to develop an automatic feature selection system for each exercise. Additionally, due to patients' convalescence over time, we need to automatically accommodate our features to their progress.

Acknowledgments. This work is co-financed from the state budget by the Technology Agency of the Czech Republic under the TREND Programme; project "VisioTherapy: Supporting physiotherapy treatments using computer-based movement analysis" (No. FW09020055).

References

1. Bazarevsky, V., Grishchenko, I., Raveendran, K., Zhu, T., Zhang, F., Grundmann, M.: Blazepose: On-device real-time body pose tracking. arXiv preprint arXiv:2006.10204 (2020)
2. Debnath, B., O'Brien, M., Yamaguchi, M., Behera, A.: A review of computer vision-based approaches for physical rehabilitation and assessment. Multimedia Syst. **28**, 209–239 (2022). https://doi.org/10.1007/s00530-021-00815-4
3. Dubey, S., Dixit, M.: A comprehensive survey on human pose estimation approaches. Multimedia Syst. 1–29 (2022). https://doi.org/10.1007/s00530-022-00980-0
4. Exer Labs Inc: Motion engine (2022). https://patents.google.com/patent/US20220327714A1
5. Gimigliano, F., Negrini, S., et al.: The World Health Organization: rehabilitation 2030: a call for action. Eur. J. Phys. Rehabil. Med. **53**(2), 155–168 (2017)
6. Google LLC: Physical training assistant system (2015). https://patents.google.com/patent/US9154739B1
7. He, T., Chen, Y., Wang, L., Cheng, H.: An expert-knowledge-based graph convolutional network for skeleton-based physical rehabilitation exercises assessment. IEEE Trans. Neural Syst. Rehabil. Eng. **32**, 1916–1925 (2024). https://doi.org/10.1109/TNSRE.2024.3400790
8. Kaia Health Software GmbH: Monitoring the performance of physical exercises (2022). https://patents.google.com/patent/US11282298B2

9. Müller, M., Röder, T.: Motion templates for automatic classification and retrieval of motion capture data. In: ACM SIGGRAPH/Eurographics Symposium on Computer Animation (SAC), pp. 137–146. Eurographics Association (2006)
10. Pereira, B., Cunha, B., Viana, P., Lopes, M., Melo, A.S.C., Sousa, A.S.P.: A machine learning app for monitoring physical therapy at home. Sensors **24**(1) (2024). https://doi.org/10.3390/s24010158
11. Sakoe, H., Chiba, S.: Dynamic programming algorithm optimization for spoken word recognition. IEEE Trans. Acoust. Speech Signal Process. **26**(1), 43–49 (1978). https://doi.org/10.1109/TASSP.1978.1163055
12. Sardari, S., Sharifzadeh, S., Daneshkhah, A., Nakisa, B., Loke, S.W., Palade, V., Duncan, M.J.: Artificial intelligence for skeleton-based physical rehabilitation action evaluation: a systematic review. Comput. Biol. Med. **158** (2023).https://doi.org/10.1016/j.compbiomed.2023.106835
13. Sedmidubsky, J., Elias, P., Budikova, P., Zezula, P.: Content-based management of human motion data: Survey and challenges. IEEE Access **9**, 64241–64255 (2021). https://doi.org/10.1109/ACCESS.2021.3075766
14. Senin, P.: Dynamic time warping algorithm review. Tech. Rep. Univ. Hawaii **855**(1–23), 40 (2008)
15. Silva, D.F., Giusti, R., Keogh, E., Batista, G.E.: Speeding up similarity search under dynamic time warping by pruning unpromising alignments. Data Min. Knowl. Disc. **32**, 988–1016 (2018). https://doi.org/10.1007/s10618-018-0557-y
16. Valcik, J., Sedmidubsky, J., Zezula, P.: Assessing similarity models for human-motion retrieval applications. Comput. Animation Virtual Worlds **27**(5), 484–500 (2016). https://doi.org/10.1002/cav.1674
17. Zhao, W., Reinthal, M.A., Espy, D.D., Luo, X.: Rule-based human motion tracking for rehabilitation exercises: realtime assessment, feedback, and guidance. IEEE Access **5**, 21382–21394 (2017). https://doi.org/10.1109/ACCESS.2017.2759801

Impact of the Neighborhood Parameter on Outlier Detection Algorithms

Félix Iglesias[1,2](✉), Conrado Martínez[3], and Tanja Zseby[1]

[1] Institute of Telecommunications, TU Wien, Vienna, Austria
tanja.zseby@tuwien.ac.at
[2] Le Studium Loire Valley Institute for Advanced Studies, Orléans, France
felix.iglesias@tuwien.ac.at
[3] Department of Computer Science, Universitat Politcnica de Catalunya, Catalunya, Spain
conrado@cs.upc.edu

Abstract. We study the impact and stability of the neighborhood parameter for a selection of popular outlier detection algorithms: kNN, LOF, ABOD, LoOP and SDO. We conduct a sensitivity analysis with data undergoing controlled changes related to: cardinality, dimensionality, global outliers ratio, local outliers ratio, layers of density, density differences between inliers and outliers, and zonification. Experiments reveal how each type of data variation affects algorithms differently in terms of accuracy and runtimes, and discloses the performance dependence on the neighborhood parameter. This serves not only to know how to select its value, but also for assessing accuracy robustness against common data phenomena, as well as algorithms' tolerance to adjustment variations. kNN, ABOD and SDO stand out, with kNN being the most accurate, ABOD the most suitable for both global and local outliers at the same time, and SDO the most stable in the parameterization. The findings of this work are key to understanding the intrinsic behavior of algorithms based on distance and density estimations, which remain the most efficient and reliable in anomaly detection applications.

Keywords: outlier detection · anomaly detection · k-neighborhood

1 Introduction

Outlier detection is a main branch of unsupervised learning. Its fundamental objective is to discriminate —by means of a label or a score— those dataset instances or elements[1] that differ from the rest due to their anomalous, noise, or novel nature. The applications of outlier detection are multiple and well known, addressing from cybersecurity and network intrusion detection to failure identification in industrial systems, medical anomaly diagnosis, fraud detection, process quality control, event monitoring in environmental data, etc. [15].

---

[1] For simplicity, hereafter we refer to the elements of a dataset simply as data points.

Outlier detection algorithms traditionally use a neighborhood parameter. This parameter, commonly referred to as k, defines the measures of distance or density between data points that are necessary to perform outlierness estimates. Although there are alternatives that do not require it, e.g., Isolation Forests [13], and Deep Learning approaches [16], methods relying on a neighborhood parameter are still the predominant and most accurate in general [2,5,8]. Hence, we study the dependence of the accuracy on the neighborhood parameter and the impact of its configuration on the different variations that the data may undergo. To this end, we perform sensitivity analyses on some representative algorithms: kNN, LOF, ABOD, LoOP and SDO. Our study is key to understand the intrinsic behavior of these algorithms, the effects of their adjustment, as well as their interrelation with common situations that occur in data.

The rest of the paper is organized as follows: In Sect. 2 we introduce studied algorithms and review previous work on the adjustment of the neighborhood parameter. Section 3 describes the experimental setup, while Sect. 4 discusses results. Finally, our conclusions are summarized in Sect. 5.

2 Outlier Detection Based on k-Neighborhood

Multiple methods perform distance or density estimations based on neighborhood parameters. We refer the reader to [2,5] for details and alternatives. In our study, we selected algorithms based on popularity, recent appearance and availability[2]. Moreover, we tried to make a representative choice of algorithms working under different principles. They are: **kNN** (k-Nearest Neighbors) [17], where the outlier score of a data point is the distance to its kth nearest neighbor; **LOF** (Local Outlier Factor) [1], which establishes the outlierness of a point as a relative measure of its density with regard to the densities of its k nearest neighbors (all of them); **LoOP** (Local Outlier Probabilities) [10], a variation of LOF, less sensitive to the choice of k and where scores conform to probabilistic interpretations [11]; **ABOD** (Angle-Based Outlier Detection) [12], in which the outlierness of a data point is a function of the variation of the angle between this point and any other pair of points in the dataset (in practice, implemented only on the k nearest neighbors); and **SDO** (Sparse Data Observers) [9], which is—unlike the rest on this list—based on models. The score of a point is measured as the average distance to the k (called x in [9]) closest points in the model.

The optimal k for classification has been broadly studied. Hall et al. [4] review multiple works on the subject, and claim that "application of this method [kNN classification] is inhibited by lack of knowledge about its properties, in particular, about the manner in which it is influenced by the value of k; and by the absence of techniques for empirical choice of k". There is a broad consensus that the selection of k directly depends on the data peculiarities. In this respect, Ghosh [3] emphasizes that "there is no theoretical guideline for choosing the optimum value of k. This optimum value depends on the specific data set, and

---

[2] Packages: https://github.com/yzhao062/pyod [19], https://github.com/vc1492a/PyNomaly and https://github.com/CN-TU/pysdoclust/tree/main.

it is to be estimated using the available training sample observations". In fact, the usual way to adjust k is by testing on the training data with some kind of resampling (e.g., bootstrap), later selecting the value of k that leads to the lowest misclassification error.

However, this assumes that the best k in the samples will also be the best k in the complete dataset. In [20] Zimek et al. prove how ensembles of classifiers operating on subsamples usually improve outlier detection classification. They show that a proportional increase in density between two areas implies the reduction of outlierness differences. This raises some doubt that the resampling approach is effective in finding the optimal k in a non-ensemble analysis setup.

Outlier detection can be taken as a specific case of classification (i.e., one-class classification or imbalance binary classification), where outlierness scores, when properly transformed, are interpreted as probabilities of being an outlier [11]. Up to this point, we have referred to kNN classifiers, but note that other methods will show different sensitivity in relation to the neighborhood parameter. For example, already in the original LOF proposal [1], Breunig et al. discuss the arbitrary, non-monotonic and data-dependent behavior of this parameter, which tends to be more unstable than in kNN. It is worth remembering that algorithms with minimum dependence on parameters (or low-sensitive parameters) are desirable and a main goal in new generation machine learning algorithms.

Perhaps [18] is the most recent work on finding the optimal k for outlier detection. Here, Yang et al. propose two methods—or one method with two variants: KFCS (score-based) and KFCR (rank-based)—and compare them with six previous alternatives. Our work does not attempt to find the optimal k, but rather to shed light on the sensitivity of popular outlier detection algorithms to variations in this parameter, as well as to understand what kind of change in the data implies the need for its readjustment. Hence, this paper is important to understand the nature of the algorithms studied, as well as how distance and density estimates change due to common variations of dataset characteristics.

3 Sensitivity Analysis

Algorithms are tested with sensitivity analysis on the neighborhood parameter for values: $k \in \{3, 5, 10, 20, 50\}$.[3] Other parameters take default values according to source libraries. Performances are evaluated with adjusted $P@n$ (adj_Patn), adjusted AP, and ROC-AUC—metrics described in [2]—plus the required processing time (runtime). To avoid non-deterministic deviations, each algorithm is run 5 times on each scenario and the average values are recorded. All experimental material is available for reuse and replication in the repository [6].

Data Scenarios. The baseline dataset consists of a spherical two-dimensional set of 1000 data points, with 970 inliers and 30 outliers. Inliers are confined within a sphere of radius $r_{in} = 0.1$, while outliers are distributed between spheres with

---

[3] We set a maximum value of 50 to ensure a lightweight total experimental time.

radii $r_{out_l} = 0.1$ and $r_{out_h} = 0.4$. We aim for uniform outlierness across both inlier and outlier zones, causing a sharp transition between them. This is achieved with the approach of the MDCgen data generation tool for radial clusters [7]. Taking the baseline dataset as a template, we construct 7 families of 8 datasets by varying one (or a few) parameters each time, see Table 1.

Table 1. Description of dataset configurations. $\{i \in \mathbb{N} : 0 \leq i \leq 7\}$.

Cardinality	car	from 1000 ($i = 0$) up to 99000 ($i = 7$) data points.
Dimensionality	dim	from 2 ($i = 0$) up to 345 ($i = 7$) dimensions.
Global outliers ratio	gor	from 1% ($i = 0$) to 22% ($i = 7$) of global outliers. The sphere with outliers expands proportionally to keep low density.
Local outliers ratio	lor	from 1% ($i = 0$) to 22% ($i = 7$) of local outliers. Here, we drop the sphere with outliers and turn random inliers into local outliers by removing the 15 closest neighbors and relocating them in dense areas.
Layers of density	lay	this scenario maintains a high number of outliers (47.4%) that are distributed in concentric layers of different density depending on the value of i. The total number of data points is 10000, to allow up to 9 layers of density with significant representation.
Inl.-out. density ratio	den	here, the radius of inliers increases and the radius of outliers decreases proportionally, causing the relative density to vary from approx. 129 ($i = 0$) to approx. 4 ($i = 7$).
Zonification	zon	the baseline dataset is cloned up to 8 times ($i = 7$), each time placing the new set in a distant space location (no cluster overlap), and each time expanding intra-distances by 50%.

datasets	points	dims.	outliers	r_{inl}	r_{out_l}	r_{out_h}	clusters	dens.layers
car	$1000(1+2i^2)$	2	3.0%	0.1	0.1	0.4	1	2
dim	1000	$2+i^3$	3.0%	0.1	0.1	0.4	1	2
gor	1000	2	$(1+3i)$%	0.1	0.1	$0.3 + 0.03i$	1	2
lor	1000	2	$(1+3i)$%	0.1	—	—	1	1
lay	10000	2	47.4%	0.1	0.1	1.0	1	$2+i$
den	1000	2	3.0%	$0.1+0.01i$	$0.1+0.01i$	$0.4-0.01i$	1	2
zon	$1000(1+i)$	2	3.0%	various	various	0.4	$1+i$	$2(1+i)$

4 Results

Figure 1 shows the results of the sensitivity experiments for the adjusted $P@n$. The numerical table behind the figure has been omitted due to the lack of space and also considering that, for these experiments, figures are more useful to efficiently and intuitively interpret results. For the same reason, we do not show results with adjusted AP and ROC-AUC, since they are all equivalent and the adjusted $P@n$ is the measurement that most clearly shows performance differences. All experimental results are available in our repository [6].

Algorithms. In our experiments, algorithms can be evaluated based on: (a) the tendency to show high accuracy, and (b) the variability of accuracy as a function

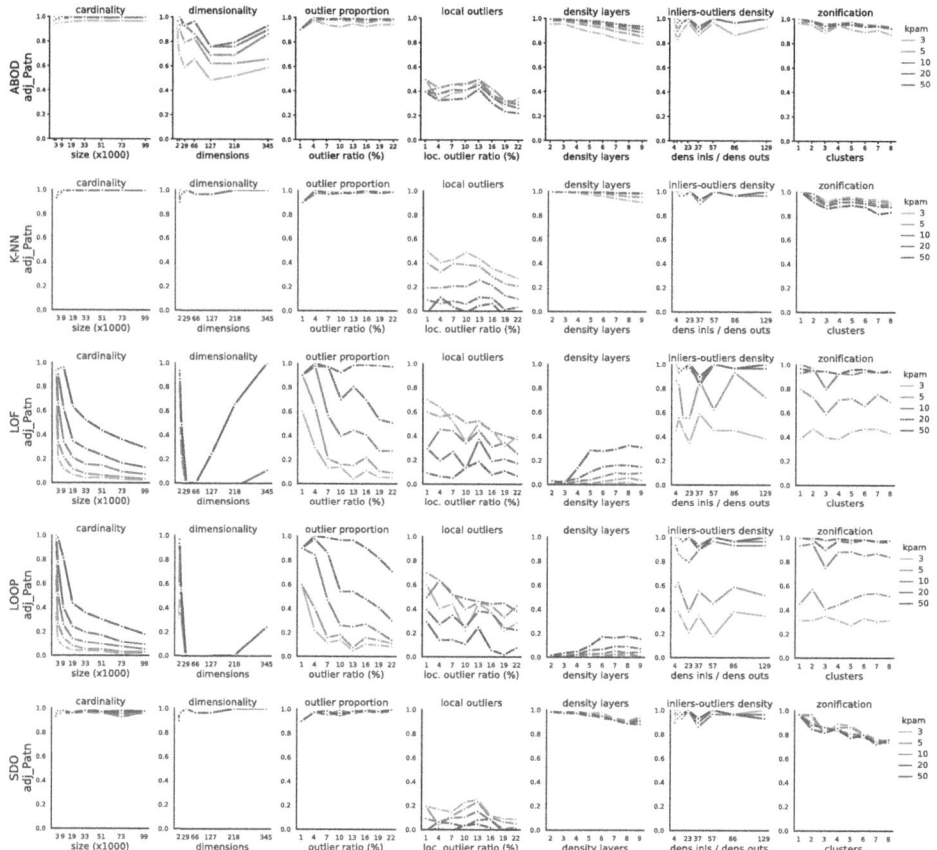

Fig. 1. Sensitivity Analysis Results (adjusted $P@n$).

of the neighborhood parameter (k or x, as appropriate, represented in Fig. 1 by kpam). In general, ABOD, kNN and SDO show better accuracy and are less sensitive to kpam variations. The difference between them can be summarized as follows: while kNN obtains the best accuracy, SDO is least affected by kpam variations. ABOD is better than kNN and SDO for detecting local outliers (loc), although dimensionality changes significantly affect it. SDO is the least suitable for local outliers. LOF and LoOP show similar behavior, both are highly sensitive to kpam. Note that these algorithms are particularly designed to detect local outliers; however, their performances are only slightly superior to ABOD for the local outlier scenario, and only when adjusted with the appropriate kpam.

Data Scenarios. In general, algorithms obtain better accuracy the higher the value of the neighborhood parameter (kpam in our experiments). This is consistent with the traditional way of selecting the best k (we call it k^*) in relation to the total number of points m: k^* should grow with m, but sublinearly, that is, $\lim_{m \to \infty} k^*(m) = \infty$ and $\lim_{m \to \infty} k^*(m)/m = 0$ [14].

The exception is the case of local outliers (lor), where k and accuracy are inversely proportional for kNN, ABOD and SDO, while in LOF and LoOP the optimal k^* values range from low to intermediate values: $3 \leq k^* \leq 10$ in our experiments. This is because a high value of k reinforces the consensus when identifying global outliers, but it has the effect of diluting the weight of the immediate environment in the case of local outliers. For instance, as k increases for kNN, the score of a local outlier in the middle of a high-density zone will tend to equal (or even be lower than) the score of an inlier in a low-density zone. LOF and LoOP, with a more refined approach for estimating locality, also suffer from a too high value for k, but in a different way. In LOF and LoOP, setting a large value for k might also include points that are not close enough, thus distorting the measurement. Therefore, estimating the local character of a point requires a certain number of references that are particular not only to the dataset, also to the point itself, a fact that complicates the adjustment of k. Also, kNN and SDO work better with low values of k in the zonification (zon) tests This is because each zone (or cluster) has a different density in both its inliers and outliers, hence outliers also acquire a local character, the key neighborhood of them being of a much larger scale than in lor.

Let us discuss now the remaining scenarios. An increase in the number of data points (car) does not require an adjustment of k. It does not affect accuracy either, except for LOF and LoOP, presumably due to the reduction in outlierness differences with higher density, as shown by Zimek et al. [20]. Dimensionality (dim) severely affects LOF and LoOP, while, for the rest, increasing dimensionality does not require adjustment of k, except in the case of ABOD. Changes in the proportion of global outliers (gor) or in the density differences between inliers and outliers (den) also neither significantly impacts accuracy nor needs adjustment of k in kNN, ABOD and SDO. Increasing density layers (lay) and increasing zonification (zon) decrease the accuracy[4] in kNN, ABOD and SDO, and benefit slightly from a readjustment of k in kNN and ABOD. Note that in lay and zon, LOF and LoOP with high k do not vary their accuracy when the number of distinct densities increases.

The arbitrary performances of LOF and LoOP are caused by their local nature. Thus, they emphasize residual random differences in areas of equal density, while they are blind to capture the global view. We recall that, except for the lor case, in our tests Ground Truths define outliers from a global perspective.

Runtimes. Figure 2 shows the runtimes of tests involving changes in dataset size, i.e., car and dim. While ABOD shows to be a solid option, with high accuracy, relatively low dependence on k and good for detecting both global and local outliers, it is the algorithm with highest computational cost, and where a higher value of k also means the strongest increase in this regard. Other algorithms show no significant impact on runtimes due to a higher value of k, even if dimensionality or cardinality increases. However, differences between LoOP and LOF when increasing data points (car) are striking, although this is caused by the implementations used, since both have theoretical equivalent complexity.

---

[4] We remind that zonification in our tests implies a larger number of distinct densities.

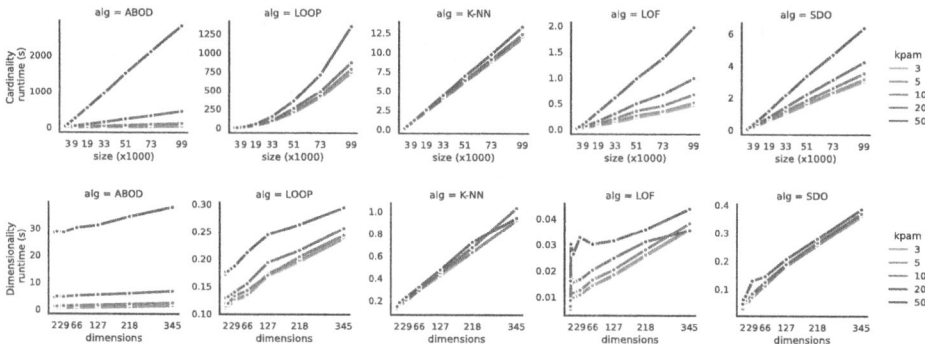

Fig. 2. Sensitivity Analysis Results (runtimes).

5 Conclusions

We have analyzed the performance of different outlier detection algorithms and their dependence on the neighborhood parameter (usually defined as k) when facing controlled changes in data properties. In summary, ABOD shows the best trade-off between global and local outlier detection, kNN the best overall accuracy, and SDO the least dependence on k. Performing similarly, LOF and LoOP have more variable accuracy and a high dependence on k. In terms of computational costs, ABOD is the algorithm most affected by large k.

Experiments reveal that, for global (or distance-based) algorithms, while a high k is generally advisable, a low k is better for detecting local outliers and when datasets show multiple zones of different density. In local (or density-based) algorithms, k is much more sensitive; a low k may fall short at defining the neighborhood, while a high k may exceed the ideal neighborhood and make the estimation be more global. In fact, results suggest that the optimal k is different for each point and requires consideration of both local and global aspects.

Our recommendation is to use two indices to express the outlierness of a point as a combination of global and local perspectives. For the global score, between kNN (lazy, instance-based) and SDO (eager, model-based), kNN is more accurate, but SDO is suitable when a data model is required and a more flexible definition of anomaly is desired (e.g., detecting novelties as anomalies, even if they appear as dense clusters). For the local score, LOF suffices. Enhanced interpretability is achieved if both scores are normalized to probabilistic estimates [11]. As a general rule, without previous knowledge, $k \in [5, 15]$ is a good default choice for LOF and kNN (or SDO) in the suggested combined approach.

Acknowledgments.. This work has been partially supported by funds from the MOTION Project (Project PID2020-112581GB-C21) of the Spanish Ministry of Science and Innovation MCIN/AEI/10.13039/501100011033, and the JUNON "Ambition Research Development Centre-Val de Loire" (ARD CVL) program.

References

1. Breunig, M.M., Kriegel, H.P., Ng, R.T., Sander, J.: LOF: Identifying density-based local outliers. In: ACM SIGMOD, pp. 93–104 (2000). https://doi.org/10.1145/335191.335388
2. Campos, G.O., Zimek, A., Sander, J., Campello, R.J., Micenková, B., Schubert, E., Assent, I., Houle, M.E.: On the evaluation of unsupervised outlier detection: measures, datasets, and an empirical study. Data Min. Knowl. Discov. **30**(4), 891–927 (2016). https://doi.org/10.1007/s10618-015-0444-8
3. Ghosh, A.K.: On optimum choice of k in nearest neighbor classification. Comput. Stat. Data Anal. **50**(11), 3113–3123 (2006). https://doi.org/10.1016/j.csda.2005.06.007
4. Hall, P., Park, B.U., Samworth, R.J.: Choice of neighbor order in nearest-neighbor classification. Ann. Stat. **36**(5), 2135–2152 (2008). https://doi.org/10.1214/07-AOS537
5. Han, S., Hu, X., Huang, H., Jiang, M., Zhao, Y.: Adbench: Anomaly detection benchmark. In: Koyejo, S., Mohamed, S., Agarwal, A., Belgrave, D., Cho, K., Oh, A. (eds.) NeurIPS 2022. Curran Assoc., Inc. (2022)
6. Iglesias, F.: Analysis of the neighborhood parameter on outlier detection algorithms—evaluation tests (2024). https://doi.org/10.48436/xvy1m-jwg83
7. Iglesias, F., Zseby, T., Ferreira, D., Zimek, A.: MDCGen: multidimensional dataset generator for clustering. J. Classif. **36**(3), 599–618 (2019)
8. Iglesias Vázquez, F., Hartl, A., Zseby, T., Zimek, A.: Anomaly detection in streaming data: A comparison and evaluation study. Expert Syst. Appl. **233**(C) (2023). https://doi.org/10.1016/j.eswa.2023.120994
9. Iglesias Vázquez, F., Zseby, T., Zimek, A.: Outlier detection based on low density models. In: IEEE International Conference on Data Mining Workshops, pp. 970–979 (2018)
10. Kriegel, H.P., Kröger, P., Schubert, E., Zimek, A.: LoOP: local outlier probabilities. In: ACM CIKM, pp. 1649–1652 (2009). https://doi.org/10.1145/1645953.1646195
11. Kriegel, H.P., Kröger, P., Schubert, E., Zimek, A.: Interpreting and unifying outlier scores. In: SIAM International Conference on Data Mining (SDM'11), pp. 13–24 (2011)
12. Kriegel, H.P., Schubert, M., Zimek, A.: Angle-based outlier detection in high-dimensional data. In: ACM SIGKDD KDD, pp. 444–452 (2008)
13. Liu, F.T., Ting, K.M., Zhou, Z.H.: Isolation forest. In: 2008 Eighth IEEE International Conference on Data Mining, pp. 413–422 (2008). https://doi.org/10.1109/ICDM.2008.17
14. Loftsgaarden, D.O., Quesenberry, C.P.: A nonparametric estimate of a multivariate density function. Ann. Math. Stat. **36**(3), 1049–1051 (1965). https://doi.org/10.1214/aoms/1177700079
15. Nassif, A.B., Talib, M.A., Nasir, Q., Dakalbab, F.M.: Machine learning for anomaly detection: a systematic review. IEEE Access **9**, 78658–78700 (2021). https://doi.org/10.1109/ACCESS.2021.3083060
16. Pang, G., Shen, C., Cao, L., Hengel, A.V.D.: Deep learning for anomaly detection: a review. ACM Comp. Surveys **54**(2) (2021). https://doi.org/10.1145/3439950
17. Ramaswamy, S., Rastogi, R., Shim, K.: Efficient algorithms for mining outliers from large data sets. SIGMOD Rec. **29**(2), 427–438 (2000)
18. Yang, J., Tan, X., Rahardja, S.: Outlier detection: How to select k for k-nearest-neighbors-based outlier detectors. Pattern Recogn. Lett. **174**(C), 112–117 (2023). https://doi.org/10.1016/j.patrec.2023.08.020

19. Zhao, Y., Nasrullah, Z., Li, Z.: PyOD: a python toolbox for scalable outlier detection. J. Mach. Learn. Res. **20**(96), 1–7 (2019). http://jmlr.org/papers/v20/19-011.html
20. Zimek, A., Gaudet, M., Campello, R.J., Sander, J.: Subsampling for efficient and effective unsupervised outlier detection ensembles. In: ACM SIGKDD KDD, pp. 428–436 (2013)

Optimizing CLIP Models for Image Retrieval with Maintained Joint-Embedding Alignment

Konstantin Schall, Kai Uwe Barthel, Nico Hezel, and Klaus Jung

Visual Computing Group, HTW Berlin, 12459 Berlin, Germany
{konstantin.schall,barthel,hezel,klaus.jung}@htw-berlin.de

Abstract. Contrastive Language and Image Pairing (CLIP), a transformative method in multimedia retrieval, typically trains two neural networks concurrently to generate joint embeddings for text and image pairs. However, when applied directly, these models often struggle to differentiate between visually distinct images that have similar captions, resulting in suboptimal performance for image-based similarity searches. This paper addresses the challenge of optimizing CLIP models for various image-based similarity search scenarios, while maintaining their effectiveness in text-based search tasks such as text-to-image retrieval and zero-shot classification. We propose and evaluate two novel methods aimed at refining the retrieval capabilities of CLIP without compromising the alignment between text and image embeddings. Through comprehensive experiments, we demonstrate that these methods enhance CLIP's performance on various benchmarks, including image retrieval, k-NN classification, and zero-shot text-based classification, while maintaining robustness in text-to-image retrieval using only one embedding per image.

Keywords: Multi-modal similarity search · Content-based image retrieval · Representations learning for general-purpose feature extraction

1 Introduction

Contrastive Language and Image Pairing (CLIP) [19] has emerged as one of the most influential developments in deep learning and similarity search in recent years. It introduces a robust joint-embedding model that excels in text-to-image similarity searches and has paved the way for pioneering work in image generation [20,21], large-scale vision foundation models [9,35,41] and other cross-modal tasks like text-to-video [28] or text-to-motion retrieval [13]. The prevalence of images captioned with text on the web and social media platforms has enabled more cost-effective and efficient collection of annotated images compared to traditional single-label datasets like ImageNet [22]. This approach has enabled the creation of datasets on an unprecedented scale, featuring up to 8 billion image-text pairs [39]. CLIP trains two neural networks concurrently: an image encoder and a text encoder, aiming to produce highly similar embeddings for corresponding image-text pairs. This training approach not only results in neural networks with more robust representations and improved generalization but also supports the development of architectures with a larger number of parameters [38].

However, since CLIP employs natural text supervision, the image encoders trained by CLIP tend to perform semantic similarity searches, which are suboptimal for instance-level content-based image retrieval scenarios. For example, one image might show a burger with chips while another displays a fast-food restaurant building; both could be captioned with "visiting a fast-food restaurant". Consequently, an image-based search using one of the images could incorrectly return the other, despite their visual differences. A method to mitigate this issue involves specifically fine-tuning CLIP-trained image encoders for general-purpose content-based image retrieval, as described in [24]. This fine-tuning allows the encoders to perform image-based similarity searches across diverse instance-based datasets from various domains, achieving high generalization without requiring domain-specific adaptations. However, this fine-tuning of only the image encoder can lead to a misalignment between the textual and visual embeddings, making the retrieval-optimized image embeddings less effective for text-to-image search scenarios. Applications that require good performance in both image-to-image and text-to-image retrieval would therefore have to maintain two separate vector sets.

This paper explores the development of joint-embedding models that perform well in both search modalities and presents two different methods to achieve a retrieval-optimized joint-embedding space. The first method involves a two-stage fine-tuning approach. Initially, the image encoder is optimized for retrieval, similar to the methods described in [24]. Subsequently, the text encoder is realigned with the newly optimized image embeddings. The second method uses several pseudo-captions per image to directly introduce the alignment process to the retrieval-optimization fine-tuning. We conduct extensive experiments to highlight the strengths and weaknesses of these approaches. Moreover, we evaluate our optimized CLIP models across various benchmarks, demonstrating that our methods not only improve the original model's performance on tasks such as image retrieval, k-NN classification, and text-based zero-shot classification but also maintain high quality in text-to-image retrieval and allow for these advantages with one embedding per image. Code and optimized model weights are accessible at: https://github.com/Visual-Computing/MCIP.

2 Related Work

2.1 Retrieval Optimizations

Deep neural networks have significantly outperformed traditional methods like SIFT [11] in nearest neighbor search, offering substantial improvements. Image embedding models now generate comprehensive representations, encapsulating information across the entire image within a single global vector. For retrieval tasks, transitioning from global average pooling to advanced techniques such as RMAC [27] or GeM [18] has been advantageous. Recent advancements have further refined these global image representations by incorporating local details from intermediate network layers [26]. Typically, these models are trained on the Google Landmarks v2 dataset [41] and evaluated against benchmarks like RParis and ROxford [17] or the Google Landmarks v2 test set. However, the effectiveness of these techniques in more generic retrieval contexts remains uncertain, as they have been primarily developed and tested within highly specific domains, where data homogeneity does not adequately represent the diversity of images encountered in general-purpose retrieval applications.

Instead of focusing on architectural modifications to improve fine-grained image detail aggregation, GPR [24] and UnED [37] take a different approach to increase image retrieval performance in applications with a wide range of image domains. Both use a large combination of publicly available datasets with a total of over 100,000 classes to re-train CLIP image encoders for the specific case of image retrieval with appropriate loss functions like the ArcMargin loss [4]. This method greatly increases the desired retrieval performance, however the fine-tuned image encoders can not be used for text-to-image similarity search anymore, since the joint-embedding space became misaligned through the fine-tuning process.

2.2 CLIP Optimizations

CLIP models have successfully been optimized in various aspects. The LAION organisation published an openly available dataset of 5 billion text-image pairs and trained a variety of large models to further improve the accuracy and performance of the original CLIP [25]. SigLIP [39] further improves CLIP's qualities by utilizing a novel Sigmoid based loss function for language and image pairing and training with the *webli* dataset [39], consisting of over 8 billion text-image pairs. Locked-Image-Tuning (LiT) [40] uses an image encoder pre-trained on a classification problem instead of a randomly initialised one and locks its weights during the CLIP training. This allows significantly faster training, since only the text-encoder has to be trained and yields better zero-shot classification performance, when the image-encoder was pre-trained with a large-scale proprietary dataset. RA-CLIP [34] proposes a superior training procedure by introducing a retrieval augmented image encoder. Prior to CLIP training, image embeddings are extracted with a pre-trained network like DinoV2 [15] to enhance the image embeddings of the trained image encoder with a retrieval augmentation module (RAM). However, none of these approaches try to optimize a CLIP models image encoder for effective image similarity search, while preserving the quality of text-based tasks like text-to-image retrieval and zero-shot classification.

3 Method

The goal of our proposed methods is to optimize CLIP-like joint-embedding models in a way that allows improved image-based similarity search with the image embeddings, while also enabling a high quality for text-to-image retrieval and text-based zero-shot classification of images. We investigate two different approaches to achieve this goal, which will be explained in detail in this chapter after some necessary background information has been provided.

3.1 Background

Contrastive Language-Image Pairing: A CLIP-like joint-embedding model usually consists of two main components: an image encoder and a text encoder and is trained with N paired instances $\{(x_i, t_i)\}_{i=1}^{N}$, where x_i is the image and t_i represents the associated text of the pair with index i. The image encoder, denoted as $f_x(x_i)$, transforms an input image x_i into an intermediate image embedding. This embedding is then projected into a shared embedding space in $\mathbb{R}^d$ with the image projector g_x, resulting in $u_i = g_x(f_x(x_i))$. Similarly, for the textual counterpart, the text encoder $f_t(t_i)$ processes an input text t_i to produce an intermediate text embedding y_i, which is then mapped to the same shared space in $\mathbb{R}^d$ by the text projector g_t, giving $v_i = g_t(f_t(t_i))$. The collections of embedding vectors are denoted by $U = [u_1, u_2, \cdots, u_N]$ and $V = [v_1, v_2, \cdots, v_N]$.

To optimize this joint embedding space, the model employs a contrastive loss function known as InfoNCE (Noise-Contrastive Estimation) [14]. This loss function aims to minimize the distance between the correct pairs of text and image embeddings in the shared space while maximizing the distance between mismatched pairs. It can be represented as

$$\mathcal{L}_{\text{InfoNCE}} = -\frac{1}{N}\sum_{i=1}^{N}\log\frac{\exp(u_i^\top v_i/\tau)}{\sum_{j=1}^{N}\exp(u_i^\top v_j/\tau)} - \frac{1}{N}\sum_{i=1}^{N}\log\frac{\exp(u_i^\top v_i/\tau)}{\sum_{j=1}^{N}\exp(u_j^\top v_i/\tau)} \quad (1)$$

where τ is a temperature parameter that scales the dot products of the exponentials and since the elements of U and V are L^2-normalized the dot products represent the cosine similarities of the embeddings.

General-Purpose Retrieval Fine-Tuning: General-Purpose Retrieval (GPR) [24] fine-tunes the image encoder $f_x(x_i)$ (the projector is discarded) with a combined single-label dataset consisting of ImageNet20k (without the ImageNet1k classes) [22], Google-Landmarks V2 [32], AliProducts [1], iNat21 [8] and VGGFaces2 [16], totaling in 22.6 million images from 168k classes. The authors introduced a benchmark specifically for general-purpose retrieval and k-NN classification formed from various domains and showed that fine-tuning a CLIP image encoder achieves the best retrieval results across multiple domains without the need of a specific domain adoption.

The average benchmark performance of the CLIP encoder can be increased by about 10% points when fine-tuning with this combined dataset and using the ArcMargin loss [4,24]. The mix of data from a wide range of domains with different class-distributions and class boundaries, e.g. narrow boundaries in the VGGFaces2 data and wide class boundaries in the ImageNet20k data parts, forces the previously semantically trained image-encoder to focus on visual features to distinguish the 168k classes of the training dataset while maintaining the generalization achieved with CLIP pre-training.

The ArcMargin loss is a modification used in deep learning to enhance the discriminative power of features extracted by neural networks, especially in narrow class-boundary tasks like face recognition. This approach involves adjusting the angle between the feature of each sample and the corresponding weights of the samples class-vector in the classification layer. The main objective is to introduce a margin between classes within the angular space, thereby improving the separation of classes. The ArcMargin loss can be expressed as follows:

$$\mathcal{L}_{\text{ArcMargin}} = -\frac{1}{N} \sum_{i=1}^{N} \log \frac{\exp(s \cdot \cos(\theta_{i,y_i} + m))}{\exp(s \cdot \cos(\theta_{i,y_i} + m)) + \sum_{j \neq y_i} \exp(s \cdot \cos(\theta_{i,j}))} \quad (2)$$

where N is the number of samples in the batch, y_i denotes the true-class index for the ith sample, and $\theta_{i,k}$ refers to the angle between the feature of the ith sample and the weight vector of the class at index k. The parameter s represents a scaling factor applied to the cosine of these angles, and m is the margin added to the angle of the true class. $\cos(\theta_{y_i} + m)$ effectively pushes the decision boundary away from the feature vectors of the true class, enhancing model robustness.

3.2 Two-Stage Fine-Tuning

The first proposed method includes two separate fine-tuning stages. The first stage optimizes the image-encoder in the same way as in the previously described GPR fine-tuning approach using the same combination of training data and the ArcMargin loss function. The fine-tuned image-encoder produces embeddings that capture more visual information of the images and performs better across a wide range of image-based similarity search tasks. Since [24] has shown that image-to-image retrieval performance benefits from discarding g_x, we do the same for a fair comparison.

Next, a re-alignment fine-tuning of the image and text embeddings is introduced with a image-text pair dataset (Fig. 1). Since the first stage only fine-tuned the image-encoder f_x, in the second step the image-projector g_x, the text-encoder f_t, and the text-projector g_t have to be re-trained to achieve an optimal alignment in the joint embedding space of $u \in \mathbb{R}^d$ and $v \in \mathbb{R}^d$. Similar to the work presented in LiT [40], f_x is locked during this training phase, i.e. the weights are not updated during back-propagation. This allows to extract all image embeddings prior to the re-alignment fine-tuning, which significantly reduces the memory load during training and enables the usage of large batch-sizes, which is beneficial for contrastive text-image training [19,39]. In this stage, the InfoNCE loss (Eq. 1) is used. This method is denoted as 2SFT (two-stage fine-tuning) in the upcoming experiment section. The first stage will be called GPR-FT (general-purpose retrieval fine-tuning) and the second stage Re-A (re-alignment).

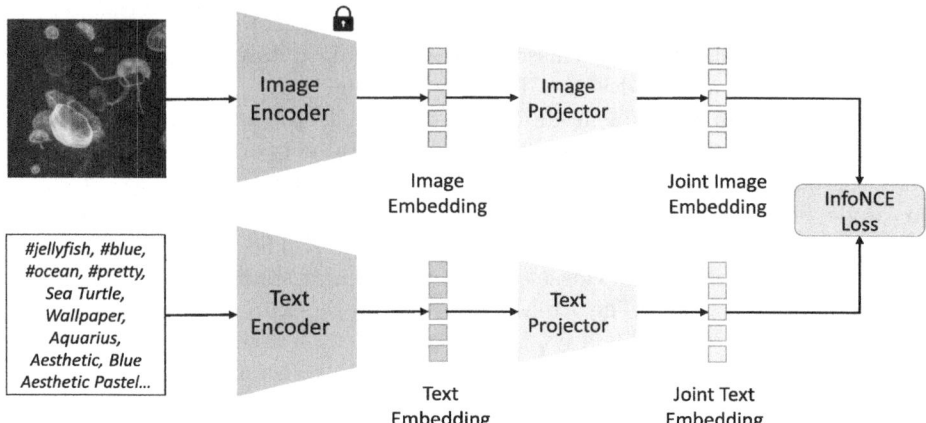

Fig. 1. Re-alignment fine-tuning: The fine-tuned image-encoder is locked and the other three parts of the model are trained to re-align the text embeddings with the image embeddings. An image-captions paired dataset is used in this step.

3.3 General-Purpose Retrieval Fine-Tuning with Multi-caption Image Pairings

In contrast to the first method, we now try to incorporate the text-image pairing approach directly to the general-purpose retrieval fine-tuning with the goal to create an joint-embedding space with optimal image-retrieval while maintaining the alignment of the image and text embeddings. However, since the fine-tuning with the ArcMargin loss is performed using a single-class per image dataset without any available text-captions, the CLIP loss cannot be used in this fine-tuning strategy directly.

Pseudo-Captioning: We propose to employ a pseudo-captioning pipeline that utilizes a large collection of image-text pairs, specifically the Conceptional 12M dataset (CC12M) [3]. First, we extract image-embeddings for all images of the GPR training dataset with a pre-trained CLIP model and second, we extract text-embeddings for all captions available in CC12M. Next, we perform a similarity search with each of the image-embeddings to retrieve the k-nearest-neighbors from the text-embedding collection with $k = 10$. Similar to previous work [33], we subsequently discard all retrieved captions with a cosine similarity smaller than 0.27, which was chosen empirically and only includes the most fitting captions. This operation leads to a single-class-multi-caption-image paired collection, consisting of N elements $\{(x_i, y_i, \{t_{i,1}, t_{i,2}, \ldots, t_{i,C_i}\})\}_{i=1}^{N}$, with x_i being the image, y_i being the true-class label and $\{t_{i,1}, t_{i,2}, \ldots, t_{i,C_i}\}$ being the collection of $C_i \leq k$ pseudo-captions. Figure 2 shows the generated captions for three images from one example class of the training dataset. It is apparent, that even though the pseudo-captions are noisy and repetitive, the combination of multiple captions manages to describe the image content and has low intra-class variance.

1. A handsome Tree Swallow in May. (0.33)
2. A tree swallow perching on a piece of wood (0.30)
3. A tree swallow keeps a watchful eye, Lido Beach Passive Nature Area (0.29)
4. Tree swallow peeking out from the hole of a nesting box (0.26)
5. The new neighbors (tree swallows) seem nice. (0.26)

1. A handsome Tree Swallow in May. (0.31)
2. Bird on a Wire: Tree Swallow (0.30)
3. A swallow high on a power wire, Bodie SHP, <PERSON> (0.29)
4. Adaptations of a Tree Swallow' 'Bird on a Wire: Tree Swallow (0.28)
5. May Jasper the tree swallow (0.26)

1.A handsome Tree Swallow in May. (0.34)
2.A tree swallow perching on a piece of wood (0.32)
3.A tree swallow keeps a watchful eye, Lido Beach Passive Nature Area (0.31)
4. Tree Swallow perched on a branch (0.29)
5. This Tree Swallow was nesting in a hollow in an aspen tree (0.28)
6. A tree swallow keeps a watchful eye, Lido Beach Passive Nature Area (0.26)

Fig. 2. Generated pseudo-captions for three images from one example class of the GPR-FT training set. Even though the texts are redundant, the combination of several captions succeeds in describing the content of the images.

ArcMargin Loss for Multi-Caption Image Pairs: To make use of the generated single-label-multi-caption-image tuples during the general-purpose image retrieval-finetuning, we propose an extension of the ArcMargin loss that utilizes the pseudo-captions to maintain joint-embedding alignment during the image-encoder optimization. The ArcMargin loss function as given in Eq. 2 can be extended to the Multi-Caption-ArcMargin (MCArcMargin) loss to enable training with multiple captions per image as follows:

$$\mathcal{L}_{\text{MCArcMargin}} = -\frac{1}{N}\sum_{i=1}^{N}\frac{1}{C_i}\sum_{c=1}^{C_i}\log\frac{\exp(s\cdot\cos(\theta_{i,t_{i,c}}+m))}{\exp(s\cdot\cos(\theta_{i,t_{i,c}}+m))+\sum_{t\neq t_{i,c}}\exp(s\cdot\cos(\theta_{i,t}))} \quad (3)$$

where N is the number of images in the batch, C_i is the number of captions generated for the image at index i, $\theta_{i,t_{i,c}}$ is the angle between the image-embedding u_i and the text-embedding $v_{i,c}$ (the cth generated caption of sample i) and $\theta_{i,t}$ refers to the angle between the feature of the ith sample and the text-embedding of caption t in the batch other than $t_{i,c}$. m and s are the same hyper-parameters as in the original ArcMargin loss function (Eq. 2). This loss function ensures that the angles between an image-embedding and the text-embeddings of the generated captions for this image are small and enforces a larger angle for the text-embeddings of other images in the batch.

Additionally, we also compute the standard ArcMargin loss for the class-labels, akin to the previous approach, setting the final loss objective to:

$$\mathcal{L} = \lambda_1\mathcal{L}_{\text{ArcMargin}} + \lambda_2\mathcal{L}_{\text{MCArcMargin}} \quad (4)$$

where λ_1 and λ_2 are two hyper-parameters that allow to modify the weight of the respective loss function.

In order to maintain the joint-embedding alignment, the image-encoder and the image-projector are simultaneously fine-tuned, while the text-encoder and

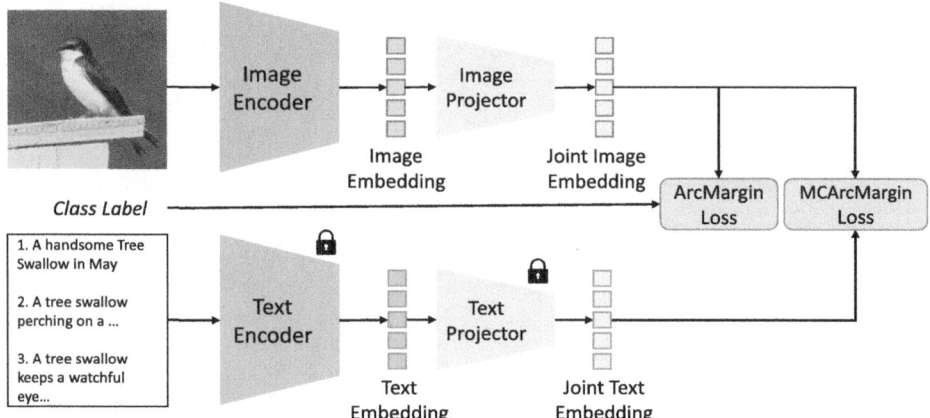

Fig. 3. Fine-tuning with multi-caption image pairing: This method utilizes the generated pseudo-captions to the train the image encoder and projector while keeping the text encoder and projector locked to maintain alignment of the joint-embedding space in a single fine-tuning run. In addition to the single-label training with the ArcMargin loss, the introduced multi-caption-ArcMargin loss is used.

text-projector remain locked. This method is further denoted as MCIP (Multi-Caption-Image-Pairing) and is depicted in Fig. 3.

4 Experiments

This section presents the experiments to evaluate our two proposed methods. We report results for the best learning rate, which varies for each of the evaluated models. For the MCIP approach, we use $\lambda_1 = 0.5$ and $\lambda_2 = 0.5$.

4.1 Evaluation

To evaluate the qualities in domain-independent similarity search, we introduce a benchmark that consists of several evaluation sets for each of the following tasks:

Content-Based Image Retrieval: Starting with content-based image retrieval, we evaluate our methods with GPR1200 [23], a challenging evaluation set, consisting of 1200 categories from various domains. It has been shown that the GPR1200 mean-average-precision (mAP) has a very high correlation to the average of nine different evaluation sets [24]. Additionally, we add the INSTRE dataset [31], an evaluation set with mixed domains, the CUB200 dataset [29] and the ROxford dataset [17] on the medium setting. We apply the respective evaluation protocol for each dataset and report mAP values for INSTRE, GPR1200 and ROxford and Recall@1 values for CUB200.

k-NN Classification: The main advantage of k-NN classification over linear-probing is that the model has not have to be fine-tuned to handle the specific class-distribution of an evaluation set and can directly be used by extracting the image-embeddings of the training split from a dataset. For each test-image, the k nearest neighbors are then simply retrieved from previously extracted training-data embeddings and the predicted class is the one with the most occurrences in the retrieved neighbors collection. We set $k = 21$ and test on ImageNet1k-Validation (IN-V) [22], ImageNet-Sketch (IN-S) [30], ImageNet-Adversarial (IN-A) [7], and FGVCAircraft (FGVCA) [12], each reflecting unique challenges in image classification. Accuracy is reported for each dataset.

Zero-Shot Classification: For zero-shot classification, which is crucial where labeled images are scarce, we parse class names into prompts and compute cosine similarities between image and text embeddings. We test on the same datasets used in k-NN classification plus ObjectNet (ON) [2], evaluating the classification accuracy.

Text-to-Image Retrieval: We evaluate the ability to match images to textual descriptions on Flickr30k (F30k) [36] and MS-COCO-Caption (COCO) [10], reporting Recall@5, since it is closer to real-world applications compared to Recall@1.

4.2 Results

We begin by examining the impact of each individual step in the two-stage fine-tuning approach (GPR-FT + Re-A), as well as the multi-caption-image-pairing (MCIP) fine-tuning, both with and without an additional re-alignment step. We compare these results to the baseline model, which is the original OpenAI ViT-L@336 [19], a vision transformer [5] that processes images with a resolution of 336 × 336 pixels. The first part of Table 1 shows the results for each of the introduced evaluation sets.

The model achieves the highest scores among the tasks of image-to-image retrieval and k-NN classification after the GPR fine-tuning, but, as expected, achieves the lowest evaluation scores across the text-based tasks in this state. Since only the image encoder was fine-tuned with a discarded image-projector, the image-embeddings are not aligned with the image-projector that has to be used to perform the text-based evaluations, which leads to significantly lower results. The re-alignment fine-tuning with the 12 million image-text pairs from the CC12M dataset leads to a strong improvement in all of the text-based tasks (Zero-Shot and T2I), however, the image-based evaluation scores are slightly reduced compared to the previous model state.

Next, the model was trained with the introduced multi-caption-image-pairing approach (MCIP). It achieves higher scores across all of the evaluation tasks compared to the two-stage-fine-tuning method after a single training run. However, it can be observed that the performance is further slightly increased with an additional re-alignment fine-tuning step. This is performed as described in Fig. 1 with the small adjustment of additionally locking the image projector, since this

Table 1. **I2I**: Image-to-Image retrieval (mAP/R@1), **k-NN** classification (accuracy), $k = 21$, **Zero-Shot** classification (accuracy), **T2I**: Text-to-Image retrieval (R@5). First section: Results for each individual step of our proposed methods when applied to the original OpenAI CLIP model (ViT-L@336). Second section: Results for the Datacomb trained ViT-L by LAION. Last section: Results for the SigLIP ViT-SO400M model.

Model	GPR-FT	MCIP	Re-A	I2I GPR1200	INSTRE	CUB200	ROxford	k-NN IN-V	IN-S	IN-A	FGVCA	Zero-shot IN-V	IN-S	IN-A	ON	T2I Flickr30k	MS-COCO	AVG
OpenAI				76.0	77.5	77.0	43.4	78.2	52.7	46.4	51.3	76.6	61.0	77.6	72.0	89.0	61.7	67.2
OpenAI	✓			87.1	85.0	88.2	67.3	84.2	60.5	61.2	57.8	74.5	60.8	76.2	74.5	84.9	57.2	68.4
OpenAI	✓	✓		86.7	84.4	87.9	66.4	83.8	60.4	60.9	57.1	77.2	61.5	77.9	74.6	89.8	64.6	73.8
OpenAI	✓		✓	87.0	84.8	88.0	67.4	84.1	60.4	61.0	57.7	77.0	61.1	78.3	74.5	90.1	65.6	74.1
OpenAI	✓	✓	✓	87.0	84.8	88.0	67.4	84.1	60.4	61.0	57.7	77.3	61.6	78.4	74.6	90.2	66.1	74.2
LAION				77.2	80.2	82.2	50.8	79.5	61.0	43.8	59.3	79.2	68.0	69.6	74.3	91.6	70.3	70.5
LAION		✓	✓	87.5	82.1	90.2	72.3	83.8	65.8	61.5	63.1	80.2	68.2	75.2	79.7	91.7	69.8	76.5
SigLIP				78.2	84.1	82.4	51.3	85.2	67.1	59.7	71.8	83.1	74.5	82.5	76.9	93.5	76.7	76.2
SigLIP		✓	✓	89.0	85.1	91.2	70.5	86.4	70.2	68.7	72.3	82.9	74.7	84.4	80.3	95.0	75.3	80.4

part was also trained in the MCIP step. These results show that the MCIP+Re-A fine-tuning leads to better results compared to the naive two-stage-fine-tuning, however the increase in average performance is rather low (73.8 vs 74.2). MCIP requires the generation of pseudo-captions and therefore has a slightly higher computational overhead.

In the second and third part of Table 1 we evaluated our methods with two additional models. First, a Vit-L variant that was trained with the Datacomp dataset [6], even though smaller in size, it achieves better evaluation results compared to a ViT-H, trained with LAION5B [25]. Second, the ViT-SO400M, a high-performance vision transformer where the number of parameters was optimized to 400 million [39]. This model was trained with the webli dataset [39] and with the Sigmoid-Language-Image-Pairing loss function (SigLIP) and reaches the highest scores in our evaluation in the baseline state. Since our MCIP method overall performed better than our 2SFT, we only report results for the multi-caption-image-pairing fine-tuning + Re-A. While our proposed method leads to a significant increase in both image-based similarity search tasks, text-based zero-shot classification was only improved with the ViT-L model for all datasets and only for some with the ViT-SO400M model. Interestingly, the text-to-image retrieval quality was increased on the Flickr30k dataset and slightly decreased on the MS-COCO evaluation set after fine-tuning with our MCIP approach. Nevertheless, this experiments show that our method is able to optimize several CLIP-like models for image-retrieval while also maintaining or even improving the performance in text-based tasks like zero-shot classification and text-to-

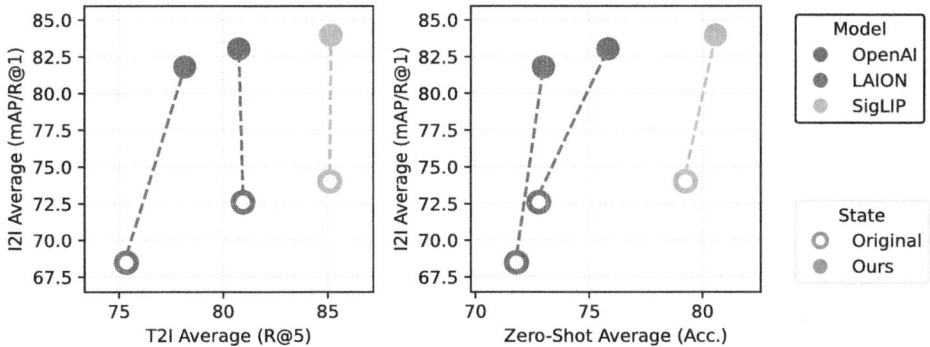

Fig. 4. Highlighted results for three different models fine-tuned with our proposed MCIP approach. In all three cases image-based similarity search could significantly be improved while maintaining text-to-image retrieval qualities and additionally improving zero-shot classification capabilities.

image retrieval. Figure 4 shows the results of image-to-image search compared to text-to-image search and zero-shot classification respectively.

On a single NVIDIA A100 GPU the OpenAI model needs about 20 h for the GPR-FT step and the additional Re-A fine-tuning takes 8 h to converge in the case of the first method. For the second method, the pseudo-caption generation takes 24 h, the MCIP fine-tuning 22 h and the Re-A step only 2 h. Once the pseudo-captions have been extracted, the MCIP approach therefore converges slightly faster compared to the two-stage fine-tuning.

5 Conclusions

In this paper, we addressed the challenge of optimizing CLIP models for image retrieval while preserving the joint-embedding alignment, which is crucial for effective text-based search tasks. Through a detailed exploration of two novel methodologies, we demonstrated how fine-tuning CLIP models can enhance image-based similarity search capabilities without compromising text-to-image retrieval and zero-shot classification performance.

Our first proposed method, the two-stage fine-tuning (2SFT) approach, involves initially fine-tuning the image encoder for improved retrieval, followed by a re-alignment of the text encoder. This method, while effective, showed a slight reduction in text-based tasks performance after re-alignment. The second method, which integrates multi-caption-image pairing (MCIP) during the retrieval optimization phase, resulted in slightly better overall performance. However, since this method involves the creation of pseudo-captions the additional computation time might not justify the small gain in retrieval quality over the two-stage fine-tuning approach.

We validated our methods on various models, including the original OpenAI CLIP ViT-L@336, a Datacomp trained ViT-L model, and the SigLIP ViT-SO400M model. Our experiments revealed significant improvements in image retrieval benchmarks like GPR1200 and INSTRE, while maintaining or even enhancing the performance in zero-shot classification and text-to-image retrieval tasks. The results confirm that our proposed methodologies not only optimize CLIP models for diverse image retrieval scenarios but also maintain the robustness required for text-based search tasks. This dual optimization is critical for real-world applications where both modalities are often required to work seamlessly together. The code and optimized model weights for our experiments are available at: https://github.com/Visual-Computing/MCIP.

References

1. Alibaba, T.: CVPR 2020 Aliproducts Challenge: Large-Scale Product Recognition (2020). https://tianchi.aliyun.com/competition/entrance/231780/information
2. Barbu, A., Mayo, D., Alverio, J., Luo, W., Wang, C., Gutfreund, D., Tenenbaum, J., Katz, B.: Objectnet: a large-scale bias-controlled dataset for pushing the limits of object recognition models. In: Wallach, H., Larochelle, H., Beygelzimer, A., d'Alché-Buc, F., Fox, E., Garnett, R. (eds.) Advances in Neural Information Processing Systems, vol. 32. Curran Associates, Inc. (2019)
3. Changpinyo, S., Sharma, P., Ding, N., Soricut, R.: Conceptual 12M: pushing web-scale image-text pre-training to recognize long-tail visual concepts. In: CVPR (2021)
4. Deng, J., Guo, J., Xue, N., Zafeiriou, S.: Arcface: additive angular margin loss for deep face recognition. In: CVPR (2019)
5. Dosovitskiy, A., Beyer, L., Kolesnikov, A., Weissenborn, D., Zhai, X., Unterthiner, T., Dehghani, M., Minderer, M., Heigold, G., Gelly, S., Uszkoreit, J., Houlsby, N.: An Image is Worth 16x16 Words: Transformers for Image Recognition at Scale. CoRR (2020)
6. Gadre, S.Y., Ilharco, G., Fang, A., Hayase, J., Smyrnis, G., Nguyen, T., Marten, R., Wortsman, M., Ghosh, D., Zhang, J., Orgad, E., Entezari, R., Daras, G., Pratt, S., Ramanujan, V., Bitton, Y., Marathe, K., Mussmann, S., Vencu, R., Cherti, M., Krishna, R., Koh, P.W., Saukh, O., Ratner, A., Song, S., Hajishirzi, H., Farhadi, A., Beaumont, R., Oh, S., Dimakis, A., Jitsev, J., Carmon, Y., Shankar, V., Schmidt, L.: Datacomp: In Search of the Next Generation of Multimodal Datasets (2023). https://arxiv.org/abs/2304.14108
7. Hendrycks, D., Zhao, K., Basart, S., Steinhardt, J., Song, D.: Natural Adversarial Examples (2019). https://doi.org/10.48550/ARXIV.1907.07174
8. Horn, G.V., Aodha, O.M., Song, Y., Cui, Y., Sun, C., Shepard, A., Adam, H., Perona, P., Belongie, S.J.: The inaturalist species classification and detection dataset. In: CVPR (2018)
9. Jia, C., Yang, Y., Xia, Y., Chen, Y., Parekh, Z., Pham, H., Le, Q.V., Sung, Y., Li, Z., Duerig, T.: Scaling up visual and vision-language representation learning with noisy text supervision. CoRR abs/2102.05918 (2021). https://arxiv.org/abs/2102.05918
10. Lin, T., Maire, M., Belongie, S.J., Bourdev, L.D., Girshick, R.B., Hays, J., Perona, P., Ramanan, D., Doll'a r, P., Zitnick, C.L.: Microsoft COCO: Common Objects in Context. CoRR abs/1405.0312 (2014). http://arxiv.org/abs/1405.0312
11. Lowe, D.G.: Distinctive image features from scale-invariant keypoints. Int. J. Comput. Vision **60**(2), 91–110 (2004). https://doi.org/10.1023/B:VISI.0000029664.99615.94

12. Maji, S., Kannala, J., Rahtu, E., Blaschko, M., Vedaldi, A.: Fine-grained visual classification of aircraft. Tech. Rep. (2013)
13. Messina, N., Sedmidubsky, J., Falchi, F., Rebok, T.: Text-to-motion retrieval: towards joint understanding of human motion data and natural language. In: Proceedings of the 46th International ACM SIGIR Conference on Research and Development in Information Retrieval, pp. 2420–2425. SIGIR '23, Association for Computing Machinery, New York, NY, USA (2023). https://doi.org/10.1145/3539618.3592069
14. van den Oord, A., Li, Y., Vinyals, O.: Representation Learning with Contrastive Predictive Coding (2019). https://arxiv.org/abs/1807.03748
15. Oquab, M., Darcet, T., Moutakanni, T., Vo, H., Szafraniec, M., Khalidov, V., Fernandez, P., Haziza, D., Massa, F., El-Nouby, A., Assran, M., Ballas, N., Galuba, W., Howes, R., Huang, P.Y., Li, S.W., Misra, I., Rabbat, M., Sharma, V., Synnaeve, G., Xu, H., Jegou, H., Mairal, J., Labatut, P., Joulin, A., Bojanowski, P.: Dinov2: Learning Robust Visual Features Without Supervision (2024). https://arxiv.org/abs/2304.07193
16. Parkhi, O.M., Vedaldi, A., Zisserman, A.: Deep face recognition. In: British Machine Vision Conference (2015)
17. Radenovic, F., Iscen, A., Tolias, G., Avrithis, Y., Chum, O.: Revisiting Oxford and Paris: large-scale image retrieval benchmarking. In: CVPR (2018)
18. Radenovic, F., Tolias, G., Chum, O.: Fine-tuning CNN image retrieval with no human annotation. IEEE Trans. Pattern Anal. Mach. Intell. (2019)
19. Radford, A., Kim, J.W., Hallacy, C., Ramesh, A., Goh, G., Agarwal, S., Sastry, G., Askell, A., Mishkin, P., Clark, J., Krueger, G., Sutskever, I.: Learning transferable visual models from natural language supervision. In: Meila, M., Zhang, T. (eds.) Proceedings of the 38th International Conference on Machine Learning, ICML 2021, 18–24 July 2021, Virtual Event. Proceedings of Machine Learning Research, vol. 139, pp. 8748–8763. PMLR (2021). http://proceedings.mlr.press/v139/radford21a.html
20. Ramesh, A., Pavlov, M., Goh, G., Gray, S., Voss, C., Radford, A., Chen, M., Sutskever, I.: Zero-Shot Text-to-Image Generation (2021)
21. Rombach, R., Blattmann, A., Lorenz, D., Esser, P., Ommer, B.: High-Resolution Image Synthesis with Latent Diffusion Models (2021)
22. Russakovsky, O., Deng, J., Su, H., Krause, J., Satheesh, S., Ma, S., Huang, Z., Karpathy, A., Khosla, A., Bernstein, M., Berg, A.C., Fei-Fei, L.: Imagenet large scale visual recognition challenge. Int. J. Comput. Vision (IJCV) (2015)
23. Schall, K., Barthel, K.U., Hezel, N., Jung, K.: Gpr1200: A benchmark for general-purpose content-based image retrieval. In: Multi-media Modeling: 28th International Conference, MMM 2022, Phu Quoc, Vietnam, June 6–10, 2022, Proceedings, Part I, pp. 205–216. Springer-Verlag, Berlin, Heidelberg (2022). https://doi.org/10.1007/978-3-030-98358-1_17
24. Schall, K., Barthel, K.U., Hezel, N., Jung, K.: Improving Image Encoders for General-Purpose Nearest Neighbor Search and Classification, pp. 57–66. ICMR '23, Association for Computing Machinery, New York, NY, USA (2023). https://doi.org/10.1145/3591106.3592266
25. Schuhmann, C., Beaumont, R., Vencu, R., Gordon, C., Wightman, R., Cherti, M., Coombes, T., Katta, A., Mullis, C., Wortsman, M., Schramowski, P., Kundurthy, S., Crowson, K., Schmidt, L., Kaczmarczyk, R., Jitsev, J.: Laion-5b: An Open Large-Scale Dataset for Training Next Generation Image-Text Models (2022). https://doi.org/10.48550/ARXIV.2210.08402

26. Shao, S., Chen, K., Karpur, A., Cui, Q., Araujo, A., Cao, B.: Global Features Are All You Need for Image Retrieval and Reranking (2023)
27. Tolias, G., Sicre, R., Jégou, H.: Particular object retrieval with integral max-pooling of CNN activations. In: ICLR (Poster) (2016)
28. Vadicamo, L., Arnold, R., Bailer, W., Carrara, F., Gurrin, C., Hezel, N., Li, X., Lokoc, J., Lubos, S., Ma, Z., Messina, N., Nguyen, T.N., Peska, L., Rossetto, L., Sauter, L., Schöffmann, K., Spiess, F., Tran, M.T., Vrochidis, S.: Evaluating performance and trends in interactive video retrieval: insights from the 12th VBS competition. IEEE Access **12**, 79342–79366 (2024). https://doi.org/10.1109/ACCESS.2024.3405638
29. Wah, C., Branson, S., Welinder, P., Perona, P., Belongie, S.: The Caltech-UCSD Birds-200-2011 dataset. Tech. Rep. CNS-TR-2011-001. California Institute of Technology (2011)
30. Wang, H., Ge, S., Lipton, Z., Xing, E.P.: Learning robust global representations by penalizing local predictive power. In: Advances in Neural Information Processing Systems, pp. 10506–10518 (2019)
31. Wang, S., Jiang, S.: Instre: A New Benchmark for Instance-Level Object Retrieval and Recognition. TOMM (2015)
32. Weyand, T., Araujo, A., Cao, B., Sim, J.: Google landmarks dataset v2—a large-scale benchmark for instance-level recognition and retrieval. In: CVPR (2020)
33. Wu, W., Timofeev, A., Chen, C., Zhang, B., Duan, K., Liu, S., Zheng, Y., Shlens, J., Du, X., Gan, Z., Yang, Y.: Mofi: Learning Image Representations from Noisy Entity Annotated Images (2024). https://arxiv.org/abs/2306.07952
34. Xie, C.W., Sun, S., Xiong, X., Zheng, Y., Zhao, D., Zhou, J.: Ra-clip: retrieval augmented contrastive language-image pre-training. In: Proceedings of the IEEE/CVF Conference on Computer Vision and Pattern Recognition (CVPR), pp. 19265–19274 (2023)
35. Yang, J., Li, C., Zhang, P., Xiao, B., Liu, C., Yuan, L., Gao, J.: Unified Contrastive Learning in Image-Text-Label Space (2022)
36. Young, P., Lai, A., Hodosh, M., Hockenmaier, J.: From image descriptions to visual denotations: new similarity metrics for semantic inference over event descriptions. Trans. Assoc. Comput. Linguist. **2**, 67–78 (2014). https://doi.org/10.1162/tacl_a_00166
37. Ypsilantis, N.A., Chen, K., Cao, B., Lipovský, M., Dogan-Schönberger, P., Makosa, G., Bluntschli, B., Seyedhosseini, M., Chum, O., Araujo, A.: Towards universal image embeddings: a large-scale dataset and challenge for generic image representations. In: Proceedings of the IEEE/CVF International Conference on Computer Vision (ICCV), pp. 11290–11301 (2023)
38. Zhai, X., Kolesnikov, A., Houlsby, N., Beyer, L.: Scaling Vision Transformers. CoRR (2021). https://arxiv.org/abs/2106.04560
39. Zhai, X., Mustafa, B., Kolesnikov, A., Beyer, L.: Sigmoid Loss for Language Image Pre-training (2023)
40. Zhai, X., Wang, X., Mustafa, B., Steiner, A., Keysers, D., Kolesnikov, A., Beyer, L.: Lit: Zero-Shot Transfer with Locked-Image Text Tuning (2022). https://arxiv.org/abs/2111.07991
41. Zhang, H., Zhang, P., Hu, X., Chen, Y.C., Li, L.H., Dai, X., Wang, L., Yuan, L., Hwang, J.N., Gao, J.: Glipv2: Unifying Localization and Vision-Language Understanding. arXiv preprint arXiv:2206.05836 (2022)

Bayesian Estimation Approaches for Local Intrinsic Dimensionality

Zaher Joukhadar[1]($\boxtimes$), Hanxun Huang[1], Sarah Monazam Erfani[1], Ricardo J. G. B. Campello[2], Michael E. Houle[1,3], and James Bailey[1]

[1] University of Melbourne, Melbourne, VIC, Australia
{zjoukhadar,curtis.huang1,sarah.erfani,baileyj}@unimelb.edu.au
[2] University of Southern Denmark, Odense, Denmark
campello@imada.sdu.dk
[3] New Jersey Institute of Technology, Newark, NJ, USA
michael.houle@njit.edu

Abstract. Local Intrinsic Dimensionality (LID) is a measure of data complexity in the vicinity of a query point. In this work, we address the problem of estimating LID from a Bayesian perspective by establishing a theoretical framework that derives the distribution of LID given a data sample. Using this framework, we develop new LID estimators that can outperform the Maximum Likelihood Estimator (MLE) in certain contexts. The framework also provides a convenient way to incorporate prior LID knowledge through informative priors. Additionally, we demonstrate how to aggregate multiple LID distributions in a Bayesian manner using logarithmic pooling. We conduct a variety of experiments, demonstrating that a Bayesian approach to LID is effective with a small number of nearest neighbors and when incorporating informative priors. We also show that in deep neural networks, MLE produces highly volatile LID estimates, whereas a Bayesian approach that incorporates prior LID information smoothes and reduces the variance of these estimates.

Keywords: Local intrinsic dimensionality · LID · Bayesian · Estimation

1 Introduction

Intrinsic dimensionality (ID) assesses the effective number of dimensions in which a data manifold resides, either globally (GID) or locally (LID). GID utilizes all data points to estimate the number of dimensions of the dataset as a whole, whereas LID assesses the ID in the vicinity of a designated query point.

When working with LID, estimation is a fundamental consideration and a key challenge. Several approaches have been proposed, with one popular method based on extreme value statistics being the Maximum Likelihood Estimator (MLE) [1,10,17]. It utilizes a k-nearest neighbor (k-NN) approach to estimate the local intrinsic dimensionality. However, one challenging aspect of using a

k-nearest neighbor approach is the choice of k. To maintain locality, a smaller neighborhood is desirable, but this would require a smaller sample size, which can result in higher variance. Conversely, when using a large value of k, the locality assumption is compromised, which may lead to less accurate estimation.

Another challenge is estimating LID for use in regularizing deep representations [11] when training a Deep Neural Network (DNN). Conventional estimators such as MLE can suffer from high volatility due to neighborhoods derived from small sized mini-batches [18,19], since obtaining the deep representations for the entire dataset is computationally expensive. Mini-batches may be augmented with representations from previous mini-batches, derived from earlier states of the model using a memory bank [24]; if so, samples that end up in the k-NN set might not be truly local, thereby increasing the variance of the LID estimates and rendering them less reliable.

In this work, we explore Bayesian approaches for estimating LID. Bayesian methods offer advantages over frequentist approaches, particularly with small sample sizes, making them more effective and 'local.' Additionally, Bayesian techniques can incorporate prior information in a principled manner, enabling more accurate LID re-estimation over time. Our main contributions are:

- We develop a Bayesian framework for LID estimation, exploring both informative and non-informative priors. This approach yields new point estimators and insights into existing ones.
- We demonstrate Bayesian aggregation of multiple LID distributions using logarithmic pooling.
- Experiments show that Bayesian estimation outperforms MLE with a small number of nearest neighbors, especially when incorporating informative priors, making it ideal for integration into machine learning training.

2 Related Work

Dimensional models are classified into two types: parametric and expansion models. Parametric models, such as Principal Component Analysis (PCA) [15], offer a global measure of data complexity by projecting data onto a lower-dimensional space. Some parametric models, such as Local PCA, Gaussian Mixture Models (GMMs), and Locally Linear Embedding (LLE), also capture local structures. Expansion models, on the other hand, measure intrinsic dimensionality locally around a query point by evaluating the growth rate of data points with increasing distance. These models are practical and efficient, requiring fewer data points. A key example is the expansion dimension, later generalized to continuous distance distributions by Houle [9]. This Local Intrinsic Dimensionality (LID) model was later shown to be strongly associated with Extreme-Value Theory (EVT) [10].

LID has proven useful in various applications. Ma et al. [18] showed that adversarial samples have distinct LID signatures, aiding in adversarial detection. Huang et al. [11] used LID to prevent dimensional collapse in self-supervised learning. LID has also improved training with noisy labels [19] and been effective in outlier detection [2], material pattern identification [25], and distinguishing zones in granular media [23].

Houle [10] provides a theoretical foundation for LID and its Maximum Likelihood Estimator (MLE) [1]. Many other estimators have been proposed, including an angle based approach whose formulation makes use of a beta distribution [22].

To the best of our knowledge, no previous work has explored incorporating prior knowledge of intrinsic dimensionality (ID) into ID estimation. In the context of DNN training, an ID calculated for a sample at one epoch could inform its estimation at another. A Bayesian framework offers a principled way to integrate this prior knowledge. Denti et al. [6] examined Two-NN estimation [7]—which also can be viewed as an estimator of LID [1]—from a Bayesian perspective using a conjugate Gamma prior distribution. In contrast, we consider both informative and non-informative priors with a different likelihood formulation.

3 Background on LID

Let F be a real-valued function that satisfies $F(0) = 0$ and is non-zero over some open interval containing $r \in \mathbb{R}$, $r \neq 0$.

Definition 1. ([10]) The *intrinsic dimensionality of F at r* is defined as follows, whenever the limit exists:

$$\text{IntrDim}_F(r) \triangleq \lim_{\epsilon \to 0} \frac{\ln\left(F((1+\epsilon)r)/F(r)\right)}{\ln((1+\epsilon)r/r)}.$$

Theorem 1. ([10]) *If F is continuously differentiable at r, then*

$$\text{ID}_F(r) \triangleq \frac{r \cdot F'(r)}{F(r)} = \text{IntrDim}_F(r).$$

We will be particularly interested in functions F that satisfy the conditions of a cumulative distribution function (CDF). Let $\mathbf{q}$ be a location of interest within a data domain $\mathcal{S}$ for which the distance measure $d : \mathcal{S} \times \mathcal{S} \to \mathbb{R}_{\geq 0}$ has been defined. To any generated sample $\mathbf{s} \in \mathcal{S}$, we associate the distance $\text{dist}(\mathbf{q}, \mathbf{s})$; in this way, a *global* distribution that produces the sample $\mathbf{s}$ can be said to induce the random value $\text{dist}(\mathbf{q}, \mathbf{s})$ from a *local* distribution of distances taken with respect to $\mathbf{q}$. The CDF $F(r)$ of the local distance distribution is simply the probability of the sample distance lying within a threshold r—that is, $F(r) \triangleq \Pr[\text{dist}(\mathbf{q}, \mathbf{s}) \leq r]$.

To characterize the local intrinsic dimensionality in the vicinity of location $\mathbf{q}$, we consider the limit of $\text{ID}_F(r)$ as the distance r tends to 0. Regardless of whether F satisfies the conditions of a CDF, we denote this limit by

$$\text{ID}_F^* \triangleq \lim_{r \to 0^+} \text{ID}_F(r).$$

Henceforth, when we refer to the local intrinsic dimensionality (LID) of a function F, or of a point $\mathbf{q}$ whose induced distance distribution has F as its CDF, we will take 'LID' to mean the quantity ID_F^*.

Work in [1,10] has shown how to formulate likelihoods for (distance) distributions where LID is treated as a parameter of the distribution: in [10], the LID representation theorem is formulated, and in [1], the CDF of the distance distribution is shown to correspond to a reciprocal transformation $X = \frac{1}{Y}$, where Y is Pareto distributed. This transformed distribution is known under various names, including inverse Pareto distribution, reverse Pareto distribution and power function distribution:

$$F(r|r \leq w) = \left(\frac{r}{w}\right)^{\text{ID}_F^*},$$

where w is the upper bound or threshold for the distance r. Under suitable conditions as $r, w \to 0$, we have

$$F'(r|r \leq w) = \frac{\text{ID}_F^*}{w}\left(\frac{r}{w}\right)^{\text{ID}_F^* - 1}. \tag{1}$$

In the presentation that follows, we may write $F(r)$ instead of $F(r|r \leq w)$ for the CDF, or $F'(r)$ instead of $F'(r|r \leq w)$ for the associated PDF. When the distributional CDF is understood, we will denote its theoretical LID value by ID^* and its estimates by $\widehat{\text{ID}^*}$. The symbols X and r may be used interchangeably to represent distances measured from a reference location $\mathbf{q}$.

Using maximum likelihood estimation (MLE) techniques, for a data sample of k distances $\{X_1, \ldots, X_k = w\}$ to a reference location, one can derive the well-known Hill estimator [1,8,17] from the density function in Eq. 1:

$$\widehat{\text{ID}^*}_{\text{MLE}} = -\frac{k}{\sum_{i=1}^{k} \ln \frac{X_i}{w}}. \tag{2}$$

4 A Bayesian Estimation Approach for LID

In the following analysis, we will treat ID^* as a random variable and w as fixed. Bayesian estimation of the ID^* parameter is based on the use of Bayes' rule:

$$\Pr(\text{ID}^* | D) = \frac{\Pr(D | \text{ID}^*) \Pr(\text{ID}^*)}{\Pr(D)}.$$

- $\Pr(\text{ID}^*)$: a prior distribution for the ID^* parameter, either informative (based on previous information or expert knowledge) or non-informative (minimizes assumptions or when no prior information is available).
- $\Pr(D)$: the probability of the data, which acts as a normalizing constant.
- $\Pr(\text{ID}^* | D)$: the posterior distribution of LID, given the data. The posterior encapsulates all information about the ID^* value after observing the data.
- $\Pr(D | \text{ID}^*)$: probability of observing the k-NN samples $D = \{X_1, \ldots, X_k = w\}$ given the value of ID^*. The likelihood function for one observation is $\ell(\text{ID}^* | r) \triangleq F'(r)$, and assuming independence of the samples D,

$$\Pr(D | \text{ID}^*) = \prod_{i=1}^{k} \ell(\text{ID}^* | X_i) = \left(\frac{\text{ID}^*}{w}\right)^k \exp\left[(\text{ID}^* - 1) \sum_{i=1}^{k} \ln \frac{X_i}{w}\right]. \tag{3}$$

Looking next at the prior, we analyze the situation when using a well-known uninformative prior for the distribution of ID*, the Jeffreys prior [14], which is equal to the square root of the Fisher information I_w. The asymptotic Fisher information was derived in [11] and is $I_w = \frac{1}{\text{ID}^{*2}}$, leading to the prior $\pi_J(\text{ID}^*) \propto \frac{1}{\text{ID}^*}$. The Jeffreys prior is designed to be non-informative and neutral across different parameterizations, making it robust when prior knowledge is limited. This prior naturally assigns higher probabilities to smaller values of ID*, subtly favoring lower values before considering the data. However, while the actual data distribution might deviate from the assumptions underlying the Jeffreys prior, the prior remains a sound choice due to its minimal assumptions and ability to remain objective, making it appropriate for our analysis.

$$\pi_J(\text{ID}^*) \triangleq \sqrt{I_w(\text{ID}^*)} = \frac{1}{\text{ID}^*}. \tag{4}$$

The data probability $\Pr(D) = \int_0^\infty \Pr(D \mid \text{ID}^*) \Pr(\text{ID}^*) \, d\text{ID}^*$ is a normalizing factor to ensure that $\Pr(\text{ID}^* \mid D)$ integrates to 1. With $\Pr(\text{ID}^*) = \pi_J(\text{ID}^*)$, this is

$$p_J(D) \triangleq \int_0^\infty \left(\frac{\text{ID}^*}{w}\right)^k \exp\left[(\text{ID}^* - 1) \sum_{i=1}^k \ln \frac{X_i}{w}\right] \frac{1}{\text{ID}^*} \, d\text{ID}^*$$

$$= \frac{1}{w^\alpha} \int_0^\infty (\text{ID}^*)^{\alpha - 1} e^{-\beta(\text{ID}^* - 1)} \, d\text{ID}^*, \text{ where } \alpha = k \text{ and } \beta = -\sum_{i=1}^k \ln \frac{X_i}{w}$$

$$= \frac{e^\beta}{w^\alpha} \int_0^\infty (\text{ID}^*)^{\alpha - 1} e^{-\beta \text{ID}^*} \, d\text{ID}^*$$

$$= -\frac{e^\beta}{w^\alpha \beta^\alpha} \Big[\Gamma(\alpha, \beta\,\text{ID}^*)\Big]_0^\infty,$$

where $\Gamma(\cdot, \cdot)$ is the upper incomplete Gamma function

$$= \frac{e^\beta \Gamma(\alpha)}{w^\alpha \beta^\alpha}, \text{ where } \Gamma(\cdot) \text{ is the Gamma function.} \tag{5}$$

We now derive the posterior distribution for the choice of Jeffreys prior, using Bayes' rule, incorporating our derivations from (5), (4), and (3).

$$f_J(\text{ID}^* \mid D) \triangleq \frac{\Pr(D \mid \text{ID}^*) \pi_J(\text{ID}^*)}{p_J(D)}$$

$$= \frac{w^\alpha \beta^\alpha}{e^\beta \Gamma(\alpha)} \left(\frac{\text{ID}^*}{w}\right)^k \exp\left[(\text{ID}^* - 1) \sum_{i=1}^k \ln \frac{X_i}{w}\right] \frac{1}{\text{ID}^*}$$

$$= \frac{w^\alpha \beta^\alpha}{e^\beta \Gamma(\alpha)} \left(\frac{\text{ID}^*}{w}\right)^\alpha e^{-\beta(\text{ID}^* - 1)} \frac{1}{\text{ID}^*}$$

$$= \frac{\beta^\alpha}{\Gamma(\alpha)} (\text{ID}^*)^{\alpha - 1} e^{-\beta \text{ID}^*}.$$

Putting these steps together yields the following theorem.

Theorem 2. *Suppose we have a data sample $D = \{X_1, \ldots, X_k\}$ and are using the Jeffreys prior for ID^*. Then the posterior is a $\mathrm{Gamma}(\alpha, \beta)$ distribution, where $\alpha = k$ is the shape parameter and $\beta = -\sum_{i=1}^{k} \ln \frac{X_i}{X_k}$ is the rate parameter.*

Table 1. Bayesian point estimators derived from the Jeffreys posterior distribution $f_J(\mathrm{ID}^* \mid D) = \mathrm{Gamma}(\alpha, \beta)$. ($\psi$ is the digamma function.)

Name	Loss function	Distribution statistic	Resulting $\widehat{\mathrm{ID}^*}_{\mathrm{Bayes}}$	Corresponds to
Quadratic loss	$(\mathrm{ID}^* - \widehat{\mathrm{ID}^*})^2$	Expected value	$\frac{\alpha}{\beta} = \frac{k}{-\sum_{i=1}^{k} \ln \frac{X_i}{w}}$	MLE (Eq. 2)
Absolute loss	$\lvert \mathrm{ID}^* - \widehat{\mathrm{ID}^*} \rvert$	Median	Numerically solve	–
Stein's loss [13]	$\frac{\widehat{\mathrm{ID}^*}}{\mathrm{ID}^*} - \ln \frac{\widehat{\mathrm{ID}^*}}{\mathrm{ID}^*} - 1$	Harmonic mean	$\frac{\alpha-1}{\beta} = \frac{k-1}{-\sum_{i=1}^{k} \ln \frac{X_i}{w}}$	KL divergence between distributions associated with $\widehat{\mathrm{ID}^*}$ and ID^* [3]
Brown's loss [4]	$(\ln \mathrm{ID}^* - \ln \widehat{\mathrm{ID}^*})^2$	Geometric mean	$\frac{\exp \psi(\alpha)}{\beta} = \frac{\exp \psi(k)}{-\sum_{i=1}^{k} \ln \frac{X_i}{w}}$	Squared Fisher Rao Metric [11]
Maximum a Posteriori	$\arg\max_{\mathrm{ID}^*} f(\mathrm{ID}^* \mid D)$	Mode	$\frac{\alpha-1}{\beta} = \frac{k-1}{-\sum_{i=1}^{k} \ln \frac{X_i}{w}}$	Same estimator as for Stein's loss

4.1 Bayesian Point Estimation

Theorem 2 can be used to obtain a point estimate of the LID value. This requires the specification of a loss function $L(\mathrm{ID}^*, \widehat{\mathrm{ID}^*})$, where ID^* is the true intrinsic dimensionality and $\widehat{\mathrm{ID}^*}$ is the point estimate. The point estimate $\widehat{\mathrm{ID}^*}_{\mathrm{Bayes}}$ is obtained by minimizing the expected loss, which is given by the following equation:

$$\widehat{\mathrm{ID}^*}_{\mathrm{Bayes}} = \arg\min_{\widehat{\mathrm{ID}^*}} \int_0^\infty L(\mathrm{ID}^*, \widehat{\mathrm{ID}^*}) f(\mathrm{ID}^* \mid D) \, d\mathrm{ID}^* \ . \qquad (6)$$

By adopting different loss functions and solving the integral in (6), various estimators of ID^* can be derived. The following theorem presents the Bayesian estimators obtained by applying different loss functions to a gamma distribution for ID^* (e.g., using the Jeffreys prior).

Theorem 3. *When the posterior distribution $\Pr(\mathrm{ID}^* \mid D)$ is a $\mathrm{Gamma}(\alpha, \beta)$ distribution, Bayesian point estimation applied to the loss functions listed in Table 1 lead to the estimators shown in the table. Each corresponds to a different summary statistic of the posterior.*

Proof. We illustrate with the estimator $\widehat{\mathrm{ID}^*}_{\mathrm{QL}}$ derived using the quadratic loss (QL) where $L(\mathrm{ID}^*, \widehat{\mathrm{ID}^*}) = (\mathrm{ID}^* - \widehat{\mathrm{ID}^*})^2$; the other cases follow similarly. To find the minimum for $\widehat{\mathrm{ID}^*}$, we differentiate Eq. 6 with respect to $\widehat{\mathrm{ID}^*}$, and set the result to be equal to zero.

$$0 = \int_0^\infty 2(\mathrm{ID}^* - \widehat{\mathrm{ID}^*}) f(\mathrm{ID}^* \mid D) \, \mathrm{d} \mathrm{ID}^*$$

$$0 = \widehat{\mathrm{ID}^*} \int_0^\infty f(\mathrm{ID}^* \mid D) \, \mathrm{d} \mathrm{ID}^* - \int_0^\infty \mathrm{ID}^* f(\mathrm{ID}^* \mid D) \, \mathrm{d} \mathrm{ID}^*$$

$$0 = \widehat{\mathrm{ID}^*} - \int_0^\infty \mathrm{ID}^* f(\mathrm{ID}^* \mid D) \, \mathrm{d} \mathrm{ID}^*$$

$$\widehat{\mathrm{ID}^*} = \int_0^\infty \mathrm{ID}^* f(\mathrm{ID}^* \mid D) \, \mathrm{d} \mathrm{ID}^* ,$$

which is the expected value of $f(\mathrm{ID}^* \mid D)$. For the gamma distribution, this corresponds to $\widehat{\mathrm{ID}^*}_{\mathrm{QL}} = \frac{\alpha}{\beta}$.

From Theorem 3 we observe that

- All estimators, except the non-analytic median, are an explicit function of the gamma distribution parameters α and β, with β always in the denominator.
- The differences among the various estimators are in their numerators, which are k, $k-1$, or $\exp(\psi(k))$. Noting that $k-1 < \exp(\psi(k)) < k$, we can consider these values as a 'degrees-of-freedom' adjustment of the effective sample size. The need for correction is seen with the MLE estimator (Eq. 2), which disregards the k-th distance measurement in its final term, $\ln \frac{X_k}{X_k}$.
- Using Stein's loss, the estimator relative to MLE is $\frac{\widehat{\mathrm{ID}^*}_{\mathrm{Stein}}}{\widehat{\mathrm{ID}^*}_{\mathrm{MLE}}} = \frac{k-1}{k}$. As k increases, this ratio approaches 1, reducing the difference between $\widehat{\mathrm{ID}^*}_{\mathrm{Stein}}$ and $\widehat{\mathrm{ID}^*}_{\mathrm{MLE}}$. Thus, our new estimators outperform MLE primarily when k is small, as shown in Sect. 5.1.

4.2 Gamma Prior

We now consider the posterior distribution when using a gamma distribution as the prior. The gamma prior is $\pi_G(\mathrm{ID}^*) \triangleq \frac{\beta_0^{\alpha_0}}{\Gamma(\alpha_0)} (\mathrm{ID}^*)^{\alpha_0-1} e^{-\beta_0 \mathrm{ID}^*}$. Derivation using Equation 3 leads to the likelihood function $f_G(D \mid \mathrm{ID}^*) \triangleq \left(\frac{\mathrm{ID}^*}{w}\right)^\alpha e^{-(\mathrm{ID}^*-1)\beta}$ and the normalization factor

$$p_G(D) = \int_0^\infty \left(\frac{\mathrm{ID}^*}{w}\right)^\alpha e^{-(\mathrm{ID}^*-1)\beta} \frac{\beta_0^{\alpha_0}}{\Gamma(\alpha_0)} (\mathrm{ID}^*)^{\alpha_0-1} e^{-\beta_0 \mathrm{ID}^*} \, \mathrm{d} \mathrm{ID}^* .$$

After some manipulations we arrive at

$$f_G(\mathrm{ID}^* \mid D) = \frac{(\beta_0 + \beta)^{\alpha_0 + \alpha}}{\Gamma(\alpha_0 + \alpha)} (\mathrm{ID}^*)^{\alpha_0 + \alpha - 1} e^{-(\beta_0 + \beta) \mathrm{ID}^*} . \quad (7)$$

Theorem 4. *Given a data sample $D = \{X_1, \ldots, X_k\}$, if the prior for ID^* is a $\mathrm{Gamma}(\alpha_0, \beta_0)$ distribution, then the posterior is a $\mathrm{Gamma}(\alpha_0 + \alpha, \beta_0 + \beta)$ distribution, where $\alpha = k$ is the number of data samples and $\beta = -\sum_{i=1}^{k} \ln \frac{X_i}{X_k}$.*

Corollary 1. *Using the quadratic loss, a Bayesian point estimate of ID^* with posterior distribution $\mathrm{Gamma}(2\alpha, \beta_0 + \beta_1)$ is the harmonic mean of the ID^* estimates obtained using $\mathrm{Gamma}(\alpha, \beta_0)$ and $\mathrm{Gamma}(\alpha, \beta_1)$ as priors.*

Proof. Let $\widehat{\mathrm{ID}^*}_{\mathrm{Gamma}}$, $\widehat{\mathrm{ID}^*_0}$, and $\widehat{\mathrm{ID}^*_1}$ be these respective estimates. From Theorems 3 and 4 we have that

$$\widehat{\mathrm{ID}^*}_{\mathrm{Gamma}} = \frac{2\alpha}{\beta_0 + \beta_1} = \frac{2}{\frac{\beta_0}{\alpha} + \frac{\beta_1}{\alpha}} = \frac{2}{\frac{1}{\widehat{\mathrm{ID}^*_0}} + \frac{1}{\widehat{\mathrm{ID}^*_1}}}.$$

4.3 Use of Power Prior

In cases where an informative prior incorporates historical data, such as distances to a point $\mathbf{q}$, a power prior [12] can balance the influence of this data on the posterior distribution. This Bayesian approach ensures that new LID estimates are not overly dominated by historical data. A power prior raises the likelihood to a power $0 \leq \tau \leq 1$, where a higher value of τ corresponds to higher trust in the historical data. The parameter τ acts as a temperature hyper-parameter.

Theorem 5. *When estimating ID^* on a data sample $D_0 = \{X_1, \ldots, X_k\}$, using a Jeffreys prior together with a power prior yields a $\mathrm{Gamma}(\tau\alpha_0, \tau\beta_0)$ as the posterior distribution, where $\alpha_0 = k$ is the shape parameter and $\beta_0 = -\sum_{i=1}^{k} \ln \frac{X_i}{X_k}$ is the rate parameter.*

This posterior can then be used as a prior for new data.

Corollary 2. *When estimating ID^* on a new data sample $D_1 = \{Y_1, \ldots, Y_{\alpha_1}\}$, using the posterior of Theorem 5 as a prior with D_1 yields $\mathrm{Gamma}(\tau\alpha_0 + \alpha_1, \tau\beta_0 + \beta_1)$ as the new posterior distribution, where α_1 is the size of the new data sample and $\beta_1 = -\sum_{i=1}^{\alpha_1} \ln \frac{Y_i}{Y_{\alpha_1}}$.*

4.4 Combinations of Posteriors

In this section, the results thus far show that starting from the non-informative Jeffreys prior, ID^* can be estimated through a chain of Gamma priors and posteriors. This approach is useful for combining LID estimates from multiple sample sets into a single estimate across all data. Bayesian methods like logarithmic and linear pooling can be used, with our focus here on logarithmic pooling. Given a set of densities $f_1, \ldots, f_N$ and a vector of positive weights $\mathbf{v} = v_1, \ldots, v_N$ summing to 1, logarithmic pooling is defined as follows.

Logarithmic pooling [20]: The logarithmic pooled density f_{lp} is

$$f_{\mathrm{lp}}(\mathrm{ID}^*) \triangleq t(\mathbf{v}) \prod_{i=1}^{N} (f_i(\mathrm{ID}^*))^{v_i}, \text{ where } t(\mathbf{v}) \triangleq \left(\int_0^\infty \prod_{i=1}^{N} (f_i(\mathrm{ID}^*))^{v_i} \, \mathrm{d}\mathrm{ID}^* \right)^{-1}.$$

Assuming gamma densities, we derive the following theorem.

Theorem 6. *Suppose we have $f_i = \text{Gamma}(\alpha_i, \beta_i)$ for $i = 1, \ldots, N$. Then the density f_{lp} under logarithmic pooling is a $\text{Gamma}(\sum_{i=1}^{N} v_i \alpha_i, \sum_{i=1}^{N} v_i \beta_i)$ density.*

4.5 1-Step Priors Versus Accumulative Priors

We propose two methods for incorporating prior distributions in ID* estimation, assuming a temporal scenario where LIDs are being iteratively estimated, such as during the training of a deep neural network (DNN). The first method uses a 1-step prior, relying on the most recent estimate $\widehat{\text{ID}^*}_{t-1}$ to calculate $\widehat{\text{ID}^*}_t$. The second method accumulates prior terms over time for calculating $\widehat{\text{ID}^*}_t$. Both methods use a temperature parameter τ to control prior influence.

For both methods we have $\beta_t = -\sum_{i=1}^{k} \ln(X_i^{(t)}/X_k^{(t)})$, where $\alpha_t = k$ for $t \geq 0$, and $D^{(t)} = \{X_1^{(t)}, \ldots, X_k^{(t)}\}$ are the distances to query point $\mathbf{q}$ at time step t. For $t = 0$, we use the Jeffreys prior, yielding $\widehat{\text{ID}^*}_0 = \frac{\alpha_0}{\beta_0}$.

The 1-Step method (based on Corollary 2) is given by:

$$\alpha^{1S} = (\tau + 1)k, \quad \beta_t^{1S} = \tau \beta_{t-1} + \beta_t, \quad \widehat{\text{ID}^*}_t^{1S} = \alpha^{1S}/\beta_t^{1S}, \text{ for } t \geq 1. \quad (8)$$

The Accumulative method generalizes the 1-Step method. According to Theorem 5, given a current time step t, for each previous time step $0 \leq j \leq t$, a Jeffreys prior and power prior with temperature τ^{t-j} yield a $\text{Gamma}(\tau^{t-j}\alpha_j, \tau^{t-j}\beta_j)$ posterior. These are combined across time steps using equal-weighted logarithmic pooling to produce a single posterior. This process can be seen as a decayed exponential weighting of α_j and β_j backward in time. While the assumption of mutual independence for the sequential estimation of α_j and β_j may not always hold, the method is robust under the studied conditions. The method is given by:

$$\alpha_t^{\text{Acc}} = \sum_{j=0}^{t} \tau^{t-j} \alpha_j, \quad \beta_t^{\text{Acc}} = \sum_{j=0}^{t} \tau^{t-j} \beta_j, \quad \widehat{\text{ID}^*}_t^{\text{Acc}} = \alpha_t^{\text{Acc}}/\beta_t^{\text{Acc}}, \text{ for } t \geq 1. \quad (9)$$

5 Experimental Results

5.1 Benchmarking the New LID Estimators

To evaluate our new estimators, we use 17 popular synthetic datasets from [5,21]. We compare against MLE using Mean Squared Error (MSE) and its decomposition into Bias^2 and Variance. Table 2 shows that for $k = 10$, our estimators outperform MLE across all datasets, with the Maximum a Posteriori / Stein's loss estimator achieving the best results for most.

In Sect. 4.1, we saw that our new estimators modify the 'effective' sample size such that $k - 1 < \exp(\psi(k)) < k$. This generally decreases the estimated dimensionality, with reduced expected value and reduced variance, making the estimator less sensitive to fluctuations in the denominator. Compared to MLE, our estimators reduce both bias and variance for most datasets.

We generated nine variations of the M12 isotropic multivariate Gaussian dataset with intrinsic dimensions $d \in \{2, 3, 5, \ldots, 25, 30, 40\}$, and tested the estimators with $k \in \{5, 10, 20, 40\}$. Figure 1 shows the Normalized MSE (relative to MLE) for the three proposed estimators, calculated between true (d) and estimated ID. Our Bayesian estimators excel when k is small, likely due to the greater impact of reducing the numerator relative to MLE.

5.2 Using Priors in Training

We next address the challenge of estimating LID of deep representations by demonstrating the benefits of leveraging priors using the 1-Step and Accumulative methods. In both of Eqs. 8 and 9, t represents an epoch: for the LID estimate at a given point $\mathbf{q}$ at time t, the α and β parameters of the gamma prior are calculated from those of the previous epoch(s).

We compare against MLE as a baseline. For the sake of fair comparison, we use a memory bank of deep representations as the reference set for all methods. We use MLE estimates calculated using the entire dataset as the reference set as a 'pseudo-ground-truth' LID value $\mathrm{ID}^{*\mathrm{pGT}}$. (Results using the other estimators to calculate $\mathrm{ID}^{*\mathrm{pGT}}$ are not shown, since the performances followed the same trends as when MLE was used.) Note that $\mathrm{ID}^{*\mathrm{pGT}}$ is impractical to obtain for large datasets due to the need for performing model inference for all data at every optimization step. Pseudo-code for our approach is shown in Algorithm 1. The memory bank $\mathcal{M}$ is a fixed capacity first-in-first-out (FIFO) queue, which can store 20% of the training data. We use a standard ResNet-18 model and the last

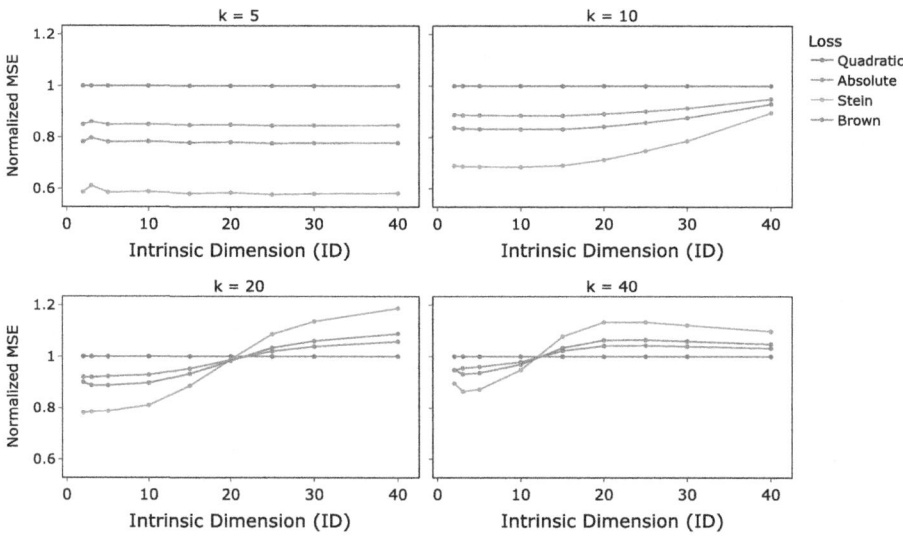

Fig. 1. Normalized MSE (with respect to the MLE estimator MSE) for the three proposed estimators using $k \in \{5, 10, 20, 40\}$ on nine variations of the M12 isotropic multivariate Gaussian dataset with intrinsic dimensions $d \in \{2, 3, 5, 10, 15, 20, 25, 30, 40\}$. Note that our quadratic loss estimator is equivalent to the baseline MLE.

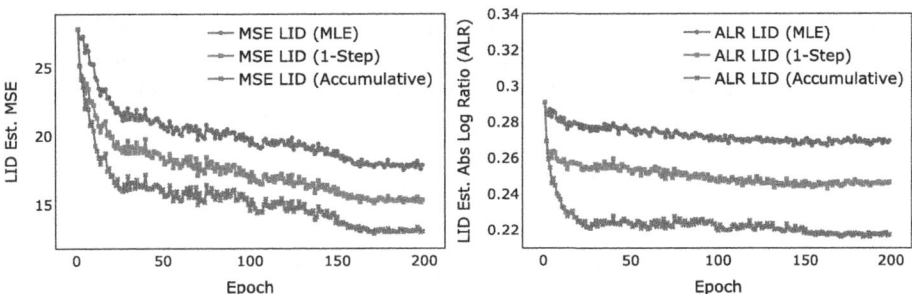

Fig. 2. MSE of LID estimates (left), and Absolute Log Ratio (ALR) $|\ln(\frac{\text{LID}^{\text{method}}}{\text{LID}^{*\text{pGT}}})|$ (right) of LID estimates in the penultimate layer (64 nodes), for the training set. LID (1-Step), LID (Accumulative), and LID (MLE) are plotted for ResNet-18 training on CIFAR-10 with $k = 20$ and $\tau = 1.0$. The final model achieves an accuracy of 92.52%. Training is done using cross entropy loss. The LID estimates are computed during training but **not** factored into the loss.

layer before the classification layer as the deep representation. We used CIFAR-10 and CIFAR-100 as datasets [16]. Since the performances were consistent for both sets, we show results only for CIFAR-10. We use the mean-squared error (MSE) between $\widehat{\text{ID}}_{\mathcal{B}}^{*}$ estimated with the memory bank and $\text{ID}_{\mathcal{B}}^{*\text{pGT}}$ calculated against the whole dataset for each mini-batch $\mathcal{B}$, as shown in Lines 7 and 9 in Algorithm 1. For all experiments, we use a batch size of 256. The MSE for each epoch is obtained by averaging over each mini-batch.

Algorithm 1 LID estimation in stochastic gradient descent algorithm
---
1: **Input:** Memory bank $\mathcal{M}$, total number of Epochs E, number of optimization steps per epoch J, a neural network g, training set $\{x_i\}_{i=1}^{N} \in \mathcal{D}$
2: **for** e to E **do**
3:     **for** J **times do**
4:         $\mathcal{B} = \text{sample}(\mathcal{D})$ ▷ Randomly sample a batch of data points without replacement
5:         $\mathcal{Z}_{\mathcal{B}} = \{g(x_b) \mid x_b \in \mathcal{B}\}$ ▷ Obtain representation $\mathcal{Z}_{\mathcal{B}}$ for the current batch
6:         $\text{push}(\mathcal{Z}_{\mathcal{B}}, \mathcal{M})$ ▷ Update memory bank $\mathcal{M}$ by pushing $\mathcal{Z}_{\mathcal{B}}$ into the queue
7:         $\widehat{\text{ID}}_{\mathcal{B}}^{*} = \text{estimation}(z_b, \mathcal{M}) \; \forall z_b \in \mathcal{Z}_{\mathcal{B}}$ ▷ Estimation using MLE, 1-Step and Accumulative methods
8:         $\mathcal{Z}_{\mathcal{D}} = \{g(x_i) \mid x_i \in \mathcal{D}\}$ ▷ Obtain representation $\mathcal{Z}_{\mathcal{D}}$ for entire training set
9:         $\text{ID}_{\mathcal{B}}^{*\text{pGT}} = \text{estimation}(z_b, \mathcal{Z}_{\mathcal{D}}) \; \forall z_b \in \mathcal{Z}_{\mathcal{B}}$ ▷ Estimation using MLE, 1-Step and Accumulative methods
10:         $\text{train}(g, \mathcal{B})$ ▷ training g with gradient descent for one optimization step
11:     **end for**
12: **end for**

Figure 2 shows MSE for the 3 methods during training of ResNet18 on CIFAR-10. Across all 3 methods, both Bayesian LID variants outperform MLE.

Analysis of LID Across Class Examples: We analyze LID estimates for samples from the CIFAR-10 Class 4 (Deer); other classes followed similar trends. We calculated LID estimates of each sample in this class at each epoch and plotted their mean and variance.

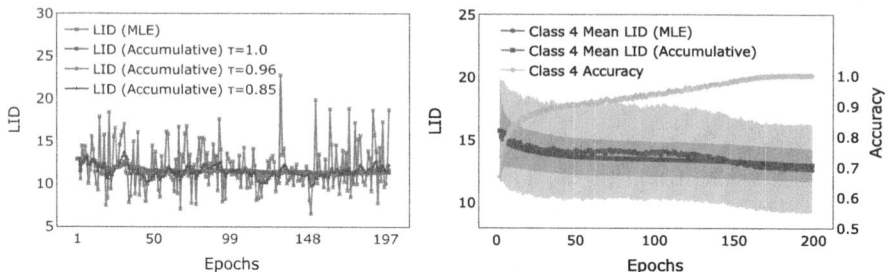

Fig. 3. *Left:* LID estimates using MLE and Accumulative estimators for a single example from Class 4 (Deer) of CIFAR-10 over epochs, from the same training run as Fig. 2. *Right:* Mean and variance of LID estimates for examples of Class 4 over epochs, along with the accuracy for that class. All examples are from the training set.

In Fig. 3 we see that the Accumulative method provides a more consistent and stable LID estimate compared to the MLE, which exhibits significant volatility over epochs and does not converge to a consistent LID value toward the end of training. This volatility is likely due to the neighbors being selected from a small reference subset (the mini-batch) with distance distributions very different from those of the closest neighbors from the whole dataset. Additionally, these mini-batches incorporate stale data from previous model epochs, due to the use of a memory bank of past representations.

The right side of Fig. 3 shows the mean and variance of LID estimates for all examples of Class 4 (Deer) across training epochs, alongside the accuracy for this class. Figure 4 shows the LID distribution for Class 4 at epoch 200 using both estimation methods. We see substantially higher variance for the MLE method, compared to the Accumulative method.

However, since we do not have a ground truth LID for this data (CIFAR-10 deep representations), we cannot verify bias reduction as we did with the synthetic data in Table 2. Nonetheless, the ability to maintain a mean estimate comparable to the MLE method while reducing variance is suggestive of an improvement in the reliability and consistency of LID estimation using the accumulative approach.

Table 2. MSE, Bias^2 (* indicates negative bias), Variance, and Avg. Est. ID for new estimators. Dataset size = 10,000, $k = 10$ nearest neighbors. d is the intrinsic dimension, m is the representational dimension. Best results in **boldface**

Dataset	d	m	Quad.(MLE) Avg est. ID	MSE Bias^2 Var	Abs. Loss Avg.est. ID	MSE Bias^2 Var	Max a-p.(Stein) Avg.est. ID	MSE Bias^2 Var	Brown Loss Avg.est. ID	MSE Bias^2 Var
M1: Sphere	10	11	13.48	42.47 12.09 30.38	13.03	37.59 9.18 28.40	12.13	**29.15 4.54 24.61**	12.81	35.34 7.89 27.45
M2: Affine space	3	5	4.20	4.55 1.45 3.11	4.06	4.04 1.13 2.91	3.78	**3.13 0.61 2.52**	3.99	3.80 0.99 2.81
M3: Fused figures	4	6	5.61	8.55 2.60 5.95	5.43	7.60 2.05 5.56	5.05	**5.92 1.11 4.82**	5.34	7.16 1.80 5.37
M4: Nonlinear manifold	4	8	5.56	7.65 2.43 5.21	5.38	6.76 1.91 4.87	5.00	**5.23 1.00 4.22**	5.28	6.36 1.64 4.71
M5b: 2D Helix	2	3	2.88	2.23 0.77 1.45	2.79	1.97 0.62 1.36	2.59	**1.53 0.35 1.18**	2.74	1.86 0.55 1.31
M6: Nonlinear manifold	6	36	8.57	19.35 6.61 12.74	8.29	17.14 5.25 11.91	7.71	**13.26 2.94 10.32**	8.15	16.11 4.62 11.50
M7: Swiss-Roll	2	3	2.82	2.03 0.68 1.35	2.73	1.79 0.53 1.26	2.54	**1.38 0.29 1.09**	2.68	1.68 0.46 1.22
M8: Curved Manifold	12	72	19.52	129.33 56.60 72.72	18.88	115.28 47.29 67.98	17.57	**89.94 31.05 58.90**	18.56	108.68 42.99 65.69
M9: Affine Space	20	20	21.86	81.20 3.45 77.76	21.13	73.97 1.28 72.70	19.67	**63.09 0.11* 62.99**	20.77	70.84 0.60 70.25
M10a: 10D hypercube	10	11	13.11	38.19 9.64 28.55	12.67	33.82 7.13 26.69	11.79	**26.35 3.22 23.13**	12.46	31.82 6.03 25.79
M10b: 17D Hypercube	17	18	20.12	75.50 9.72 65.78	19.45	67.50 6.01 61.49	18.11	**54.50 1.23 53.28**	19.12	63.92 4.50 59.42
M10c: 24D Hypercube	24	25	26.45	118.58 5.99 112.60	25.57	107.72 2.47 105.26	23.80	**91.23 0.04* 91.20**	25.14	103.00 1.29 101.71
M11: Moebius band	2	3	2.83	2.07 0.69 1.36	2.75	1.84 0.54 1.28	2.56	**1.42 0.31 1.11**	2.70	1.73 0.49 1.23
M12: Iso. MV Gaussian	20	20	24.28	113.33 18.33 95.02	23.48	100.90 12.09 88.82	21.85	**80.39 3.43 76.96**	23.08	95.30 9.47 85.83
M13a: 2D S-curve	2	3	2.84	2.07 0.69 1.36	2.75	1.84 0.54 1.28	2.56	**1.42 0.31 1.11**	2.70	1.73 0.49 1.23
Mn1: High-ID nonlinear manifold [5]	18	72	20.44	71.97 5.93 66.04	19.76	64.83 3.09 61.74	18.39	**53.65 0.15 53.49**	19.42	61.68 2.02 59.65
Mn2: High-ID Nonlinear Manifold [5]	24	96	25.91	109.88 3.65 106.24	25.05	100.42 1.11 99.32	23.32	**86.52 0.46* 86.06**	24.63	96.36 **0.39** 95.97
win-draw-lose				17-0-0		17-0-0		17-0-0		17-0-0

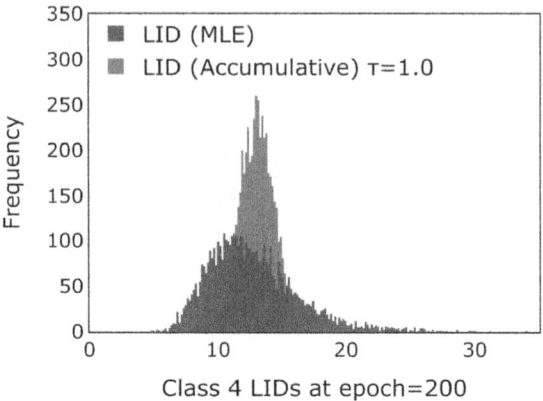

Fig. 4. Class 4 LID distribution

6 Conclusion

We have introduced a Bayesian framework for Local Intrinsic Dimensionality (LID) estimation. It allows the incorporation of prior information, leading to the derivation of several LID estimators and methods for aggregating multiple LID distributions. This also enables the development of two novel estimation strategies suitable for learning processes. We have seen that that Bayesian-based estimators can outperform Maximum Likelihood Estimators (MLE) when a small number of nearest neighbors is used. Furthermore, they can help stabilize LID estimates within deep neural network training using mini-batches.

References

1. Amsaleg, L., Chelly, O., Furon, T., Girard, S., Houle, M.E., Kawarabayashi, K., Nett, M.: Extreme-value-theoretic estimation of local intrinsic dimensionality. DMKD (2018)
2. Anderberg, A., Bailey, J., Campello, R.J.G., Houle, M.E., Marques, H., Radovanović, M., Zimek, A.: Dimensionality-aware outlier detection: theoretical and experimental analysis. In: (SDM24) (2024)
3. Bailey, J., Houle, M.E., Ma, X.: Local intrinsic dimensionality, entropy and statistical divergences. Ent. (2022). https://doi.org/10.3390/e24091220
4. Brown, L.: Inadmissibility of the usual estimators of scale parameters in problems with unknown location and scale parameters. Ann. Math. Stat. **39**(1), 29–48 (1968)
5. Campadelli, P., Casiraghi, E., Ceruti, C., Rozza, A.: Intrinsic dimension estimation: relevant techniques and a benchmark framework. Math. Probl. Eng. **2015**, 1–21 (2015)
6. Denti, F., Doimo, D., Laio, A., Mira, A.: The generalized ratios intrinsic dimension estimator. Sci. Rep. **12**(1) (2022)
7. Facco, E., d'Errico, M., Rodriguez, A., Laio, A.: Estimating the intrinsic dimension of datasets by a minimal neighborhood info. Sci. Rep. (2017)

8. Hill, B.M.: A simple general approach to inference about the tail of a distribution. **3**(5), 1163–1174 (1975)
9. Houle, M.E.: Dimensionality, discriminability, density and distance distributions. In: ICDMW13, pp. 468–473 (2013)
10. Houle, M.E.: Local intrinsic dimensionality I: an extreme-value-theoretic foundation for similarity applications. In: SISAP, pp. 64–79 (2017)
11. Huang, H., Campello, R.J.G.B., Erfani, S.M., Ma, X., Houle, M.E., Bailey, J.: LDReg: Local dimensionality regularized self-supervised learning. ICLR 24 . 10.48550/arXiv.2401.10474
12. Ibrahim, J.G., Chen, M., Gwon, Y., Chen, F.: The power prior: theory and applications. Stat. Med. **34**(28), 3724–3749 (2015)
13. James, W., Stein, C.: Estimation with quadratic loss (1992). https://doi.org/10.1007/978-1-4612-0919-5_30
14. Jeffreys, H.: An invariant form for the prior probability in estimation problems. Proc. R. Soc. Lond. A **186**(1007), 453–461 (1946)
15. Jolliffe, I.T.: Principal Component Analysis. Springer (2002)
16. Krizhevsky, A.: Learning multiple layers of features from tiny images (2009)
17. Levina, E., Bickel, P.J.: Maximum likelihood estimation of intrinsic dimension. In: NeurIPS (2004)
18. Ma, X., et al: Characterizing adversarial subspaces using local intrinsic dimensionality. In: ICLR (2018)
19. Ma, X., Wang, Y., Houle, M.E., Zhou, S., Erfani, S.M., Xia, S., Wijewickrema, S.N.R., Bailey, J.: Dimensionality-driven learning with noisy labels. In: ICML (2018)
20. Neyman, E., Roughgarden, T.: From proper scoring rules to max-min optimal forecast aggregation. In: EC '21, p. 734. ACM (2021)
21. Rozza, A., Lombardi, G., Ceruti, C., Casiraghi, E., Campadelli, P.: Novel high intrinsic dimensionality estimators. Mach. Learn. (2012)
22. Thordsen, E., Schubert, E.: ABID: angle based intrinsic dimensionality-theory and analysis. Inf. Syst. **108**, 101989 (2022)
23. Tordesillas, A., Zhou, S., Bailey, J., Bondell, H.: A representation learning framework for detection and characterization of dead versus strain localization zones from pre-to post-failure. Gra, Mat (2022)
24. Wu, Z., Xiong, Y., Yu, S.X., Lin, D.: Unsupervised feature learning via nonparametric instance discrimination. In: CVPR (2018)
25. Zhou, S., Tordesillas, A., Pouragha, M., Bailey, J., Bondell, H.: On local intrinsic dimensionality of deformation in complex materials. Nat. Sci. Rep. **11**(10216) (2021). https://doi.org/10.1038/s41598-021-89328-8

Towards Personalized Similarity Search for Vector Databases

Marek Mahrík[1], Matúš Šikyňa[1](✉), Vladimir Mic[2], and Pavel Zezula[1]

[1] Faculty of Informatics, Masaryk University, Brno, Czech Republic
492727@mail.muni.cz, {xsikyna,zezula}@fi.muni.cz
[2] Department of Computer Science, Aarhus University, Aarhus, Denmark
v.mic@cs.au.dk

Abstract. The importance of similarity search has become prominent in the fast-evolving vector databases, which apply content embedding techniques on complex data to produce and manage large collections of high-dimensional vectors. Processing of such data is only possible by using a similarity function for storage, structure, and retrieval. However, if multiple users access the collection, their views on similarity can differ as similarity, in general, is subjective and context-dependent. In this article, we elaborate on the problem of a similarity search engine implementation, where users use a common index but search with personalised views of similarity, implemented by a possibly different similarity model. Specifically, we define a foundational theoretical framework and conduct experiments on real-life data to confirm the viability of such an approach. The experiments also indicate future research directions needed to propose and implement an effective and efficient personalised similarity search engine.

Keywords: Similarity search · Personalized similarity · Vector databases

1 Introduction

Current AI models produce vector embeddings that carry semantic information capable of performing context-based reasoning. The concept of similarity is central to this process, given its critical role in both learning and teaching. Such vector embedding models produce high-dimensional vectors, the collections of which are structured into partitions of a vector database index to facilitate efficient similarity querying. Typically, vector databases utilise Euclidean distance to structure the index and identify the most similar vectors in response to a given query vector.

This work was supported by the research grants (VIL50110) from VILLUM FONDEN, and by the Open Calls for Security Research 2023–2029 (OPSEC) program granted by the Ministry of the Interior of the Czech Republic under No. VK01010147 - Automated digital data forensics lab for complex crime detection.

Table 1. Notation used throughout this paper

$D; X \subseteq D$	domain of the searched vectors; searched dataset
$\vec{q} \in \Theta \subseteq D$	query vector $\vec{q}$ and the set of tested query vectors Θ from D
$\vec{x}, \vec{y} \in D$	vectors from the domain of the searched vectors
δ	the dimensionality of vectors in the domain D
$\mathbf{A}$	positive semi-definite matrix used in the Mahalanobis distance function
$d_E; d_E(\vec{x}, \vec{y})$	Euclidean distance function; the Euclidean distance of $\vec{x}$ and $\vec{y}$
d_M	Mahalanobis distance function
$d_M(\vec{x}, \vec{y}, \mathbf{A})$	the Mahalanobis distance of $\vec{x}$ and $\vec{y}$ computed with matrix $\mathbf{A}$
$s_{M,E}(\mathbf{A})$	the scaling factor defining relation between functions d_E and d_M (Eq. 1)
r_M	search range given for Mahalanobis distance function d_M
r_E	search range defined for d_E by r_M and scaling factor $s_{M,E}(\mathbf{A})$ (Eq. 3)
$r_{\text{Opt}E}(\vec{q})$	smallest search range defined in the Euclidean space which for given $\vec{q}$ returns a superset of $R_{d_M,r_M}(\vec{q})$ (Eq. 5)
$Q(\vec{q}, <f>, <r>)$	the range query with distance function $<f>$ and search range $<r>$
$R_{<f>,<r>}(\vec{q})$	the (precise) answer of the range query $Q(\vec{q}, <f>, <r>)$
$\mathcal{P}_E$	the precision of answer $R_{d_E,r_E}(\vec{q})$ with respect to $R_{d_M,r_M}(\vec{q})$
$\mathcal{P}_{\text{Opt}E}$	the precision of answer $R_{d_E,r_{\text{Opt}E}(\vec{q})}(\vec{q})$ with respect to $R_{d_M,r_M}(\vec{q})$
$\#_E$	the number of vectors in $R_{d_E,r_E}(\vec{q})$ for a given $\vec{q}$
$\#_{\text{Opt}E}$	the number of vectors in $R_{d_E,r_{\text{Opt}E}(\vec{q})}(\vec{q})$ for a given $\vec{q}$

Undoubtedly, current AI models, trained on billions of sample objects, are capable of producing high-quality content embeddings. However, once these vectors are structured in an index using a fixed distance measure (typically, the Euclidean one), the process of partitioning and filtering remains the same for all database users. Such conditions invariably persist until a global, presumably costly, reorganisation is performed. This approach is in sharp contrast with human perception of similarity, which is subjective and context-dependent [13,24]. Contemporary approaches to similarity indexing include but are not limited to the metric space [26], LSH [6], graph-based (e.g., HNSW [15]), and quantization-based (e.g., OPQ [9]) concepts.

In this paper, we investigate the feasibility of a search engine architecture where a single similarity index accelerates query execution, even when personalised (i.e. modified or different) similarity measurements are used. We first review the related work in Sect. 2. Next, we outline the principal idea in Sect. 3 and introduce the theoretical foundations of a personalised similarity search engine in Sect. 4. In Sect. 5, the correctness of the theory and feasibility of the proposal are verified by experiments on contemporary data. Through this approach, we aim to better understand the design potential and identify its limitations and research challenges. The effort should lead to the proposal of an effective and efficient personalised search system. The specific findings are summarised in Sect. 6, and the paper concludes in Sect. 7.

2 Related Work

There have been many significant efforts to incorporate personalisation into search systems. One such approach is to utilise relevance feedback [19]. Traditional relevance feedback methods require explicit user feedback, which imposes a significant burden on users by requiring them to actively indicate the relevance of search results [12]. As opposed to explicit feedback, methods based on implicit feedback gather data such as past searches and click history [8,17,21] to improve the ranking of search results tailored to individual users. For instance, Shen et al. [20] explore short-term user search context given by previous queries for query expansion.

While implicit feedback focuses on individual user behaviour, another approach to enhancing personalisation is through collaborative filtering. Sugiyama et al. [22] derive users' preferences from their browsing history while utilising modified collaborative filtering to predict missing term weights in user profiles. Moreover, mining the search behaviour of similar users can further improve personalisation [23,25].

Deep learning techniques have shown significant potential in capturing complex patterns and user preferences. Ma et al. [14] presented time-aware LSTM architectures to model the evolving query intent and document interest over time, capturing both short-term preferences and long-term user interests. PSSL [27] utilises contrastive sampling methods to generate paired self-supervised data from users' query logs to improve the personalised results.

Personalisation can also be used for browsing. Bartolini et al. [2] presented PIBE, an adaptive image browsing system that allows users to customise a hierarchical organisation of images without the need for costly global reorganisations.

Our approach is different as it considers personalised instances of Mahalanobis distance functions adapted by users' feedback to capture the subjective similarity perception. Moreover, we leverage the relationship between Euclidean and Mahalanobis distances to personalise search results using an efficient Euclidean-based indexing structure without the need for global index reorganisation.

3 Principal Idea and the Approach

The idea of the proposed personalised similarity search system is based on the following two paradigms: (1) the *metric learning* [3], where individual subject similarity views are modelled by instances of the class of Mahalanobis distances, and (2) the *lower bounding relationship* between the Euclidean and the Mahalanobis distances: given two vectors, their Euclidean distance is always smaller or equal to the scaled Mahalanobis distance [5].

We assume that the index of the database is constructed from vectors typically extracted using the *deep learning* technology, which utilises the best possible reference knowledge available. After the index construction by the Euclidean distance, users start querying using the Mahalanobis distance function with the identity square matrix **A** – this distance function equals the Euclidean.

Each user owns a specific matrix **A**, initially the identity matrix. While searching, users provide feedback on result satisfaction, which is used to modify matrix **A**, reflecting the personalised views of users by establishing proper correlations between specific pairs of dimensions. The database index, structured with the Euclidean distance function, still executes queries using the Euclidean distances. Provided a user's matrix is not the identity matrix, the index needs to retrieve larger result sets so that the nearest neighbours defined by the Mahalanobis distances are included. Depending on the query type, this is achieved by using a proper extension, i.e., a larger search range or a larger number of nearest neighbours.

However, the problem is the size of the extension - too small extension does not guarantee full inclusion, i.e., it can decrease the recall as some nearest neighbours defined by the Mahalanobis distances can be missing, and too large extension provides a bulky result, i.e., it decreases the precision as many not-qualifying objects are also retrieved by the index.

From a technical point of view, we assume the *filter and refine* principle – the search engine first efficiently filters out most of the not-qualifying objects using the Euclidean distances and retrieves a superset of the answer, which is consequently refined by the Mahalanobis distances specified for individual users by their personalised matrix **A**. According to our experiments, the computation of Mahalanobis distances is approximately 2.5 times slower than the computation of Euclidean distances. It motivates us to investigate the search range that provides 100 per cent recall with the maximum possible precision.

In summary, we investigate the feasibility of a similarity search engine architecture – see the general model in Fig. 1 – where individual users see the vector database (i.e., *Similarity index*) and express their queries through a personalised instance of the Mahalanobis distance function (i.e., the *View metric*) while the database is structured and the index executes queries using the Euclidean distance measure (i.e., the *Index metric*). Performed experiments aim to identify the advantages and limitations of such a solution.

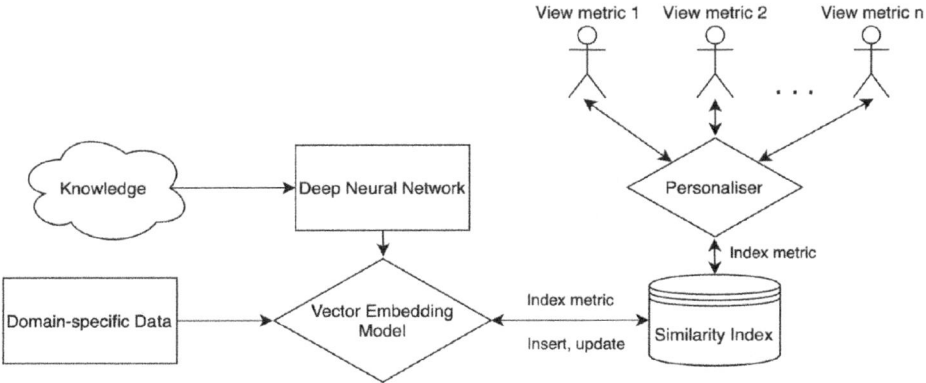

Fig. 1. Personalized similarity search system architecture

4 Theoretical Foundations

A summary of notation used in the paper is in Table 1. We use the Euclidean distance $d_E(\vec{x}, \vec{y})$, defined for two δ-dimensional vectors $\vec{x}, \vec{y}$ as the Euclidean norm of their difference:

$$d_E(\vec{x}, \vec{y}) = \|\vec{x} - \vec{y}\|.$$

The Mahalanobis distance $d_M(\vec{x}, \vec{y}, \mathbf{A})$ of vectors $\vec{x}$ and $\vec{y}$ is defined for any positive semi-definite matrix $\mathbf{A}$:

$$d_M(\vec{x}, \vec{y}, \mathbf{A}) = \sqrt{(\vec{x} - \vec{y})^T \mathbf{A} (\vec{x} - \vec{y})}.$$

Having all eigenvalues λ_j of $\mathbf{A}$, article [5] defines the lower bound on the Mahalanobis distance given by the Euclidean distance:

$$\frac{d_M(\vec{x}, \vec{y}, \mathbf{A})}{\sqrt{min_j(\lambda_j)}} \geq d_E(\vec{x}, \vec{y}) \tag{1}$$

Value $s_{M,E}(\mathbf{A}) = 1/\sqrt{\min_j\{\lambda_j\}}$ is the **scaling factor** between the Mahalanobis and Euclidean distances. Consequently, if the scaling factor $s_{M,E}(\mathbf{A})$ is 1, i.e., the smallest eigenvalue of matrix $\mathbf{A}$ is 1, then the Mahalanobis distance is lower-bounded by the Euclidean distance: $d_M(\vec{x}, \vec{y}, \mathbf{A}) \geq d_E(\vec{x}, \vec{y})$.

Having the domain D of δ-dimensional vectors, we define the goal of the personalised similarity search in a dataset $X \subseteq D$ as follows. Having a query vector $\vec{q} \in D$ and matrix $\mathbf{A}$, we want to evaluate a range query $Q(\vec{q}, d_M, r_M)$, i.e., to retrieve the *Mahalanobis answer set* $R_{d_M, r_M}(\vec{q})$:

$$R_{d_M, r_M}(\vec{q}) = \{\vec{x} \mid d_M(\vec{x}, \vec{q}, \mathbf{A}) \leq r_M\} \subseteq X \tag{2}$$

Due to the property given by Eq. 1, range r_E defined as:

$$r_E = r_M \cdot s_{M,E}(\mathbf{A}) \tag{3}$$

defines the range query $Q(\vec{q}, d_E, r_E)$ that provides the Euclidean answer set $R_{d_E, r_E}(\vec{q})$:

$$R_{d_E, r_E}(\vec{q}) = \{\vec{x} \mid d_E(\vec{x}, \vec{q}) \leq r_E\} \subseteq X \tag{4}$$

that contains the Mahalanobis answer set formalised by Eq. 2 as its subset.

Notice that r_E only depends on r_M and matrix $\mathbf{A}$ according to Eq. 3, i.e., is independent of a query vector $\vec{q}$. Therefore, besides range r_E, we define the *optimum range* $r_{\text{Opt}E}(\vec{q})$ for a given $\vec{q}$ as the minimum Euclidean range such that the Euclidean answer set $R_{d_E, r_{\text{Opt}E}(\vec{q})}(\vec{q})$ contains Mahalanobis answer set $R_{d_M, r_M}(\vec{q})$, i.e.:

$$\begin{aligned} R_{d_E, r_{\text{Opt}E}(\vec{q})}(\vec{q}) = \{\vec{x} \mid d_E(\vec{x}, \vec{q}) \leq r_{\text{Opt}E}(\vec{q})\} \supseteq R_{d_M, r_M}(\vec{q}) \\ \wedge \ \forall r < r_{\text{Opt}E}(\vec{q}) : R_{d_E, r}(\vec{q}) \not\supseteq R_{d_M, r_M}(\vec{q}) \end{aligned} \tag{5}$$

Then, the Euclidean range r_E expresses the maximum $r_{\text{Opt}E}(\vec{q})$ over all possible δ-dimensional vectors $\vec{q}$ [5,10]. Finally, we emphasise that:

- for any $\vec{q} \in D$: $R_{d_M,r_M}(\vec{q}) \subseteq R_{d_E,r_{\text{Opt}E}(\vec{q})}(\vec{q}) \subseteq R_{d_E,r_E}(\vec{q})$,
- range $r_{\text{Opt}E}(\vec{q})$ depends on $\vec{q}$, and inherently on r_M and $\mathbf{A}$ (see Eq. 5),
- r_E only depends on r_M and $s_{M,E}(\mathbf{A})$ (see Eq. 3),
- different matrices $\mathbf{A}$ can have the same scaling factor $s_{M,E}(\mathbf{A})$ as it is given just by the smallest eigenvalue of $\mathbf{A}$.

The first bullet implies the recalls 1 of the Euclidean answer sets $R_{d_E,r_{\text{Opt}E}(\vec{q})}(\vec{q})$ and $R_{d_E,r_E}(\vec{q})$ concerning the Mahalanobis answer set $R_{d_M,r_M}(\vec{q})$.

The *precision* of each of the Euclidean answer sets is consequently defined:

$$\mathcal{P}_E = \frac{|R_{d_M,r_M}(\vec{q})|}{|R_{d_E,r_E}(\vec{q})|} \tag{6}$$

and

$$\mathcal{P}_{\text{Opt}E} = \frac{|R_{d_M,r_M}(\vec{q})|}{|R_{d_E,r_{\text{Opt}E}(\vec{q})}(\vec{q})|} \tag{7}$$

Notice they differ just in the search ranges r_E and $r_{\text{Opt}E}(\vec{q})$.

5 Concept Verification

In this section, we experimentally verify the validity of theory from Sect. 4 by investigating the relations between ranges r_M, $r_{\text{Opt}E}(\vec{q})$, and r_E. We also investigate the precisions $\mathcal{P}_{\text{Opt}E}$ and $\mathcal{P}_E$ of the search with ranges $r_{\text{Opt}E}(\vec{q})$ and r_E, respectively, to reveal whether the usage of these ranges leads to small answer sets $R_{d_E,r_E}(\vec{q})$ and $R_{d_E,r_{\text{Opt}E}(\vec{q})}(\vec{q})$ which are necessary for efficient refinement with the Mahalanobis distance function d_M.

Dataset We use a set of 226,778 images that were randomly selected from the *Profiset* image collection [4]. These images are represented by 768-dimensional vector embeddings generated by the CLIP ViT-L/14 model [18]. The vectors are normalised to unit vectors and form the searched dataset X.

Learning Personalized Matrices In this article, we address the similarity search purely on a geometric level, i.e., we abstract from the semantics of original images. Conversely, our goal is to investigate the relations of the Euclidean search ranges r_E (Eq. 3), $r_{\text{Opt}E}(\vec{q})$ (Eq. 5), and Mahalanobis search range r_M. Specifically, the contributions of the indexing with Euclidean distances to the searching with the Mahalanobis distances are proportional to the size of the answer set $R_{d_E,r_E}(\vec{q})$, or possibly $R_{d_E,r_{\text{Opt}E}(\vec{q})}(\vec{q})$ – the smaller, the better. Besides the number of vectors in the Euclidean answer sets, we also report their precisions $\mathcal{P}_E$ and $\mathcal{P}_{\text{Opt}E}$.

The focus on a geometric level of image representations allows us to skip user relevance feedback and simulate it with various matrices **A** learnt for the Mahalanobis distance function. We learn matrices **A** with two *metric learning* techniques, namely *Information Theoretic Metric Learning (ITML)* [7] and *Sparse Distance Metric Learning (SDML)* [16]. Both techniques to learn **A** use pairs of vectors that are marked either as *similar* or *dissimilar*. The learning process aims to minimise the Mahalanobis distances of similar vectors and maximise the distances of dissimilar vectors at the same time.

We get the similar and dissimilar pairs of vectors in two different ways denoted as *random* and *nearest*. The random selection means that we work with n vector pairs selected in random from the dataset. In the nearest selection, we control a mutual closeness of vectors $(\vec{x}, \vec{y})$ in n pairs: vector $\vec{x}$ is always selected at random, but $\vec{y}$ is chosen randomly from a set of 1000 nearest neighbours of a given $\vec{x}$ from the dataset, as defined by Euclidean distances. In both cases, we randomly assess whether vectors in each pair are mutually similar or dissimilar. Consequently, approximately half of the pairs (i.e., $n/2$) are denoted as similar.

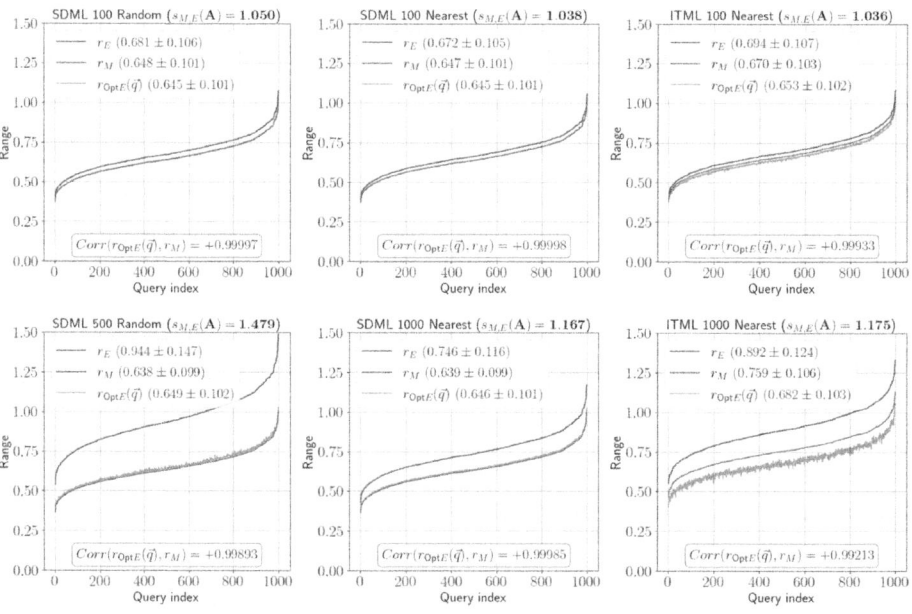

Fig. 2. Ranges r_M, $r_{\text{Opt}E}(\vec{q})$, and r_E, their averages and standard deviations for six learned matrices **A**. The x-axis expresses query indices – query vectors $\vec{q}$ are sorted to make r_M (i.e., the green curve) non-decreasing. Scaling factors $s_{M,E}(\mathbf{A})$ are at the top of each plot, and the Pearson correlations $Corr(r_{\text{Opt}E}(\vec{q}), r_M)$ are at the bottom.

We have investigated about 30 matrices **A** and picked 6 representatives for a presentation in the article. All the results we have observed are consistent with the presented findings. The matrices are denoted "<method> n <selection>" where <method> is either *SDML* or *ITML*, n is the number of vector pairs used in **A** learning, and <selection> denotes the way how the vector pairs were sampled, i.e., either *Random* or *Nearest*.

Matrices *SDML 100 Nearest* and *ITML 100 Nearest* have almost the same scaling factor $s_{M,E}(\mathbf{A})$, and we demonstrate sizes of the corresponding Euclidean answer sets created as the super sets of the Mahalanobis answer sets with these **A**. In total, we present results for 6 matrices where 3 matrices were learnt with 100 vector pairs (i.e., *SDML 100 Random*, *SDML 100 Nearest*, and *ITML 100 Nearest*), and the other 3 were learnt in the same way just with more vector pairs (*SDML 500 Random*, *SDML 1000 Nearest*, and *ITML 1000 Nearest*). This selection of matrices helps us to demonstrate the consequences of an increasing number of learning vector pairs. We conduct all experiments with 1000 query vectors selected randomly from the dataset as the query set Θ. These query vectors were not used to learn any **A**.

Lower Bounding Relationship Verification For the following experiments, we need to choose the Mahalanobis search ranges r_M. We consider each $\vec{q} \in \Theta$ independently, and compute $d_M(\vec{q}, \vec{x}, \mathbf{A})$ for all $\vec{x} \in X$. For each $\vec{q}$, we define $r_M = d_M(\vec{q}, \vec{x_{100}}, \mathbf{A})$ which is the Mahalanobis distance of $\vec{q}$ to its 100th nearest neighbour $\vec{x_{100}}$ from X. These ranges r_M define the Mahalanobis answer sets $R_{d_M, r_M}(\vec{q})$ which each contains 100 vectors[1].

The following experiments verify the key result of the theoretical Sect. 4, i.e., the relation of the answer sets defined by Eqs. 2, 4, and 5:

$$R_{d_M, r_M}(\vec{q}) \subseteq R_{d_E, r_{\text{OptE}}(\vec{q})}(\vec{q}) \subseteq R_{d_E, r_E}(\vec{q}) \tag{8}$$

We start with the Mahalanobis ranges r_M, and compute the optimum range $r_{\text{OptE}}(\vec{q})$ for each $\vec{q}$ as the maximum Euclidean distance $d_E(\vec{x}, \vec{q})$, $\vec{x} \in R_{d_M, r_M}(\vec{q})$. The optimum range $r_{\text{OptE}}(\vec{q})$ enables us to compute the Euclidean answer set $R_{d_E, r_{\text{OptE}}(\vec{q})}(\vec{q})$ to confirm that it is the superset of the Mahalanobis answer set $R_{d_M, r_M}(\vec{q})$, i.e.:

$$\forall \vec{q} \in \Theta \;:\; R_{d_M, r_M}(\vec{q}) \subseteq R_{d_E, r_{\text{OptE}}(\vec{q})}(\vec{q}) \tag{9}$$

Therefore, the recall of $R_{d_E, r_{\text{OptE}}(\vec{q})}(\vec{q})$ concerning $R_{d_M, r_M}(\vec{q})$ is always 1.

We continue with computations of ranges r_E according to Eq. 3 and confirm that $r_{\text{OptE}}(\vec{q}) \leq r_E$ for each $\vec{q}$ and **A**. This is the result consistent with Eq. 5. For each $\vec{q} \in \Theta$, both ranges $r_{\text{OptE}}(\vec{q})$ and r_E apply to the same Euclidean space. The smaller search range thus implies a smaller answer set, formally:

$$\forall \vec{q} \in \Theta \;:\; R_{d_E, r_{\text{OptE}}(\vec{q})}(\vec{q}) \subseteq R_{d_E, r_E}(\vec{q}) \tag{10}$$

---

[1] We did not observe any equality $d_M(\vec{q}, \vec{x_{100}}, \mathbf{A}) = d_M(\vec{q}, \vec{x_{101}}, \mathbf{A})$ for any $\vec{q} \in \Theta$.

Eqs. 9 and 10 are thus witnesses of Eq. 8, which we want to demonstrate. Finally, the recall of $R_{d_E,r_E}(\vec{q})$ concerning $R_{d_M,r_M}(\vec{q})$ is also always 1, as a consequence of the transitivity of Eq. 9 and Eq. 10.

We have conducted all these experiments with all 30 matrices **A** and a query set Θ with 1,000 vectors. All the results have been consistent with the claims above. We report in detail the results for 6 selected matrices **A** in Fig. 2 and Table 2. Curves in Fig. 2 illustrate the ranges r_E (in blue), r_M (in green), and $r_{\text{Opt}E}(\vec{q})$ (in red). Each plot is made for a different matrix **A**, and the scaling factors $s_{M,E}(\mathbf{A})$ are presented at the top of each plot. The values of the ranges are on the y-axis, and the x-axis depicts the index of the query vector $\vec{q}$. The queries are ordered to make ranges r_M (i.e., the green curves) non-decreasing.

Table 2. Scaling factors $s_{M,E}(\mathbf{A})$; Averages of precisions $\mathcal{P}_E, \mathcal{P}_{\text{Opt}E}$; Averages of retrieved vector counts $\#_E$ and $\#_{\text{Opt}E}$. Dataset X contains 226,778 vectors.

Learned matrix **A**	$s_{M,E}(\mathbf{A})$	Avg. $\mathcal{P}_E$	Avg. $\#_E$	Avg. $\mathcal{P}_{\text{Opt}E}$	Avg. $\#_{\text{Opt}E}$
SDML 100 Random	1.050	0.410	317 ± 263	0.982	102 ± 2
SDML 500 Random	1.479	0.008	$108{,}025 \pm 82{,}569$	0.887	115 ± 23
SDML 100 Nearest	1.038	0.506	228 ± 110	0.990	101 ± 2
SDML 1000 Nearest	1.167	0.120	$4{,}110 \pm 10{,}828$	0.965	104 ± 5
ITML 100 Nearest	1.036	0.310	497 ± 528	0.804	128 ± 23
ITML 1000 Nearest	1.175	0.012	$70{,}681 \pm 69{,}508$	0.408	321 ± 235

Besides theoretically justified results, mainly $r_{\text{Opt}E}(\vec{q}) \leq r_E$, Fig. 2 compares ranges r_M (green curves) and $r_{\text{Opt}E}(\vec{q})$ (red curves). In all plots, the ranges are strongly correlated – see the Pearson correlations $Corr(r_{\text{Opt}E}(\vec{q}), r_M)$ at the bottom of each plot in Fig. 2 that are from $+0.992$ to $+0.99998$. This is a very promising observation that should be further addressed to try to efficiently estimate the optimum search range $r_{\text{Opt}E}(\vec{q})$ from r_M.

Precisions and Sizes of Euclidean Answer Sets We investigate the precisions $\mathcal{P}_{\text{Opt}E}$ and $\mathcal{P}_E$ (defined by Eq. 7 and Eq. 6), as well as the sizes of Euclidean answer sets $R_{d_E, r_{\text{Opt}E}(\vec{q})}(\vec{q})$ and $R_{d_E,r_E}(\vec{q})$ (Eq. 5 and Eq. 4) to reveal the viability of the proposal. Contributions of the indexing with Euclidean distances to the searching with the Mahalanobis distances are proportional to the size of the answer set $R_{d_E,r_E}(\vec{q})$, or possibly $R_{d_E, r_{\text{Opt}E}(\vec{q})}(\vec{q})$. Specifically, at least the optimum range $r_{\text{Opt}E}(\vec{q})$ should provide a sufficiently small answer set for each $\vec{q}$ to make the proposal viable. If also answer sets $R_{d_E,r_E}(\vec{q})$ are small enough, then even better, as the search range r_E is directly given by the Mahalanobis range r_M by Eq. 3.

For each matrix **A** and each query vector $\vec{q} \in \Theta$, we calculate $\mathcal{P}_E, \mathcal{P}_{\text{Opt}E}$, and numbers of vectors in corresponding Euclidean answer sets denoted as $\#_E$ and $\#_{\text{Opt}E}$. We remind that each Mahalanobis answer set $R_{d_M,r_M}(\vec{q})$ contains

100 vectors. Averages of the measured values over query vectors $\vec{q} \in \Theta$ are in Table 2. Specifically, there are: (1) the scaling factors $s_{M,E}(\mathbf{A})$, (2) the average precisions $\mathcal{P}_E$ and $\mathcal{P}_{\text{Opt}E}$, and (3) average numbers of vectors $\#_E$ and $\#_{\text{Opt}E}$ with standard deviations, all for 6 matrices $\mathbf{A}$. We only discuss the sizes of the retrieved answer sets $\#_{\text{Opt}E}$ and $\#_E$ in detail because the precisions are related to them. We also remind the dataset size of 226,778 vectors.

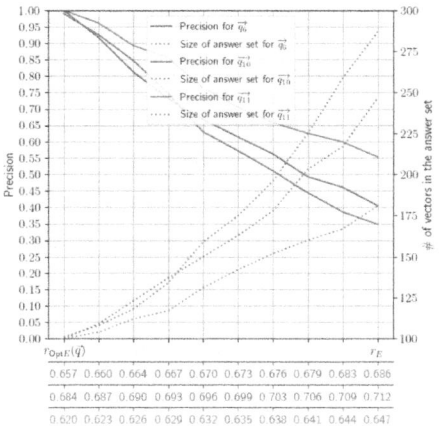
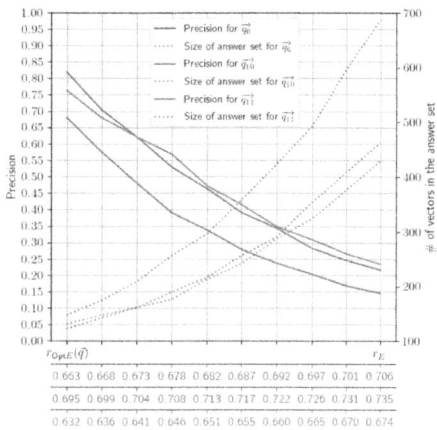

(a) SDML 100 Nearest, $s_{M,E}(\mathbf{A}) = 1.038$ (b) ITML 100 Nearest, $s_{M,E}(\mathbf{A}) = 1.036$

Fig. 3. Precision of range queries with three selected query vectors $\vec{q}$ (correspond to colours) evaluated with two matrices $\mathbf{A}$ (correspond to sub-figures). The x-axis depicts the search ranges r that cover the interval $[r_{\text{Opt}E}(\vec{q}), r_E]$ for each $\vec{q}$. The primary y-axis expresses the precision of range queries $Q(\vec{q}, d_E, r)$ and applies to the solid lines. The secondary y-axis depicts the number of vectors in the Euclidean answer set $R_{d_E, r}(\vec{q})$ and applies to the dashed lines.

Search ranges r_E provide answer sets with just hundreds of vectors in case of all matrices learned with 100 vector pairs, i.e., *SDML 100 Random, SDML 100 Nearest* and *ITML 100 Nearest* – see the fourth column in Table 2. While these are very promising results, experiments with matrices learnt with more vector pairs suggest a very quick growth of the answer set $R_{d_E, r_E}(\vec{q})$ – see all other matrices in the fourth column in Table 2. We thus conclude that the conditions leading to small answer sets $R_{d_E, r_E}(\vec{q})$ should be properly investigated since, in some cases, an index built with Euclidean distances and queried with range r_E provides small answer sets $R_{d_E, r_E}(\vec{q})$ that can be efficiently refined even with an expensive Mahalanobis distance function. But sometimes, Euclidean search with range r_E returns by orders of magnitude larger answer set than desired – see the matrices learned with 1000 or 500 vector pairs in Table 2.

Optimum search range $r_{\text{Opt}E}(\vec{q})$ leads to the retrieval of just slightly more than 100 candidates that need to be refined by expensive Mahalanobis distances, especially when the *SDML* technique to learn $\mathbf{A}$ is used (see the last column in

Table 2). While range $r_{OptE}(\vec{q})$ clearly leads to smaller answer sets than r_E, it cannot be directly derived, so its efficient estimation from r_M forms a challenge for future work. Overall, Table 2 illustrates the very good viability of the proposal and reveals the future challenges.

Figure 3 illustrates the search ranges for three representative query vectors $\vec{q} \in \Theta$ depicted in three different colours. Curves show how for each given $\vec{q}$ and an increasing Euclidean search range from $r_{OptE}(\vec{q})$ to r_E decreases the precision of the Euclidean answer set from $\mathcal{P}_{OptE}$ to $\mathcal{P}_E$. Correspondingly, the number of vectors in the Euclidean answer set increases. Figure 3a and 3b differ only in the matrix $\mathbf{A}$. Specifically, they depict the results for the same triplet of query vectors $\vec{q}$. Even though both examined matrices, *SDML 100 Nearest* in Fig. 3a and *ITML 100 Nearest* in Fig. 3b have nearly the same scaling factors $s_{M,E}(\mathbf{A})$ (1.038 and 1.036, respectively), the sub-figures demonstrate a different influence of matrices on different query vectors $\vec{q}$.

6 Final Discussion and Future Research Objectives

The experiments reveal the following facts:

- The theory is valid and the Euclidean search with ranges $r_{OptE}(\vec{q})$ and r_E always results in recall 1 with respect to the Mahalanobis search with range r_M. Both ranges can be effectively used for retrieving the superset of the Mahalanobis answer set $R_{d_M, r_M}(\vec{q})$, which can be further refined using Mahalanobis distance d_M.
- Range $r_{OptE}(\vec{q})$ is usually much smaller than r_E and the recall is still 1.
- Ranges $r_{OptE}(\vec{q})$ and r_M are surprisingly strongly correlated in our experiments; specifically, we observed the Pearson correlations around +0.999 in cases of 5 matrices out of 6 examined, and +0.992 for the last one.
- Experiments suggest that for a given matrix learning $<method>$ and $<selection>$, the precisions $\mathcal{P}_E$ and $\mathcal{P}_{OptE}$ decrease with increasing number of vector pairs n used for the $\mathbf{A}$ learning, which also increases the scaling factor $s_{M,E}(\mathbf{A})$.
- Given two different matrices $\mathbf{A}$ with nearly the same scaling factor $s_{M,E}(\mathbf{A})$, the precision $\mathcal{P}_E$ can be distinct even for the same query vector $\vec{q}$ with the search range r_E derived from r_M according to Eq. 3.
- Precision $\mathcal{P}_{OptE}$ is usually much higher than precision $\mathcal{P}_E$ because $r_{OptE}(\vec{q})$ is usually much lower than range r_E even when the scaling factor $s_{M,E}(\mathbf{A})$ is constant.
- Scaling factor $s_{M,E}(\mathbf{A})$ much depends on the way how the matrix $\mathbf{A}$ for the Mahalanobis distance function is learnt.

To develop a personalised similarity search engine, the following challenges need to be properly explored:

- The search range r_E guaranteed by Eq. 3 to cover the Mahalanobis answer set for given matrix $\mathbf{A}$ is usually too large. Given an extreme Pearson correlation

between Mahalanobis ranges r_M and optimum Euclidean ranges $r_{\text{Opt}E}(\vec{q})$, a heuristic to estimate $r_{\text{Opt}E}(\vec{q})$ from r_M should be proposed. The heuristic could be *approximate*, i.e., without guarantees on its correctness.
- We need theories and paradigms which would lead to a proper definition of the personalised matrix **A**. This metric learning should be fast, simple, but effective from the user's point of view. At the same time, it should produce matrices **A** with a small scaling factor $s_{M,E}(\mathbf{A})$ as they still can lead to small Euclidean answer sets $R_{d_E,r_E}(\vec{q})$.
- Range r_E is linearly dependent on the scaling factor $s_{M,E}(\mathbf{A})$, but the number of vectors in the Euclidean answer set $R_{d_E,r_E}(\vec{q})$ is not. In practice, the k-NN queries are preferred over the range queries because they do not require knowledge of proper search ranges. Estimating the proper extension of the k value to efficiently execute the k-NN queries according to the schema proposed in this article is thus challenging. Moreover, also the position of a specific query vector $\vec{q}$ should be considered [1,11].
- Real-life applications deal with divergence and plenty of subjective similarity perceptions as well as with the diversity of searched datasets. This opens a space for different search engine architectures and implementations. Performance models of such systems would allow for convenient decisions that would best suit the specifics of applications.

7 Conclusion

Traditional research in similarity searching follows the effectiveness or efficiency complementary objectives, possibly both. To respect the subjectivity and individual views of users, we have elaborated on the personalised similarity search engine for vector databases. We have formalised a paradigm and tested the feasibility on a real-life dataset. Encouraging results indicate research directions that must be explored to create a truly usable personalised search system.

References

1. Amsaleg, L., Chelly, O., Furon, T., Girard, S., Houle, M.E., Kawarabayashi, K.i., Nett, M.: Estimating local intrinsic dimensionality. In: Proceedings of the 21th ACM SIGKDD International Conference on Knowledge Discovery and Data Mining, pp. 29–38. KDD '15 (2015)
2. Bartolini, I., Ciaccia, P., Patella, M.: Adaptively browsing image databases with PIBE. Multim. Tools Appl. **31**(3), 269–286 (2006)
3. Bellet, A., Habrard, A., Sebban, M.: Metric Learning. Synthesis Lectures on Artificial Intelligence and Machine Learning (2015)
4. Budíková, P., Batko, M., Zezula, P.: Evaluation platform for content-based image retrieval systems. In: Research and Advanced Technology for Digital Libraries—International Conferences on Theory and Practice of Digital Libraries, TPDL 2011, Germany. Proceedings. LNCS, vol. 6966, pp. 130–142 (2011)
5. Ciaccia, P., Patella, M.: Searching in metric spaces with user-defined and approximate distances. ACM Trans. Database Syst. **27**(4), 398–437 (2002)

6. Datar, M., Immorlica, N., Indyk, P., Mirrokni, V.S.: Locality-sensitive hashing scheme based on p-stable distributions. In: 20th Annual Symposium on Computational Geometry (SCG), pp. 253–262 (2004)
7. Davis, J.V., Kulis, B., Jain, P., Sra, S., Dhillon, I.S.: Information-theoretic metric learning. In: Machine Learning, Proceedings of the 24th International Conference (ICML 2007), Oregon, USA. ACM International Conference Proceeding Series, vol. 227, pp. 209–216 (2007)
8. Dou, Z., Song, R., Wen, J.: A large-scale evaluation and analysis of personalized search strategies. In: Proceedings of the 16th International Conference on World Wide Web, WWW 2007, Banff, Alberta, Canada, 8–12 May 2007. pp. 581–590. ACM (2007)
9. Ge, T., He, K., Ke, Q., Sun, J.: Optimized product quantization. IEEE Trans. Pattern Anal. Mach. Intell. **36**(4), 744–755 (2014)
10. Hafner, J.L., Sawhney, H.S., Equitz, W., Flickner, M., Niblack, W.: Efficient color histogram indexing for quadratic form distance functions. IEEE Trans. Pattern Anal. Mach. Intell. **17**(7), 729–736 (1995)
11. Houle, M.E.: Dimensionality, discriminability, density and distance distributions. In: 13th IEEE International Conference on Data Mining Workshops, ICDM Workshops, TX, USA, 7-10 Dec 2013, pp. 468–473 (2013)
12. Kelly, D., Teevan, J.: Implicit feedback for inferring user preference: a bibliography. SIGIR Forum **37**(2), 18–28 (2003)
13. Krenková, M., Mic, V., Zezula, P.: Similarity search with the distance density model. In: Similarity Search and Applications—15th International Conference, SISAP 2022, Bologna, Italy, 5-7 Oct 2022, Proceedings. Lecture Notes in Computer Science, vol. 13590, pp. 118–132 (2022)
14. Ma, Z., Dou, Z., Bian, G., Wen, J.: PSTIE: time information enhanced personalized search. In: CIKM '20: The 29th ACM International Conference on Information and Knowledge Management, Virtual Event, Ireland, 19–23 Oct 2020. pp. 1075–1084. ACM (2020)
15. Malkov, Y.A., Yashunin, D.A.: Efficient and robust approximate nearest neighbor search using hierarchical navigable small world graphs. arXiv (1603.09320) (2018)
16. Qi, G., Tang, J., Zha, Z., Chua, T., Zhang, H.: An efficient sparse metric learning in high-dimensional space via l_1-penalized log-determinant regularization. In: Proceedings of the 26th Annual International Conference on Machine Learning, ICML 2009, Canada. ACM International Conference Proceeding Series, vol. 382, pp. 841–848 (2009)
17. Qiu, F., Cho, J.: Automatic identification of user interest for personalized search. In: Proceedings of the 15th international conference on World Wide Web, WWW 2006, Edinburgh, Scotland, UK, 23-26 May 2006, pp. 727–736 (2006)
18. Radford, A., Kim, J.W., Hallacy, C., Ramesh, A., Goh, G., Agarwal, S., Sastry, G., Askell, A., Mishkin, P., Clark, J., Krueger, G., Sutskever, I.: Learning transferable visual models from natural language supervision. In: Proceedings of the 38th International Conference on Machine Learning, ICML 2021, Virtual Event. Proceedings of Machine Learning Research, vol. 139, pp. 8748–8763 (2021)
19. Salton, G., Buckley, C.: Improving retrieval performance by relevance feedback. J. Am. Soc. Inf. Sci. **41**(4), 288–297 (1990)
20. Shen, X., Tan, B., Zhai, C.: Implicit user modeling for personalized search. In: Proceedings of the 2005 ACM CIKM International Conference on Information and Knowledge Management, Bremen, Germany, 31 Oct–5 Nov 2005, pp. 824–831. ACM (2005)

21. Speretta, M., Gauch, S.: Personalized search based on user search histories. In: 2005 IEEE/WIC/ACM International Conference on Web Intelligence (WI 2005), Sept 19–22 2005, Compiegne, France, pp. 622–628 (2005)
22. Sugiyama, K., Hatano, K., Yoshikawa, M.: Adaptive web search based on user profile constructed without any effort from users. In: Proceedings of the 13th International Conference on World Wide Web, WWW 2004, New York, NY, USA, 17-20 May 2004, pp. 675–684. ACM (2004)
23. Teevan, J., Morris, M.R., Bush, S.: Discovering and using groups to improve personalized search. In: Proceedings of the Second International Conference on Web Search and Web Data Mining, WSDM 2009, Barcelona, Spain, 9–11 Feb 2009, pp. 15–24 (2009)
24. Tversky, A.: Features of similarity. Psychol. Rev. **84**(4), 327–352 (1977)
25. White, R.W., Chu, W., Awadallah, A.H., He, X., Song, Y., Wang, H.: Enhancing personalized search by mining and modeling task behavior. In: 22nd International World Wide Web Conference, WWW '13, Rio de Janeiro, Brazil, 13–17 May 2013, pp. 1411–1420 (2013)
26. Zezula, P., Amato, G., Dohnal, V., Batko, M.: Similarity search—the metric space approach. Adv. Database Syst. **32** (2006)
27. Zhou, Y., Dou, Z., Zhu, Y., Wen, J.: PSSL: self-supervised learning for personalized search with contrastive sampling. In: CIKM '21: The 30th ACM International Conference on Information and Knowledge Management, Virtual Event, Queensland, Australia 1–5 Nov 2021. pp. 2749–2758 (2021)

Information Dissimilarity Measures in Decentralized Knowledge Distillation: A Comparative Analysis

Mbasa Joaquim Molo[1,2(✉)], Lucia Vadicamo[2], Emanuele Carlini[2], Claudio Gennaro[2], and Richard Connor[3]

[1] Department of Computer Science, University of Pisa, Pisa, Italy
[2] Institute of Information Science and Technologies, CNR, Via G.Moruzzi 1, 56124 Pisa, Italy
joaquim.molo@phd.unipi.it,
{lucia.vadicamo,emanuele.carlini,claudio.gennaro}@isti.cnr.it
[3] School of Computer Science, University of St Andrews, St Andrews KY16 9SS, Scotland
rchc@st-andrews.ac.uk

Abstract. Knowledge distillation (KD) is a key technique for transferring knowledge from a large, complex "teacher" model to a smaller, more efficient "student" model. Although initially developed for model compression, it has found applications across various domains due to the benefits of its knowledge transfer mechanism. While Cross Entropy (CE) and Kullback-Leibler (KL) are commonly used in KD, this work investigates the applicability of loss functions based on underexplored information dissimilarity measures, such as Triangular Divergence (TD), Structural Entropic Distance (SED), and Jensen-Shannon Divergence (JS), for both independent and identically distributed (iid) and non-iid data distributions. The primary contributions of this study include an empirical evaluation of these dissimilarity measures within a decentralized learning context, i.e., where independent clients collaborate without a central server coordinating the learning process. Additionally, the paper assesses the performance of clients by comparing pairwise distillation averaging among clients to conventional peer-to-peer pairwise distillation. Results indicate that while dissimilarity measures perform comparably in iid settings, non-iid distributions favor SED and JS, which also demonstrated consistent performance across clients.

Keywords: Information dissimilarity measure · Divergence Function · Knowledge Distillation · Distributed intelligence

1 Introduction

The integration of Artificial Intelligence in edge processing has led to the emergence of an interdisciplinary field known as *Distributed Intelligence* or *Edge Intelligence*, which aims to develop systems composed of software agents, robots, sensors, and computer systems that can collaborate effectively [15,22,23]. In this

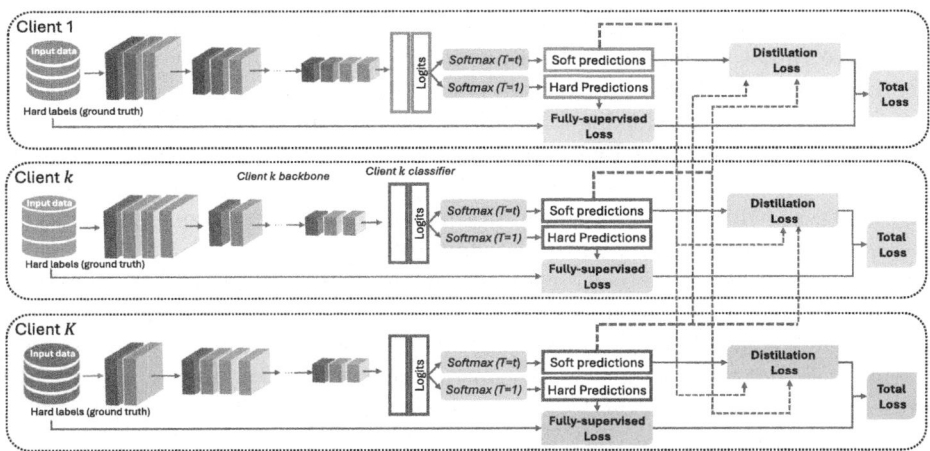

Fig. 1. KD-based decentralized network consisting of K clients, where distillation is performed using soft predictions for effective knowledge transfer.

field, *Knowledge distillation* (KD) has been employed to facilitate knowledge transfer between edge devices, enhancing the development of more efficient and accurate models [28]. KD is a machine learning technique designed to transfer knowledge from a large, complex model (the *teacher*) to a smaller, more efficient one (the *student*) [7,8,10]. In addition to its primary role in model compression, it has started to find applications in other areas, including distributed intelligence [3] and continual learning [4].

In KD-based distributed learning framework, *clients* exchange information to enhance their learning process, where each client operates both as learner and source of knowledge for other clients. These clients are part of a decentralized system where no single model acts as the central teacher. Instead, each client trains on its local dataset and shares knowledge with others. As illustrated in Fig. 1, this information exchange is achieved through a combination of two types of losses. The first loss component, indicated as "fully-supervised loss", is usually the cross-entropy (CE) with "hard" targets derived by the ground-truth labels of the input samples. The second component is the "distillation loss" designed to ensure that each learning client mimics the output of other remote clients [25]. This loss is typically implemented by comparing the probability distributions of the models involved, where one model acts as the student and others take turns serving as teachers. This encourages the student's output probabilities to closely match those of the teacher. The model's output probabilities are typically computed using a softmax layer. Adjusting the softmax temperature during training has proven to be crucial in metric learning and distillation processes. In the context of distributed intelligence, this technique is also employed to generate soft predictions for effective distillation. Hence, the distillation loss is expressed as minimizing the gap between the soft predictions of one client with respect to the soft predictions of all other clients [1,2,27].

Given that the softmax function transforms an array of logits into an array of positive values summing to 1, various information dissimilarity measures can theoretically be used to implement the distillation loss. However, in practice, it is predominantly realized using CE, in addition to Kullback-Leibler (KL) Divergence, and Mean Squared Error (MSE) [13]. These methods have been extensively studied and proven effective for knowledge transfer in diverse machine learning tasks, while a wide range of information distance functions remain unexplored in the literature related to distributed learning.

In this work, we break new ground by investigating alternative dissimilarity measures – specifically, Triangular Divergence (TD), Structural Entropic Distance (SED), and Jensen-Shannon (JS) divergence – in the context of KD for decentralized learning scenarios. Recently, the correlations among these measures and the commonly used CE have been examined in [6] for independent and identically distributed (iid) data. Our work aims to expand the understanding of how these dissimilarity measures can enhance KD techniques, particularly in settings where data distribution may vary across learning clients (with a non-iid data distribution). To the best of our knowledge, our study is the first to empirically evaluate the effectiveness of TD, SED, and JSD for KD in a decentralized learning framework, offering novel insights and expanding the potential of KD applications beyond conventional CE and KL-based approaches. Our main contributions include designing a distributed KD environment suitable for investigating the aforementioned information dissimilarity measures and examining the performance of a set of clients by comparing pairwise distillation averaging among clients to the conventional peer-to-peer pairwise distillation, considering the various information dissimilarity measures.

The rest of this article is structured as follows. Section 2 provides background and related works on knowledge distillation and the dissimilarity measures utilized. Section 3 details the fully decentralized learning model employed in our study. Section 4 presents our experimental setup, while Sect. 5 discusses the results. Section 6 provides the conclusions.

2 Background and Related Works

2.1 Information Dissimilarity Measures and Statistical Divergences

Information distance refers to a measure that quantifies the dissimilarity between two sources of information (e.g., two finite objects). This concept is distinct but also related to statistical divergences, which quantify the dissimilarity between two probability distributions. For example, some information distances, including SED, can be used to compare also probabilities, while statistical divergence can be interpreted as information distances when the source of information are probability distributions. In the following, we provide formal definitions of the divergence functions and information distances used in this paper.[1]

---

[1] Please note that, as in [6], we use the delimiter ':' as argument separator of non-symmetric divergence instead of the double bar notation '||' used in information theory.

Kullback-Leibler Divergence. The KL divergence measures the difference between two probability distributions as the amount of information lost when one distribution is used to approximate the other. Given two distributions **q** and **p**, it is defined as

$$KL(\mathbf{q}:\mathbf{p}) = \sum_{i=1}^{N} q_i \log \frac{q_i}{p_i} \qquad (1)$$

Jensen-Shannon Divergence. The JS divergence, historically introduced in [26], is a "smoothed, symmetrised" version of KL divergence and can be interpreted as the total KL divergence relative to the average distribution $\frac{\mathbf{q}+\mathbf{p}}{2}$ [21]. In this paper, we use the following definition:

$$JS(\mathbf{q},\mathbf{p}) = \frac{1}{2}\left(KL\left(\mathbf{q}:\frac{\mathbf{q}+\mathbf{p}}{2}\right) + KL\left(\mathbf{p}:\frac{\mathbf{q}+\mathbf{p}}{2}\right)\right) \qquad (2)$$

Structural Entropic Distance. SED [20] is an information-theoretic measure that compares the Shannon entropy H of two probability vectors with that of their arithmetic mean, where $H(\mathbf{p}) = -\sum_i^N p_i \ln p_i$ represents the amount of information needed to describe the probability vector $\mathbf{p} = [p_1, \ldots, p_N]$ [5]. Considering two probability vectors **p** and **q**, SED can be calculated as the ratio of the complexity of the mean vector to the geometric mean of the complexities of individual vectors:

$$SED(\mathbf{q},\mathbf{p}) = \frac{C(\frac{\mathbf{q}+\mathbf{p}}{2})}{\sqrt{C(\mathbf{q})C(\mathbf{p})}} - 1 \qquad (3)$$

where the complexity is computed as $C(\mathbf{p}) = b^{-\sum_{i=1}^{N} p_i \log_b p_i}$. The formulation in Eq. (3) gives an outcome in the range $[0,1]$, where 0 implies the two input vectors are identical, and 1 implies that they are orthogonal.

Triangular Divergence. The Triangular Divergence[2], also known as Triangular Discrimination [24] is defined as: $TD(\mathbf{q},\mathbf{p}) = \sum_{i=1}^{N} \frac{(q_i-p_i)^2}{q_i+p_i}$. Since the range of this function is $[0,2]$, in our work we use its scaled form:

$$TD(\mathbf{q},\mathbf{p}) = \frac{1}{2}\sum_{i=1}^{N} \frac{(q_i - p_i)^2}{q_i + p_i} = 1 - \sum_{i=1}^{N} \frac{2q_i p_i}{q_i + p_i} \qquad (4)$$

where the formulation in the right part of Eq. (4) is an optimized version obtained observing that $(q_i - p_i)^2 = (q_i + p_i)^2 - 4q_i p_i$ and $\sum_{i=1}^{N} p_i = \sum_{i=1}^{N} q_i = 1$.

---

[2] Note that its square root is a metric, referred to as *Triangular Distance*, *Vincze-Le Cam* distance and the symmetric *chi-squared* distance [17].

Cross Entropy. CE is a divergence measure widely used in machine learning to compare two probability distributions. It is defined as:

$$CE(\mathbf{q}:\mathbf{p}) = -\sum_{i=1}^{n} q_i \log p_i \quad (5)$$

It is worth noting that in the context of machine learning, as shown in [6], for spaces with certain properties, CE, KL, JS, TD shows very tight correlation. Specifically, if q is fixed, the perfect correlation between cross-entropy and Kullback-Leibler divergence is well-known and derives from simple algebra ($KL(\mathbf{q}:\mathbf{p}) = CE(\mathbf{q}:\mathbf{p}) - H(\mathbf{q})$) [6]. Moreover, Jensen-Shannon correlates almost perfectly with triangular divergence in almost all high-dimensional spaces [24], while cross-entropy and triangular divergence are strongly correlated when the probabilities are obtained within the softmax function (Eq. (6)) with high temperature. Note that triangular is much cheaper calculation than cross-entropy and if the correlation is very strong the latter may be used instead.

2.2 Knowledge Distillation as the Teacher-Student Approach

Knowledge distillation was initially introduced to transfer knowledge from pre-trained teacher (large) networks to student (small) networks. This involves approximating the soft output or intermediate representation of teacher networks, aiming to derive a compact and faster model [16].

Concretely, for any input data x, the teacher network generates a vector of logits $\mathbf{z}(x) = [z_1(x), \ldots, z_N(x)]$ that are turned into a probability vector $\mathbf{p}(x) = [p_1(x), \ldots, p_N(x)]$ using the softmax function: $p_i(x) = \frac{e^{z_i(x)}}{\sum_j^N e^{z_i(x)}}$. Typically, neural networks produce probability distributions with sharp peaks, which might lack informativeness. To address this, Hinton et al. [10] proposed temperature scaling in the softmax to soften these probabilities:

$$p_i(x, T) = \frac{e^{z_i(x)/T}}{\sum_j^N e^{z_j(x)/T}}, \forall i \in \{1, \ldots, N\} \quad (6)$$

where T is a hyperparameter called temperature.

In KD, both the student and the teacher generate softened probability distributions, denoted as $\mathbf{p}_S(x, T)$ and $\mathbf{p}_T(x, T)$, respectively. The student's total loss is then defined as a linear combination of a supervised student loss $\mathcal{L}_{stu}$ and a knowledge distillation loss $\mathcal{L}_{KD}$:

$$\mathcal{L} = \alpha \mathcal{L}_{stu} + (1-\alpha) \mathcal{L}_{KD} \quad (7)$$

where $\alpha \in [0, 1]$ is a hyperparemeter. Typically, $\mathcal{L}_{stu} = CE\left(\mathbf{y} : \mathbf{p}_S(x, T=1)\right)$ and $\mathcal{L}_{KD} = CE\left(\mathbf{p}_T(x, T=t) : \mathbf{p}_S(x, T=t)\right)$, with $\mathbf{y}$ being the hard labels (ground-truth). Note that the distillation loss is expressed as minimizing the gap between the output representation of the teacher and the output representation of the student.

KD-Based Distributed Learning. Recent research has explored KD for decentralized learning [28]. While much of this work focuses on a central teacher supervising student model training, there is a growing interest in fully decentralized settings where multiple clients collaborate to share knowledge without relying on a central authority.

Kim et al. [13] explored the role of the temperature hyperparameter in KD, showing higher temperature results in logit matching, which generally offers better generalization than label matching obtained with lower temperatures. They proposed employing MSE loss for direct logit matching. They showed that KL divergence loss stretches the second-to-last layer representations more than MSE loss and that KL divergence, especially with low temperature, is more resilient to noisy labels. Mishra et al. [18] developed EarlyLight, a method for training lightweight deep neural networks (DNNs) on edge devices using knowledge distillation from larger DNNs, considering also factors like storage, processing speed, and execution time. Molo et al. [19] proposed a knowledge distillation approach for vehicle detection using smart cameras in parking lots, where a large detector (teacher) guides smaller edge-based models (students) without additional labeled data. Their experimental results showed that students improve performance and can even surpass models trained with annotations.

Other approaches used a KD-based learning without a single teacher. Zhmoginov et al. [28] introduced Multi-Headed Distillation for distributed learning on the ImageNet dataset. This approach uses multiple model heads distilling to each other and simultaneous distillation of client model predictions and network embeddings, resulting in significantly higher accuracy than naive distillation methods. Jin et al. [12] introduced a personalized Federated Learning (FL) framework using self-KD to transfer historical personalized knowledge, balancing personalization and generalization. Similarly, Jeong et al. [11] addressed personalization challenges in FL for clients with diverse data and behaviors by proposing a KD-based algorithm to compare local models, enhancing client performance without data sharing and showing improved test accuracy, especially under non-iid data distributions.

Most works in the literature use KL divergence or CE as dissimilarity measures for distillation. However, there remains significant potential to investigate and utilize alternative dissimilarity measures, which could offer new insights into efficient knowledge transfer and performance across various learning tasks and scenarios.

3 Fully Decentralized Learning Model

In this section, we outline the decentralized learning environment used to evaluate the effectiveness of various information dissimilarity measures, introduced in Sect. 2.1, whose results are discussed in Sect. 5. The notation used is summarized in Table 1.

Table 1. Summary of notation used

Notation	Description
T	Temperature in the softmax
N	Number of classes
$(\mathcal{G}, \epsilon)$	Network of clients. $\mathcal{G}$ is the set of nodes, ϵ is the set of edges
K, k	Number of clients, Index of current client
C^k	Current client
$D^k = (X^k, y^k)$	Local annotated dataset on client k. X^k is the data, y^k are the labels
$(x, y) \in D^k$	Data sample x and the corresponding label y
Φ_k, ϕ	Set of indices of remote clients with respect to C^k, Index of a remote client
$\mathcal{M}^k = [\mathcal{M}_{h_1}^k, \mathcal{M}_{h_2}^k]$	Multi-head model held by client k
$\mathbf{w}_k = [\mathbf{w}_1^k, \mathbf{w}_2^k]$	Weight parameters of the local model of client k
$\mathcal{L}_{k,CE}$	Fully supervised Loss computed on client k
$\mathcal{L}_{k,KD}$	Distillation loss used for client k
α	Loss weight parameter

We consider a full network of K clients represented by a directed graph $(\mathcal{G}, \epsilon)$, where $\mathcal{G} = \{G^k \mid k \in K\}$ is a set of nodes and ϵ is the set of edges between the nodes. Each node G^k represents a client C^k holding a local dataset D^k composed of a pair (X^k, y^k), with $X^k = \{x_i^k\}_{i=1}^I$ representing the set of input data and $y^k = \{y_i^k\}_{i=1}^I$ the corresponding ground-truth labels. Each client C^k holds a model $\mathcal{M}^k$, which we assume to be a multi-head neural network. Specifically, the model has a backbone, which is the main body of the neural network that processes input data into a feature representation, and two heads, which take the features extracted by the backbone and perform final task-specific operations. The heads consist of a set of fully connected layers added on top of the backbone. We denote the models consisting of the backbone and the first head as $\mathcal{M}_{h_1}^k$, and the backbone and the second head as $\mathcal{M}_{h_2}^k$. The model with the first head, $\mathcal{M}_{h_1}^k$, is trained on the local distribution D^k, while the second, $\mathcal{M}_{h_2}^k$, is trained using knowledge distillation from connected clients.

The considered KD-based training procedure for this decentralized network involves training multiple clients concurrently, allowing them to share knowledge through distillation to improve overall model performance. Initially, each client's first model $\mathcal{M}_{h_1}^k$ is trained in a supervised manner until convergence with local data D^k. Then, as shown in Fig. 2, for each $k \in \{1, \ldots, K\}$, the first model from client C^k is shared with all outgoing connected clients in $\mathcal{G}$. Concurrently, client C^k receives the first head from all other incoming connected clients. This exchange enables each client to integrate knowledge from others while preserving their local data and model specialization, facilitating collaborative learning across the decentralized network. For the purposes of this study, we assume that all clients are interconnected. However, the proposed approach can be easily adapted to accommodate networks with different topologies and size.

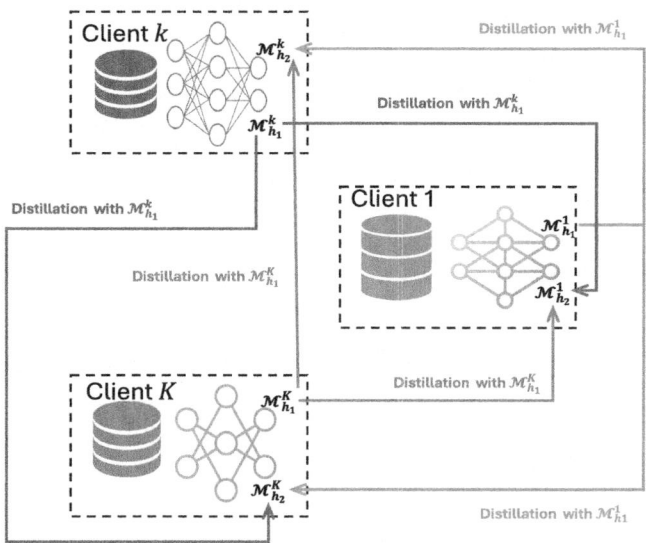

Fig. 2. KD-based decentralized network consisting of K clients, where distillation is performed using soft labels for effective knowledge transfer. In this setting, the first head of each client is communicated to the neighboring clients.

For a fixed k, we used the notation Φ_k to indicate all the indices except k. We refer C^k as the current client and $\{C^\phi \mid \phi \in \Phi_k\}$ as the remote clients. So, once the first head of the models are trained, C^k communicates $\mathcal{M}_{h_1}^k$ to all remote clients and receives the models $\{\mathcal{M}_{h_1}^\phi\}_{\phi \in \Phi_k}$ from them. The client C^k performs distillation using the available models from remote clients to train its second head $\mathcal{M}_{h_2}^k$. Specifically the parameters $\mathbf{w}_2^k$ of $\mathcal{M}_{h_2}^k$ are trained by optimizing a local total loss $\mathcal{L}_k$, which is obtained as a combination of a cross-entropy loss $\mathcal{L}_{k,CE}$ and a distillation loss $\mathcal{L}_{k,KD}$:

$$\mathcal{L}_k = \alpha \mathcal{L}_{k,CE} + (1-\alpha)\mathcal{L}_{k,KD}, \tag{8}$$

where $\alpha \in [0,1]$ is a parameter that weights the contribution of the losses with respect to the total loss. This dual-phase training approach allows each client to effectively train its local model while leveraging shared knowledge from other clients, improving generalization and performance across the network. The cross-entropy loss

$$\mathcal{L}_{k,CE} = \mathbb{E}_{(x,y) \sim D^k} \mathcal{L}_{CE}(\mathbf{w}_2^k, x, y) \tag{9}$$

is used to minimize local prediction with respect to the ground-truth labels of local data.[3]

For defining the distillation loss $\mathcal{L}_{k,KD}$ we considered two alternatives:

---

[3] Please note that $\mathcal{L}_{CE}(\mathbf{w}_2^k, x, y)$ is simply the CE dissimilarity (Eq. (5)) between the output of $\mathcal{M}_{h_2}^k$ model for the input x and the true labels y.

- **Case 1**: The *sum* of pairwise dissimilarities between the current client's soft-prediction and remote client's soft-predictions.
- **Case 2**: A distillation loss based on the dissimilarity between the current client's soft-predictions and the *average* of soft-predictions from remote clients.

Formally, let $\mathbf{p}^k(\mathbf{w}_2^k, x) = [p_1^k, p_2^k, \ldots, p_N^k]$ denote the softmax output obtained using the $\mathcal{M}_{h_2}^k$ model for the input data x, and $\mathbf{p}^\phi(x) = [p_1^\phi, p_2^\phi, \ldots, p_N^\phi]$ the softmax outputs of a remote client ϕ for the input data x (obtained using the pre-trained $\mathcal{M}_{h_1}^\phi$ model). For Case 1, we used

$$\mathcal{L}_{k,KD}(\mathbf{w}_2^k, x) = \sum_{\phi \in \Phi_k} \mathbb{E}_{x \sim X^k} f\left(\mathbf{p}^k(\mathbf{w}_2^k, x), \mathbf{p}^\phi(x)\right) \quad (10)$$

where f can be any divergence measure (e.g., CE, KL, TD, SED, JS). Since the sum of pairwise dissimilarities is used, we refer to this case as "sum" in the experiments. For Case 2, referred to as "average" in the experiments, we used

$$\mathcal{L}_{k,KD}(\mathbf{w}_2^k, x) = \mathbb{E}_{x \sim X^k} f\left(\mathbf{p}^k(\mathbf{w}_2^k, x), \frac{\sum_{\phi \in \Phi_k} \mathbf{p}^\phi(x)}{|\Phi|}\right) \quad (11)$$

Algorithm 1 summarizes the considered distillation training procedures.

4 Experimental Setup

Our analysis was conducted on a decentralized network consisting of three interconnected clients. This topology serves as a baseline evaluation, with plans for future work to extend the analysis to networks with more clients and various connectivity topologies.

We studied the effectiveness of different information dissimilarity measures (namely, CE, KL, SED, TD, JS) on distributed learning systems with different levels of data heterogeneity, ranging from scenarios where the data distribution is uniform across all clients (iid) to more extreme situations where each client focuses on its own specific tasks (non-iid). For this purpose, we used the CIFAR-10 [14] dataset and the SUN397 [29] dataset. We split the datasets into three subsets, corresponding to three clients in total.

For the CIFAR-10, the iid distribution is obtained by shuffling and evenly splitting the entire dataset, ensuring each client has different samples. For the non-iid distribution across the clients, we followed the configuration in [28]. Each client C^k receives a subset $\{\ell_i\}$ of the labels, which are designated as primary labels for C^k. Labels not included in $\{\ell_i\}$ are considered secondary for C^k. Samples for each label ℓ are distributed randomly among clients, with a higher probability ($1 + \gamma$ times greater) of being assigned to clients that have ℓ as a primary label. The parameter γ, referred to as dataset skewness, determines this distribution. When $\gamma = 0$, the data is distributed uniformly (iid), but as γ approaches infinity, samples for label ℓ are assigned exclusively to clients where ℓ is primary (non-iid). In the experiments, we used $\gamma = 15$ for CIFAR-10 and $\gamma = 10$ for SUN397.

Algorithm 1: Decentralized Training with Knowledge Distillation

Data: $(\mathcal{G}, \epsilon)$ graph representing a network of K client $\{C_1, \ldots C_K\}$
Local datasets $D^k = (X^k, y^k)$, where $X^k = \{x_i^k\}_{i=1}^I$ is set of input data and $y^k = \{y_i^k\}_{i=1}^I$ is the set of labels associated with each input, for all $k \in \{1, \ldots, K\}$.
Result: Trained model parameters for each client.
// Initialization
foreach *client* $k \in \{1, \ldots, K\}$ *in parallel* **do**
    /* Train the model with the first head, $\mathcal{M}_{h_1}^k$, until convergence. $\mathbf{w}_1^k$ are the model parameters to be updated */
    $\mathcal{M}_{h_1}^k \leftarrow$ LocalModelTraining$\left(\mathcal{L}_{CE}(\mathbf{w}_1^k, D^k)\right)$
    // Inizialize the model with the second head, $\mathcal{M}_{h_2}^k$.
    $backbone(\mathcal{M}_{h_2}^k) \leftarrow backbone(\mathcal{M}_{h_1}^k)$
    $head(\mathcal{M}_{h_2}^k)$ randomly inizialized
end
// Communication
for *each client* $k \in \{1, \ldots, K\}$ *in parallel* **do**
    $\Phi_k \leftarrow$ indices of incoming connected clients in $\mathcal{G}$ // remote client indices
    Share $\mathcal{M}_{h_1}^k$ with all outgoing connected clients in $\mathcal{G}$
    Receive $\mathcal{M}_{h_1}^\phi$ from all remote clients $\phi \in \Phi_k$
end
// Knowledge Distillation
foreach *client* $k \in \{1, \ldots, k\}$ *in parallel* **do**
    /* Train the model with the second head $\mathcal{M}_{h_2}^k$ using KD until convergence. Use the loss $\mathcal{L}_k = \alpha \mathcal{L}_{k,CE} + (1-\alpha)\mathcal{L}_{k,KD}$, where $\mathbf{w}_2^k$ are the model parameters to be updated, $\mathcal{L}_{k,KD}$ is calculated either using Eq. Eq.10 or 11 */
    $\mathcal{M}_{h_2}^k \leftarrow$ LocalModelTraining$\left(\mathcal{L}_k(\mathbf{w}_2^k, D^k)\right)$
end

Performance evaluation was conducted using 10% of the entire data distribution for both iid and non-idd datasets. For each client, we computed the accuracy of its model. In the next section, we present aggregated results, specifically the mean accuracy across the three clients.

In our implementation, we trained the three clients using independent Docker containers, each saving the model checkpoints to a shared folder. To train the second head of one client, the first heads from other remote clients are loaded from this shared folder for distillation. This choice was made to simplify the implementation and does not affect the analysis of the models' accuracy and the performance of the various losses. The study of training efficiency, including communication costs of model parameters, is left for future work.

All models are based on ResNet18 [9] and are initialized with weights pretrained on ImageNet, as provided by PyTorch. We also employ standard data augmentation techniques as recommended in the PyTorch documentation[4] for ResNet18.

For the second head, we modified the classifier of ResNet18, using two dense hidden layers with 512 and 256 neurons for CIFAR-10 and 1024 and 512 neurons for SUN397, respectively. We set the skewness parameter to 15 for CIFAR-10, and 10 for SUN397. In all cases, the batch size is set to 128. The optimizer used is SGD with an initial learning rate of 0.001, momentum of 0.9, and a weight decay of 5×10^{-4}.

We performed the distillation using various temperature T values (1, 10, and 100),[5] depending on the dataset distribution. We report the optimal temperatures for each dissimilarity measure and dataset. For CIFAR-10 in the non-iid. context, $T = 10$ provided the best results for all dissimilarity measures for both the sum and average of remote predictions. In the iid context, $T = 10$ was optimal for the sum of distillation losses and only for CE, KL, and TD in the case of the average of remote predictions. For JS and SED, $T = 1$ is used. Moreover, for the SUN397 dataset, $T = 10$ was best for CE and KL, and $T = 1$ for all other cases, both for the sum and average of remote predictions.

The code to reproduce the experiments is available at https://github.com/joaquimbasa/Distributed_KD_Information_Dissimilarity.git.

5 Results and Discussion

Consistent with the correlation findings in [6], our experiments with an iid distribution of data among three clients revealed that the various dissimilarity measures tested in the KD-loss yielded comparable results. Figure 3a and b present the average accuracy of secondary-head model $\mathcal{M}_{h_2}^k$ of clients belonging to $\mathcal{G}$ under iid data conditions on CIFAR-10 dataset, while varying the hyperparameter α. Here, $\alpha = 0$ indicates that the total loss comprises only the distillation loss, while $\alpha = 1$ indicates that no distillation from remote clients is performed, and each model is trained solely in a supervised manner using its local annotated dataset. Overall, our results indicate that KD does not significantly enhance overall accuracy when the input data is sufficient and balanced. Furthermore, all tested dissimilarity measures exhibited performance similar to CE. This observation is consistent across both cases for computing the distillation loss: using the sum of pairwise distillation losses between the current client's predictions and those of each remote client (Eq. (10)), as shown in Fig. 3a, and using the distillation loss between the current client's prediction and the average of predictions from remote clients (Eq. (11)) as shown in Fig. 3b. Based

---

[4] https://pytorch.org/vision/main/models/generated/torchvision.models.resnet101.html.

[5] As noted in [1], the best temperature is highly context-dependent, but a wide range of temperatures can be useful. They suggest using temperatures in the range of 0.1 up to 100.

on this observation, in iid settings, the choice of a dissimilarity measure may depend on implementation requirements, with a preference for computationally efficient measures such as TD. Figure 3b also demonstrates that using distillation with the average predictions of remote clients C^Φ results in similar, and in some cases slightly better, performance than the sum of pairwise losses. This approach has the added advantage of allowing the computation of a single loss instead of multiple pairwise losses, thereby reducing computational complexity.

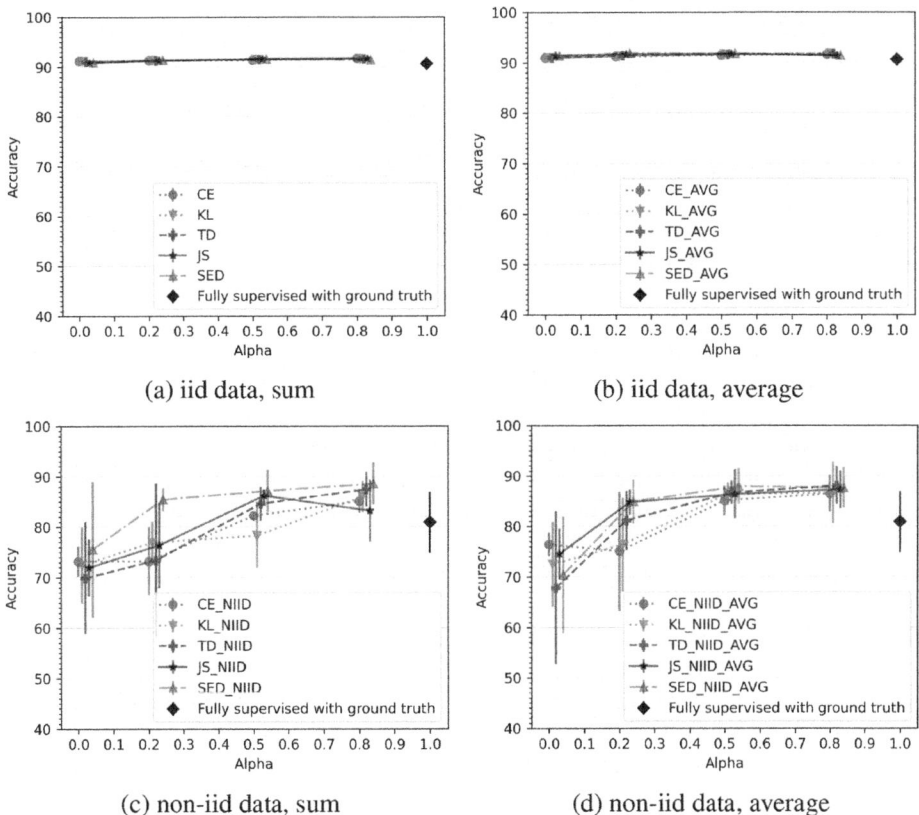

Fig. 3. CIFAR-10: Mean accuracy over three clients considering the *sum* of the distillation losses (Eq. 10) in the left-hand plots, and the the *average* of remote predictions to compute the distillation loss (Eq. 11) in the right-hand plots. Results for iid data are shown in the top row, and results for non-iid data are shown in the bottom row. The bars indicate the standard deviation of accuracy across the three clients (not visible in the iid case, where the standard deviation is less than 0.8).

In the case of non-iid distribution (Fig. 3c and d), the distillation process led to an increase in the average accuracy of the clients' models compared to the fully-supervised approach. This improvement is particularly noticeable for the value $\alpha = 0.5$. For this value, all measures show minimal variance among

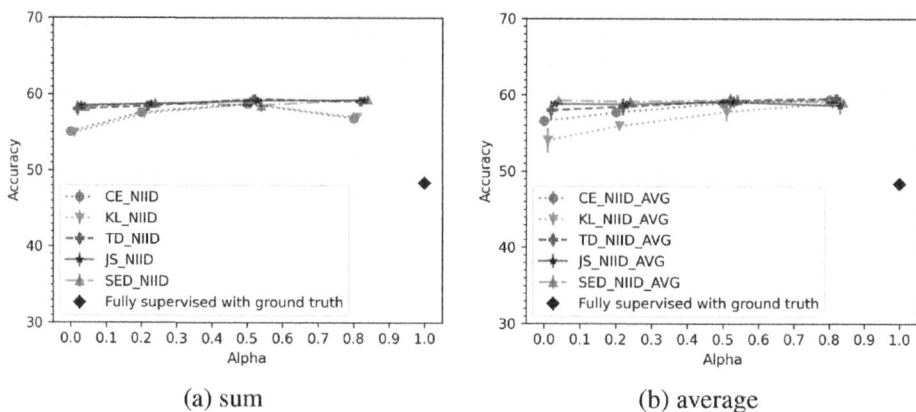

Fig. 4. SUN397 (non-iid data): Mean accuracy over three clients considering (a) the *sum* of the distillation losses (Eq. 10); (b) the *average* of C^Φ prediction to compute the distillation loss (Eq. 11). The standard deviation of accuracy across the three clients is less than 0.9 for all the plotted cases.

the three clients (as indicated by the vertical bars) except in 3c, where the KL provides a high variance compared to others. For $\alpha > 0$ values, minimal differences are observed between JS and SED when computing the distillation loss with the average of predictions generated by the remote clients, whereas CE and KL perform worse in case $\alpha = 0.2$. Furthermore, the average of the predictions obtained from remote clients, in Fig. 3d shows that for $\alpha = 0.2$, SED and JS already exhibit good performance. However, for $\alpha = 0.8$, all measures perform similarly, with KL having higher variance across clients. On the other hand, SED appears to be superior to other measures from $\alpha = 0.2$, providing minimal variance when considering the sum of distillation losses.

In addition to CIFAR-10, we also performed experiments in the non-iid scenario using the SUN397 dataset. In these experiments, adding more layers to the second head caused the model to overfit, showcasing an average accuracy of 48.33% compared to the first head, showcasing an average accuracy of 57.74% over all clients. This confirms the argument made in [28] that when the client's training data is scarce, leading to model overfitting, communication between clients can enhance generalization and improve client's performance on their private tasks. Additionally, communication between clients improves their learned representations, making them better suited for adapting to tasks from other clients.

Regarding the performance of the different dissimilarity measures, Fig. 4a and b show that CE and KL are outperformed by SED, TD, and JS distances for $\alpha = 0$ and $\alpha = 0.8$ when using the sum of distillation losses from each remote client. However, when using the average of remote predictions, the CE and KL perform worse for the values of $\alpha = 0$ and $\alpha = 0.2$. In other cases, all measures perform equally well. Notably, all measures exhibit very low variance among the three clients.

6 Conclusions

This paper empirically evaluated different information dissimilarity measures in a distributed KD setting. The core of our study was to understand the effectiveness of these measures using various data distributions. Furthermore, we used a multi-head neural network to facilitate knowledge transfer among clients, demonstrating that distance measures can significantly impact the training of distributed models using KD on non-iid data. Notably, the commonly used cross-entropy and Kullback-Leibler divergences are not always the most effective.

In future work, we plan to examine the stability of gradients (e.g., exploding or vanishing gradients) associated with the analyzed information dissimilarity measures, and evaluate the performance of the proposed distributed KD framework with a larger number of nodes and various graph topologies.

Acknowledgment. This work was partially funded by National Centre for HPC, Big Data and Quantum Computing project (EU NextGenerationEU PNRR, CUP B93C22000620006), and SUN – Social and hUman ceNtered XR (EC, Horizon Europe n. 101092612).

References

1. Agarwala, A., Pennington, J., Dauphin, Y., Schoenholz, S.: Temperature check: theory and practice for training models with softmax-cross-entropy losses. arXiv preprint arXiv:2010.07344 (2020)
2. Aguilar, G., Ling, Y., Zhang, Y., Yao, B., Fan, X., Guo, C.: Knowledge distillation from internal representations. In: Proceedings of the AAAI Conference on Artificial Intelligence, vol. 34, pp. 7350–7357 (2020)
3. Bistritz, I., Mann, A., Bambos, N.: Distributed distillation for on-device learning. Adv. Neural. Inf. Process. Syst. **33**, 22593–22604 (2020)
4. Carta, A., Cossu, A., Lomonaco, V., Bacciu, D., van de Weijer, J.: Projected latent distillation for data-agnostic consolidation in distributed continual learning. Neurocomputing 127935 (2024)
5. Connor, R.: A tale of four metrics. In: 9th International Conference on Similarity Search and Applications, SISAP 2016, pp. 210–217. Springer (2016)
6. Connor, R., Dearle, A., Claydon, B., Vadicamo, L.: Correlations of cross-entropy loss in machine learning. Entropy **26**(6) (2024)
7. Gou, J., Xiong, X., Yu, B., Du, L., Zhan, Y., Tao, D.: Multi-target knowledge distillation via student self-reflection. Int. J. Comput. Vis. **131**(7), 1857–1874 (2023)
8. Gou, J., Yu, B., Maybank, S.J., Tao, D.: Knowledge distillation: a survey. Int. J. Comput. Vis. **129**(6), 1789–1819 (2021)
9. He, K., Zhang, X., Ren, S., Sun, J.: Deep residual learning for image recognition. In: Proceedings of the IEEE Conference on Computer Vision and Pattern Recognition, pp. 770–778 (2016)
10. Hinton, G., Vinyals, O., Dean, J.: Distilling the knowledge in a neural network (2015)
11. Jeong, E., Kountouris, M.: Personalized decentralized federated learning with knowledge distillation. In: ICC 2023-IEEE International Conference on Communications, pp. 1982–1987. IEEE (2023)

12. Jin, H., Bai, D., Yao, D., Dai, Y., Gu, L., Yu, C., Sun, L.: Personalized edge intelligence via federated self-knowledge distillation. IEEE Trans. Parallel Distrib. Syst. **34**(2), 567–580 (2023)
13. Kim, T., Oh, J., Kim, N., Cho, S., Yun, S.Y.: Comparing kullback-leibler divergence and mean squared error loss in knowledge distillation. arXiv preprint arXiv:2105.08919 (2021)
14. Krizhevsky, A., Nair, V., Hinton, G.: Cifar-10 (Canadian institute for advanced research) (2009)
15. Liu, X., Yu, J., Liu, Y., Gao, Y., Mahmoodi, T., Lambotharan, S., Tsang, D.H.K.: Distributed intelligence in wireless networks. IEEE Open J. Commun. Soc. **4**, 1001–1039 (2023)
16. Luo, Y., Huang, Q., Ling, J., Lin, K., Zhou, T.: Local and global knowledge distillation with direction-enhanced contrastive learning for single-image deraining. Knowl. Based Syst. **268**, 110480 (2023)
17. Markatou, M., Chen, Y., Afendras, G., Lindsay, B.G.: Statistical distances and their role in robustness. In: New Advances in Statistics and Data Science, pp. 3–26 (2017)
18. Mishra, R., Gupta, H.P.: Designing and training of lightweight neural networks on edge devices using early halting in knowledge distillation. IEEE Trans. Mobile Comput. (2023)
19. Molo, M.J., Carlini, E., Ciampi, L., Gennaro, C., Vadicamo, L.: Teacher-student models for AI vision at the edge: a car parking case study. Proceedings Copyright **508**, 515 (2024)
20. Moss, R., Connor, R.: A multi-way divergence metric for vector spaces. In: Similarity Search and Applications: 6th International Conference, SISAP 2013, A Coruña, Spain, October 2–4, 2013, Proceedings 6, pp. 169–174. Springer (2013)
21. Nielsen, F.: On a generalization of the Jensen-Shannon divergence and the Jensen-Shannon centroid. Entropy **22**(2), 221 (2020)
22. Parker, L.E.: Distributed intelligence: overview of the field and its application in multi-robot systems. In: AAAI Fall Symposium: Regarding the Intelligence in Distributed Intelligent Systems, pp. 1–6 (2007)
23. Sahni, Y., Cao, J., Zhang, S., Yang, L.: Edge mesh: a new paradigm to enable distributed intelligence in internet of things. IEEE Access **5**, 16441–16458 (2017)
24. Topsoe, F.: Some inequalities for information divergence and related measures of discrimination. IEEE Trans. Inf. Theory **46**(4), 1602–1609 (2000)
25. Tung, F., Mori, G.: Similarity-preserving knowledge distillation. In: Proceedings of the IEEE/CVF International Conference on Computer Vision, pp. 1365–1374 (2019)
26. Wong, A.K., You, M.: Entropy and distance of random graphs with application to structural pattern recognition. IEEE Trans. Pattern Anal. Mach. Intell. **5**, 599–609 (1985)
27. Yang, Z., Zeng, A., Li, Z., Zhang, T., Yuan, C., Li, Y.: From knowledge distillation to self-knowledge distillation: a unified approach with normalized loss and customized soft labels. In: Proceedings of the IEEE/CVF International Conference on Computer Vision, pp. 17185–17194 (2023)
28. Zhmoginov, A., Sandler, M., Miller, N., Kristiansen, G., Vladymyrov, M.: Decentralized learning with multi-headed distillation. In: Proceedings of the IEEE/CVF Conference on Computer Vision and Pattern Recognition, pp. 8053–8063 (2023)
29. Zhou, B., Lapedriza, A., Xiao, J., Torralba, A., Oliva, A.: Learning deep features for scene recognition using places database. Adv. Neural Inf. Process. Syst. **27** (2014)

An Empirical Evaluation of Search Strategies for Locality-Sensitive Hashing: Lookup, Voting, and Natural Classifier Search

Malte Helin Johnsen and Martin Aumüller[(✉)]

IT University of Copenhagen, Copenhagen, Denmark
malte.h.johnsen@gmail.com, maau@itu.dk

Abstract. Approximate nearest neighbor search in high-dimensional metric spaces is crucial in modern data science pipelines. Efficient search algorithms often rely on partitioning the metric space. To find the approximate nearest neighbors of a query point, a candidate set is constructed based on the points that belong to the same part in the partition, and the closest points among these candidates are identified via a bruteforce search. Hyvönen et al. (JMLR, 2024) argue that viewing this problem as a multi-class labeling problem suggests that this traditional method is not optimal. Instead, they propose a "natural classifier" search strategy that incorporates the true labels of the candidate points, demonstrating faster searches and smaller candidate sets for the same accuracy for tree-based space partitioning methods. This paper explores the natural classifier and other search strategies for partitioning based on locality-sensitive hashing. We propose a new strategy that offers more precise control over the balance between performance and quality. Our analysis highlights the trade-offs between these methods, providing insights into optimizing search efficiency in various contexts.

Keywords: Approximate Nearest Neighbor Search · Locality-Sensitive Hashing · Similarity Search · Search Strategies

1 Introduction

Locality-sensitive hashing (LSH) as introduced by Indyk and Motwani [28] is a prominent technique in the context of similarity search in high-dimensional metric spaces. Given access to a locality-sensitive hash function h that maps each point x in the metric space to some value in a range R, one can partition a given dataset $\mathbf{X}$ by applying h to each data point $x \in \mathbf{X}$, thus *partitioning the metric space*. To find similar points in $\mathbf{X}$ to a query point q, one applies the very same hash function and checks all the points that have the hash value $h(q)$, traditionally by storing the partition of $\mathbf{X}$ using a hash table.

The main advantage of LSH-based approaches to similarity search problems is the ability to design theoretically sound algorithms that solve diverse search problems. However, since the rise of graph-based nearest neighbor search with methods such as HNSW [33], ONNG [29], and DiskANN [38], its suitability for efficient high-dimensional nearest neighbor search in a practical context must be re-evaluated. For example, the state-of-the-art evaluation of *billion-scale approximate nearest neighbor search* (ANNS) provided by Manohar et al. in [34] disregards one of the most popular LSH libraries, FALCONN [3], as too inefficient to even survive the "warmup comparison" on million-scale datasets.

What could be a missing ingredient for LSH? As also shown in [34], graph-based ANNS algorithms are able to retrieve a high-quality candidate set with only very few distance comparisons between the query point and points from the dataset $\mathbf{X}$, while both LSH and clustering-based approaches require several orders of magnitude more candidates to achieve the same quality. In an exciting development for tree-based ANNS, Hyvönen et al. [25,26] provide the following case *against* the search strategies commonly applied by these partition-based techniques. They base their argument on viewing ANNS as a multi-label classification problem: Given a set $\mathbf{X}$ of data points with ground truth labels reflecting the individual k nearest neighbors, the goal of ANNS is to predict the labels for an unknown point $q \in \mathcal{X}$. They argue that if this was the task, then returning the points colliding with q under the hash function would not be an optimal strategy for this labeling problem. Instead, one should return a majority of *the labels of these points*! Independently of this work, this labeling strategy was also considered theoretically by Efremenko et al. [21] for LSH.

In this paper, which is based on the first author's Master's thesis [30], we consider how these developments translate into practical improvements for LSH. After reviewing the basic definitions and approaches in Sect. 2, we define three different search strategies in Sect. 3. In addition to the standard lookup search and the natural classifier proposed in [25], we also consider the approach "voting search" that counts collisions and only considers points that have been seen a "sufficient number of times" [14,27]. In that section, we also propose a variant of the natural classifier that allows for easier-to-set parameters to control the quality of the result. In Sect. 4 we experimentally evaluate these search strategies for LSH index structures in the context of million-scale ANNS. Using the concept of local-intrinsic dimensionality [2,22], we evaluate how the difficulty of individual queries as considered in [5] impact the benefit of the natural classifier. We compare our results to the random projection forest baseline considered by Hyvönen et al. [25].

In a nutshell, the natural classifier improves the performance of LSH for datasets of low to medium difficulty, but is outperformed by voting search for medium to high difficulty datasets, in particular in the high recall regime. Our analysis further provides insights into the robustness of LSH-based ANNS for different index parameter choices using voting and natural classifier search, which both provide more robustness for index parameters that are usually considered crucial for the performance of LSH. In particular in the low space regime, where only very few repetitions are available, the natural classifier provides compelling performance.

1.1 Related Work

Approaches to ANNS in High-dimensional Data. Approximate nearest neighbor search algorithms usually follow a partition-based approach, such as tree (e.g., kd-trees [9], cover trees [10,24], and random projection trees [16,31]), hashing (e.g., LSH [28] and LSF [12]), and clustering (e.g., (hierarchical) k-means [20,39]) approaches. Non-partition-based approaches include graph-based approaches such as HNSW [33] and DiskANN [38], among many others. In standardized benchmarks such as [4] and [36], graph and clustering-based approaches provide the fastest empirical running times in billion-scale settings. See also the recent survey [6].

Search Strategies for LSH. LSH approaches usually come with two parameters: L, the number of repetitions and K, the number of hash functions which are compounded for each repetition. These parameters are explained in detail in Sect. 2.3. Many papers suggest different strategies for tuning these parameters in practical settings, see for example, [19,37]. Paulevé et al. [35] provide an overview over different hash family choices and pooling strategies. Lv et al. [32] suggested a multi-probing approach to improve the space-efficiency of LSH-based data structures, in which multiple "close buckets" in a single table are checked for candidate points. The LSH forest [8] provides another adaptive search strategy, which was later successfully applied in [7]. Instead of multi-probing, a trie-like data structure is used in these works to efficiently find candidate points.

2 Preliminaries

2.1 Problem Definition

Let $\mathcal{X}$ be a metric space equipped with distance function dist. Given a dataset $\mathbf{X} \subseteq \mathcal{X}$, a data point $x \in \mathcal{X}$, and an integer $k \geq 1$, we let $k\mathrm{NN}(x)$ be the set of the k closest points to x in $\mathbf{X}$. The task in *approximate nearest neighbor search* (ANNS) is to build an index $\mathcal{I}$ over $\mathbf{X}$ that supports approximate search queries. Given a query point $q \in \mathcal{X}$ and an integer $k \geq 1$, the search returns a tuple $k\mathrm{ANN}(q) := (x'_1, \ldots, x'_k) \in \mathbf{X}^k$ of k distinct data points. The quality of that solution is measured by its *recall* against the ground truth $k\mathrm{NN}(q)$. Following [4], the recall is defined as $|\{x \in k\mathrm{ANN}(q) \mid \mathrm{dist}(q,x) \leq \mathrm{dist}(q,x_k)\}|/k$, where x_k denotes the k-th most distant neighbor to q in $k\mathrm{NN}(q)$.

2.2 Partition-Based Approximate Nearest Neighbor Search

A common category of approaches to solving ANNS are the so-called partition-based approaches, which rely on searching multiple partitions of $\mathcal{X}$, where $\mathbf{X} \subseteq \mathcal{X}$. A partition P is defined as a set $\{P_1, P_2, \ldots\}$ where each part P_i is a mutually disjoint subset of $\mathcal{X}$, such that $P_i \cap P_j = \emptyset$ whenever $i \neq j$, and $\bigcup_{P_i \in P} P_i = \mathcal{X}$. Standard partition-based solutions to k-ANNS and specifically the solutions explored in this paper follow a common multi-step approach, a model of which

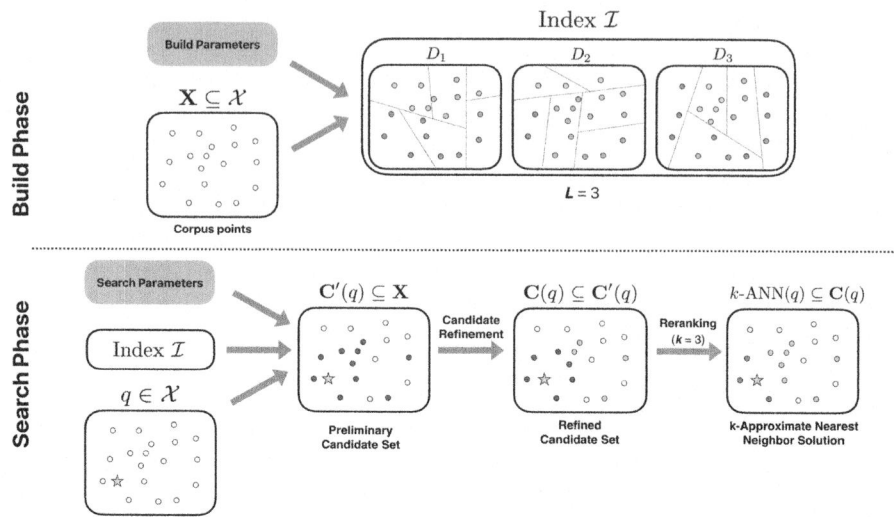

Fig. 1. Overview of partition-based k-ANNS.

is depicted in Fig. 1. At the highest level of abstraction, these approaches can be divided into a *build-phase* and a *search-phase*. In the build phase, an indexing structure is constructed based on a set of build parameters. The index structure is typically an ensemble of L data structures $\mathcal{I} = \{D_1, ..., D_L\}$, where each data structure D_i, induces a partition on $\mathcal{X}$. Once built, a single data structure $D_i \in \mathcal{I}$ supports the query operation query(D_i, q), for a point $q \in \mathcal{X}$, which returns a set of points $Q_i \subseteq \mathbf{X}$ belonging to the same part as q.

Once the build phase is completed, $\mathcal{I}$ can be used in a search phase, where given a set of search parameters and a query vector q, it is searched to return a solution, $k\text{ANN}(q)$. In this phase, each data structure in the index is queried and the returned sets of points are used to construct a preliminary candidate set $\mathbf{C}'(q)$ with $\mathbf{C}'(q) = \bigcup_{D_i \in L} \text{query}(D_i, q)$. Based on a score function, which assigns a score to each point in $\mathbf{C}'(q)$, and a set of supplied search parameters, in the candidate refinement process $\mathbf{C}'(q)$ is trimmed to a set of candidate points $\mathbf{C}(q) \subseteq \mathbf{C}'(q)$. In the final re-ranking step, a brute-force search is carried out considering only points in $\mathbf{C}(q)$, to determine the final set $k\text{ANN}(q)$. Therefore any point $x \in \mathbf{X}$ belongs to $k\text{ANN}(q)$ iff. $x \in \mathbf{C}(q)$. The combination of score function and refinement condition is called a *search strategy*. The combination of an index structure and a search strategy make up a *k-ANNS algorithm*.

2.3 Locality-Sensitive Hashing

This paper focuses on using locality-sensitive hashing [28] to partitioning the space $\mathcal{X}$. Following the definition of Charikar [11, Definition 1], a family $\mathcal{H}$ of hash functions mapping from $\mathcal{X}$ to a range R is called locality-sensitive if

Algorithm 1 Lookup Search

1: **procedure** LOOKUPSEARCH($\mathcal{I}, q$)
2:     $\mathbf{C}(q) \leftarrow \emptyset$;
3:     **for** $D_i \in \mathcal{I}$ **do**
4:         $Q_i \leftarrow$ query(D_i, q)
5:         **for** $x \in Q_i$ **do**
6:             Add x to $\mathbf{C}(q)$
7:     **return** bruteForceSearch($\mathbf{C}(q), q$)

Algorithm 2 Voting Search

1: **procedure** VOTINGSEARCH($\mathcal{I}, q, \tau$)
2:     $\mathbf{C}(q) \leftarrow \emptyset$; score $\leftarrow (0, \ldots, 0)$
3:     **for** $D_i \in \mathcal{I}$ **do**
4:         $Q_i \leftarrow$ query(D_i, q)
5:         **for** $x \in Q_i$ **do**
6:             Increment score(x) by 1
7:             **if** score$(x) = \tau$ **then**
8:                 Add x to $\mathbf{C}(q)$
9:     **return** bruteForceSearch($\mathbf{C}(q), q$)

the collision probability $\Pr_{h \in \mathcal{H}}(h(x) = h(y))$ is monotonically decreasing in the distance of $x, y \in \mathcal{X}$. Assume in the following that we have access to such a family $\mathcal{H}$.

The LSH index structure is an ensemble structure in which each data structure indexes $\mathbf{X}$ using $\mathcal{H}$. It requires the user to provide two parameters $K, L \geq 1$. A single data structure D_i consists of a hash table T_i, and a compound hash function g_i which is used to index points to buckets in T_i. The compound hash function g_i consists of a K-tuple of hash functions $g_i = (h_1, ..., h_K)$, where each h_j is drawn independently and uniformly at random from $\mathcal{H}$. To build D_i, all points $x \in \mathbf{X}$ are indexed according to the compound hash function, such that $g_i(x) = (h_1(x), ..., h_K(x)) \in R^K$, and are inserted in the corresponding bucket in the hash table T_i. A data structure $D_i \in \mathcal{I}$ is subsequently queried by hashing a query point $q \in \mathcal{X}$ according to the same compound hash function and returning the set of points $\{x \in \mathbf{X} \mid g_i(x) = g_i(q)\}$ which belong to the same bucket as q in the table T_i. The index $\mathcal{I}$ consists of L data structures $D_1, \ldots, D_L$ that are built independently. Thus, the parameter K controls the number of hash functions used for each compound hash function g_i, while the parameter L controls the number of data structures in $\mathcal{I}$. Preprocessing the dataset $\mathbf{X}$ consisting of n points requires $O(nKL)$ hash value evaluations and $O(nL)$ insertions into the hash table. To carry out a search, the query point's $O(KL)$ hash values have to be evaluated, and all points that fall into the same bucket as the query point have to be collected. Using linearity of expectation, we expect $O\left(L \cdot \sum_{x \in X} \Pr(h(x) = h(q))^K\right)$ points to collide with the query over all L data structures.

In this work, we let both K and L be parameters set by the user after an initial investigation of the data set characteristics. There exists techniques that speed up hash function evaluation by means of tensoring and pooling [13] and that automatically find good parameters for given space constraints [1,7].

Algorithm 3 Natural Classifier Search

```
1:  procedure NATURALCLASSIFIERSEARCH(I, q, τ)
2:      C(q) ← ∅; score ← (0,..., 0)
3:      for D_i ∈ I do
4:          Q_i ← query(D_i, q)
5:          for x ∈ Q_i do
6:              for y ∈ kNN(x) do
7:                  increment score(y) by 1/(|Q_i|·L)
8:                  if score(y) ≥ τ and y ∉ C(q) then
9:                      Add y to C(q)
10:     return bruteForceSearch(C(q), q)
```

3 Search Strategies

This section will describe the different search strategies that will be carried out on an index $\mathcal{I}$ containing L data structures $D_1, \ldots, D_L$. Each search strategy gets as input $\mathcal{I}$ and the query point $q \in \mathcal{X}$, but might also require additional user-set parameters.

Lookup Search. Lookup search (Algorithm 1) is the standard search operation. It carries out the query operation on each D_i. All points retrieved in this way form the candidate set $\mathbf{C}(q)$, from which the k closest points to q in $\mathbf{C}(q)$ are returned.

Voting Search. Voting search [27] expects a user-set parameter τ that defines the *voting threshold*. Abstractly, it carries out a lookup search to form the initial candidate multiset $\mathbf{C}'(q)$. $\mathbf{C}(q)$ is then formed by all points that appear at least τ times in $\mathbf{C}'(q)$. An efficient implementation that keeps track of the element count is given as Algorithm 2. Voting search can be seen as a variant of confirmation sampling by Christiani et al. [14]. They prove that if a neighbor can be sampled from a data structure that promises that the nearest neighbor is the most likely point to be sampled, a voting-based approach provides strong guarantees on the result quality.

Natural Classifier Search. Algorithm 3 describes natural classifier search, as first introduced by Hyvönen et al. in [26]. As before, it receives an additional parameter τ to be set by the user, where $0 \leq \tau \leq 1$. For each data point x that falls into the same part of the partition as the query point q, it increments the score of each point in $kNN(x)$, i.e., in particular x itself. The increment is both weighted by the number of repetitions, but also by the size of the part Q_i that contains q. The candidate set $\mathbf{C}(q)$ is then made up of all points $x \in \mathbf{X}$ that achieve a score of at least τ. In contrast to voting search, the natural classifier includes the ground truth neighbors and thus can lead to far larger $\mathbf{C}(q)$ for the same setting of K and L. A main point of our investigation is how K, L, and τ influences the observed quality of the ANN result. The natural classifier can be seen as an application of the *query expansion* technique discussed by Chum et al. [15,23], in which the results of a query are used as new query points, to

Table 1. Overview of dataset instances. n = dataset size, m = query set size.

Name	Dim.	n	m	metric	avg. NN dist	median LID
FASHION-MNIST	784	60,000	10,000	Euclidean	917.90	13.8
SIFT	128	1,000,000	10,000	Euclidean	187.75	21.9
GLOVE	100	1,183,514	10,000	Angular	0.31	42.9

enrich the original answer to the search. We precompute $k\mathrm{NN}(x)$ for all $x \in \mathbf{X}$ in the build phase to increase performance during the search phase.

Natural Classifier Search Variants Two variants of natural classifier search were evaluated alongside the one presented above: The voting and the set size variants. The voting variant is similar to the classic natural classifier search but uses a raw count score function, that is an increment value of 1 in place of a weighted value, and therefore counts the number of times a point $x \in \mathbf{X}$ is encountered during search. It adds only points to $\mathbf{C}(q)$ which are encountered at least τ times [25].

The set size variant is a novel variant, which produces a candidate set of a user-specified size. It intends to alleviate a practical issue of the classic natural classifier. We observed that typically the points in $\mathbf{C}'(q)$ have a very large number of repeat scores, leading to a candidate refinement process that only allows for very coarse tuning of the search quality. The set size variant carries out the search exactly as Algorithm 3, but instead forms $\mathbf{C}(q)$ by using the quick-select algorithm to select the τ points with the largest score, performing this selection in $O(\mathbf{C}'(q))$ comparisons.[1] Thus, this variant allows the user to set a strict limit on the number of distance computations carried out by the algorithm [30]. This technique provides a benefit from a practitioner's point of view, as it allows a single index structure to be used for a wider recall range by allowing for a finer degree of tuning, which is also easily interpreted by the user.

4 Evaluation

Implementation Notes. All source code is written in Java 11 using the standard Java SDK libraries for key data structures such as hashmaps. We stress that the goal of this paper is to study the effect of different search strategies in the context of LSH. Its results are aimed to inform whether the effort of creating a performance-oriented implementation in a low-level programming language is warranted with the proposals made in [26]. To allow for an evaluation of the performance of methods described in [26], we re-implemented their tree data structures using the same methodology. For further details and considerations on implementation choices see [30].

---

[1] The quick-select implementation uses a partitioning scheme based on the "Dutch National Flag Problem" approach as described in [18], chosen for its robustness against repeat elements.

Datasets. We make use of three standard datasets used in benchmarking ANNS implementations [4] depicted in Table 1. We used the same split between data and query points as provided in [4] to improve comparability. As observed in [5], the different LID values provide confidence in the diversity of the dataset difficulty.

Choice of Hash Function. As highlighted in Table 1, we conduct experiments both for Euclidean and for Angular distance. Datar et al. [17] define the following hash function for Euclidean distance for a data point $x \in \mathbb{R}^d$: $h(x) = \left\lfloor \frac{\boldsymbol{a} \cdot x + b}{r} \right\rfloor$, where $x, \boldsymbol{a} \in \mathbb{R}^d$, the components of $\boldsymbol{a}$ are i.i.d values drawn from the normal distribution $\mathcal{N}(0, 1)$, $r \in \mathbb{R}$ is a user-set parameter, and b is uniformly drawn from the range $[0, r]$. The intuition behind this family of locality-sensitive hash functions is that through the inner product $\boldsymbol{a} \cdot x$, any vector x is projected onto the real line and shifted by a value of b. By the division of the value r combined with the floor function, the real line is cut into equi-width segments which each map to a value in $\mathcal{Z}$, corresponding to a bucket. As mentioned in [17] and to avoid exhaustive parameter searches, we set the value r to 4 times the average distance of the nearest neighbor in **X** for the given query workload. For angular distance, we can simplify the hash function to output a single bit and can avoid user-set parameters. Given a random $\boldsymbol{a} \sim \mathcal{N}(0, 1)^d$, set $h(x) = 1$ if $\boldsymbol{a} \cdot x > 0$ and 0 otherwise [11].

Evaluation Setup. Experiments were run on 2x14 core Intel Xeon E5-2690v4 (2.60 GHz) with 512GB RAM using Ubuntu 20.04.6 LTS. The experimental pipeline is inspired by the ann-benchmarks project [4] and is able to read their datasets and produce result files that can be interpreted using their evaluation scripts. Our source code is available at https://github.com/KarateMogens/ANNSearch.

Performance and Quality Metrics. To measure the quality of the solution, we compute the recall against the ground truth k nearest neighbors, as detailed in Sect. 2, using a value of $k = 10$ for all test instances. For the natural classifier variants, we use these k true nearest neighbors as labels for each point.[2] From a memory perspective, this is similar to storing k repetitions of the LSH data structure. With regard to performance measurements, we report on the average number of searches carried out per second (SPS) across the entire set of test queries for each dataset. To get more insights into the behavior of the search strategies, we estimate the *local intrinsic dimensionality* (LID) of the query points as described in [2,22], which was shown to be a good estimator for the indexing difficulty of datasets [5].

---

[2] Following suggestions in [25], we also ran experiments where each point used the 50-NN as labels, which did not significantly improve performance.

Objectives. Our experiments are tailored to answer the following questions:

(Q1) Which natural classifier variant should be used?
(Q2) How does the difficulty of the query workload influence the search performance?
(Q3) What are the parameter trends of the LSH-based k-ANNS algorithms?
(Q4) What are the performance trends of the LSH-based k-ANNS algorithms?
(Q5) How do LSH-based search strategies compare to baseline approaches?

Comparison of Natural Classifier Search Variants. Figure 2 shows a direct comparison of the performance of the three evaluated variants of the natural classifier search. In certain ranges of the FASHION-MNIST and SIFT datasets, likely due to its limited search parameter granularity, the voting natural classifier performs slightly worse than the two other variants, while performing similarly on the GLOVE dataset. It is evident from the results that no distinct best-performing variant is discernible with the evaluated parameters, thus from a performance standpoint the three variants are equally relevant. Given that the set size variant presents a qualitative improvement over the other variants by providing a tuning parameter which directly manipulates a central metric of ANNS, we recommend to use this among the evaluated natural classifier search variants. Therefore, we will also use the set size variant as a representative for the natural classifier search strategy throughout the following evaluations.

Influence of Query Difficulty. Figure 3 provides a head-to-head comparison between the search quality of lookup search (LS) and the natural classifier search (NC), visualizing the quality on a *per-query* basis. We provide density plots that

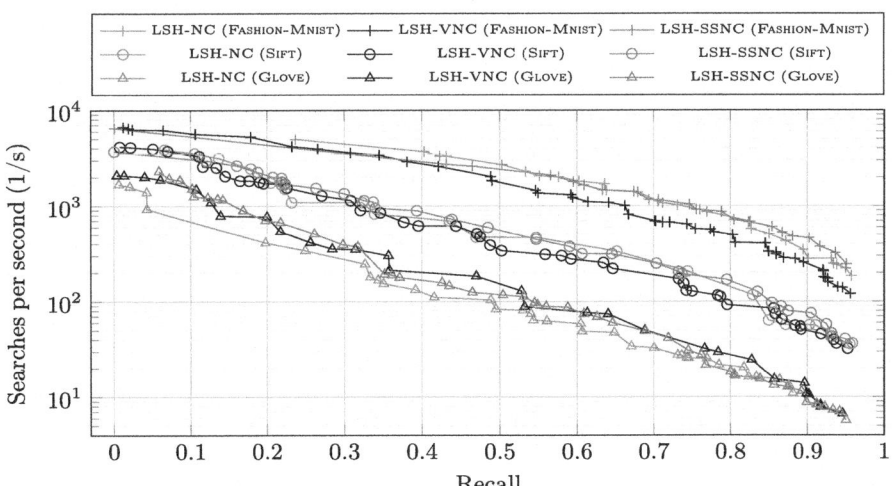

Fig. 2. Recall-Searches per second (1/s) trade-off comparison of natural classifier variants: Natural classifier (NC), voting natural classifier (VNC) and set size natural classifier (SSNC). Each search strategy is paired with the LSH index structure. Up and to the right is better.

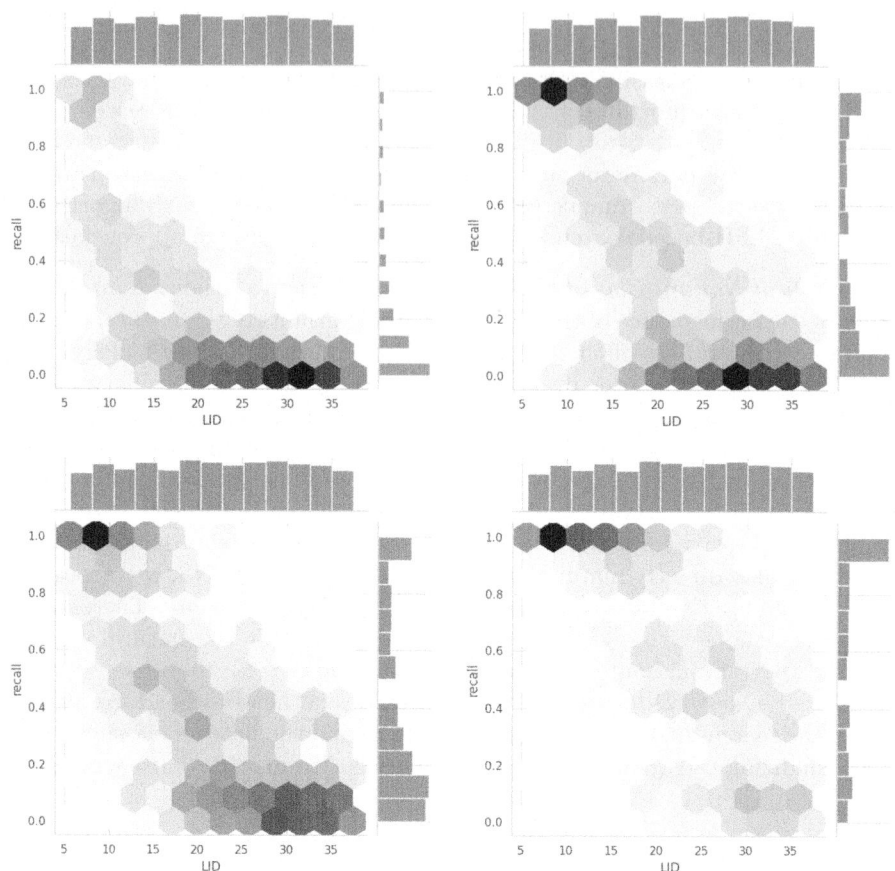

Fig. 3. Lookup search (LS) vs natural classifier search (NC) in small space settings on SIFT using $K = 20$. x-axis: estimated LID value; y-axis: recall. Heatmap visualizes the distribution of (LID, recall) values over all query points, with darker areas indicating higher density. Histogram on top displays the LID distribution of the queries, histogram on the right displays the recall distribution. Top left: LS, $L = 20$ repetitions, avg. recall: 0.18, 1835 SPS; top right: NC, $\tau = 0$, $L = 10$, avg. recall: 0.31, 816 SPS; bottom left: LS, $L = 50$, avg. recall: 0.34, 811 SPS, bottom right: NC, $\tau = 0$, $L = 40$, avg. recall: 0.60, 210 SPS.

relate recall to the LID (local intrinsic dimensionality) [22] of the query point, and we sample queries according to their LID value to get a more uniform spread of LID values, cf. [5].

In the top row of Fig. 3, we compare the results under a constrained memory setting. We compare LS using 20 repetitions, and NC using 10 repetitions, accounting for the extra memory overhead for storing the groundtruth labels. We notice that extending the results with these labels helps in particular low-LID queries, which are answered with close to perfect quality. Queries with larger

Table 2. Recall > .8, FASHION-MNIST. Lookup search (LS), Voting search (VS), Natural classifier (NC), relative running times compared to best individual time in parameter range (LS: 284.3, VS: 427.5, NC: 457.3).

K/L	20			40			80			160			320		
	LS	VS	NC	LS	VS	NC	LS	VS	NC	LS	VS	NC	LS	VS	NC
9	0.49	—	0.38	0.27	0.76	0.39	0.18	0.89	0.19	0.11	0.6	0.09	0.08	0.34	0.05
11	—	—	0.65	0.54	—	0.66	0.28	1.0	0.43	0.11	0.65	0.25	0.11	0.45	0.11
13	—	—	0.83	—	—	0.73	0.55	—	0.55	0.34	0.74	0.39	0.21	0.45	0.26
15	—	—	0.7	—	—	1.0	1.0	—	0.58	0.6	0.82	0.43	0.28	0.48	0.24
17	—	—	—	—	—	0.63	—	—	0.65	0.71	—	0.46	0.37	0.3	0.27

LID values are mostly unaffected. Allowing for 50 and 40 repetitions, respectively, easy queries are also correctly answered by LS, whereas queries with large LID remain answered with low quality. In contrast, using NC leads to much better result quality. A standard lookup search requires at least 200 repetitions to achieve the same result quality.

Robustness of Parameters. Due to space constraints, we only focus on results for FASHION-MNIST. The same trends were present for GLOVE and SIFT as well. Table 2 reports on the largest number of searches per second above an average recall of at least .8. Each running time is normalized against the fastest running time per search strategy within the showcased parameter settings.

For all variants, the best results are obtained with a moderate number of repetitions of not more than 80. For a fixed number of repetitions, lookup search improves with larger K to the point where its quality does not exceed the minimum requirement of .8. Setting the K parameter has a huge influence on the observed running time. In contrast, both the natural classifier and voting search does not necessarily improve with larger K values. For the natural classifier, the repetition count is generally lower than for the other two variants, and the difference between different K values is not as pronounced as for lookup search, except for the lowest tested K value of 9. Voting search benefits from larger repetition counts and shows robust running times for different K values for fixed L. In particular for these more difficult datasets, the natural classifier allows to obtain good recall values using little space.

LSH Search Strategy Evaluation. Figure 4 compares the performance of the three search strategies lookup search (LS), voting search (VS), and set size natural classifier (SSNC) when coupled with the LSH index structure across the three datasets FASHION-MNIST, SIFT and GLOVE. Considering first the results for FASHION-MNIST we find that LS achieves the worst performance of the three, while the SSNC achieves the best. Finally VS achieves a performance solidly between the two for the entire recall range. For a recall threshold of 0.8 the following values are achieved; LS: 284.27, VS: 489.16, SSNC: 715.18.

Turning to the SIFT dataset, LS again performs the worst of the three strategies. In the recall range up to approximately 0.55 the VS and SSNC strategies

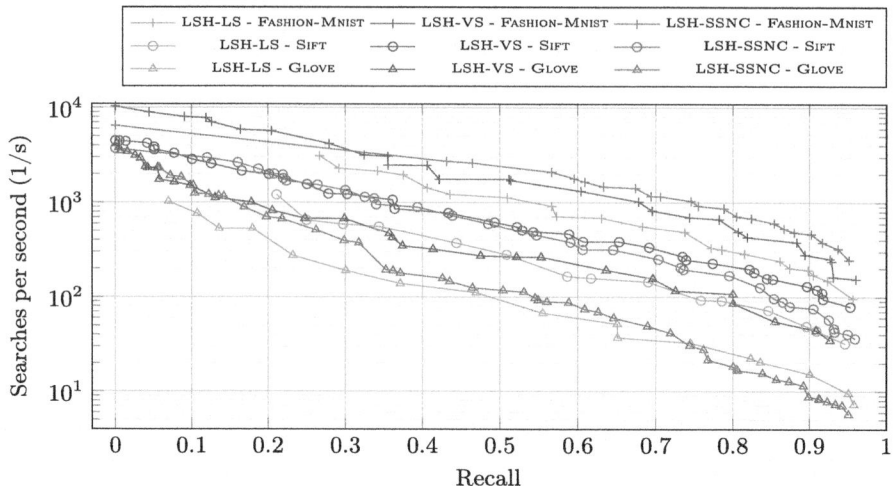

Fig. 4. Recall-Searches per second (1/s) trade-off comparison of search strategies: Lookup search (LS), voting search (VS), and set size natural classifier (SSNC). Each search strategy is paired with the LSH index structure. Up and to the right is better.

perform similarly, however in the higher recall range, especially above 0.85, VS outperforms SSNC. Finally, for the GLOVE dataset, VS outperforms the other two strategies by a wide margin in the recall range above 0.3. This dominance of VS stems from a drop in the performance of SSNC, which in the range from 0.35 to 0.75 recall degrades in performance to match LS, and in the range above 0.75 is even slightly outperformed by LS. For a recall threshold of 0.8 VS outperforms SSNC by more than a factor of 5 (LS: 22.58, VS: 107.79, SSNC: 18.44).

Our results indicate that SSNC (and thus the natural classifier search strategy in general) performs well on the easiest of the three datasets but its dominance over the other strategies degrades as the difficulty of the datasets increase, as evidenced by being outperformed by VS in the high recall range of SIFT and in all ranges of GLOVE. Our results also indicate a marked robustness of VS across increasingly difficult datasets, while performing adequately on the easiest of the three datasets, and in all cases significantly better than LS. Thus, we add nuance the findings presented in [25] which conclude the natural classifier search strategy to be a strict improvement over both the LS and VS search strategies on the datasets which they evaluate.

Comparison to Random Projection Forest (RP) Baseline. Figure 5 shows the performance of the LSH variants in context of the RP baseline that was also used in [25]. For FASHION-MNIST, the RP baseline clearly outperforms the LSH-based baselines with both VS and SSNC performing equally well. For SIFT, the RP benefits from VS over SSNC in the high recall regime, validating trends that have been observed in [25] and in our previous experiments. Still, VS on RP provides better performance in the high recall regime than LSH using VS. For

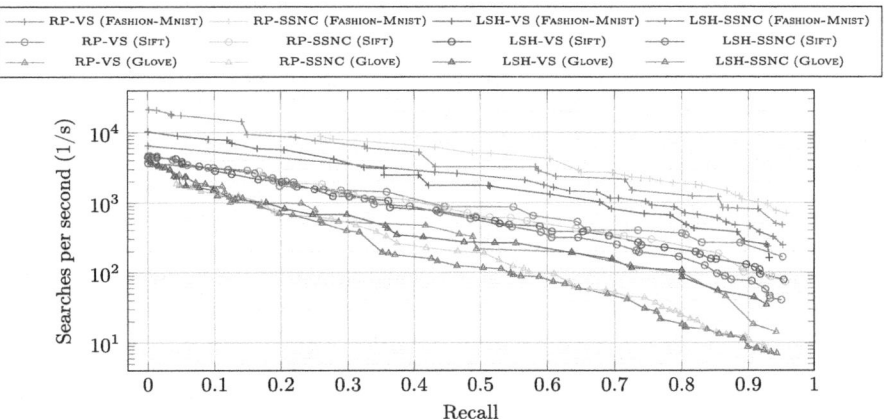

Fig. 5. Recall-Searches per second (1/s) trade-off comparison of search strategies: Voting search (VS) and set size natural classifier (SSNC) paired with random projection forest (RP) and LSH index structures. Up and to the right is better.

GLOVE, the performance of SSNC on RP decreases rapidly; in particular, SSNC shows the same trends as observed for LSH and is clearly outperformed by VS, which again adds a nuance to the findings presented in [25].

5 Discussion

This paper considered the benefits and limitations of using voting search and natural classifier search for locality-sensitive hashing. Both search strategies provide benefits over the standard lookup search strategy. We recommend the natural classifier as the method of choice for constrained memory settings in which the number of repetitions should be kept to a minimum. On the other hand, voting search provides both a more efficient and more robust alternative when more space is available, in particular for datasets with more difficult query workloads. An interesting question that we did not touch upon is how the quality of the provided labeling influences the result quality. In this work, we provided the ground truth labels through an expensive linear scan. If possible, this should be replaced in favor of some approximate labeling to improve the index construction time.

Acknowledgments. We thank the anonymous reviewers for their insightful comments and suggestions that helped us to improve this paper. This project received funding from the Innovation Fund Denmark for the project DIREC (9142-00001B).

References

1. Ahle, T.D., Aumüller, M., Pagh, R.: Parameter-free locality sensitive hashing for spherical range reporting. In: SODA (2017)
2. Amsaleg, L., Chelly, O., Furon, T., Girard, S., Houle, M.E., Kawarabayashi, K., Nett, M.: Estimating local intrinsic dimensionality. In: KDD (2015)
3. Andoni, A., Indyk, P., Laarhoven, T., Razenshteyn, I.P., Schmidt, L.: Practical and optimal LSH for angular distance. In: NIPS, pp. 1225–1233 (2015)
4. Aumüller, M., Bernhardsson, E., Faithfull, A.J.: Ann-benchmarks: A benchmarking tool for approximate nearest neighbor algorithms. Inf. Syst. **87** (2020)
5. Aumüller, M., Ceccarello, M.: The role of local dimensionality measures in benchmarking nearest neighbor search. Inf. Syst. **101**, 101807 (2021)
6. Aumüller, M., Ceccarello, M.: Recent approaches and trends in approximate nearest neighbor search, with remarks on benchmarking. IEEE Data Eng. Bull. **46**(3), 89–105 (2023)
7. Aumüller, M., Christiani, T., Pagh, R., Vesterli, M.: PUFFINN: parameterless and universally fast finding of nearest neighbors. In: ESA (2019)
8. Bawa, M., Condie, T., Ganesan, P.: LSH forest: self-tuning indexes for similarity search. In: WWW. ACM, pp. 651–660 (2005)
9. Bentley, J.L.: Multidimensional binary search trees used for associative searching. Commun. ACM **18**(9) (sep 1975)
10. Beygelzimer, A., Kakade, S.M., Langford, J.: Cover trees for nearest neighbor. In: ICML 2006. ACM
11. Charikar, M.S.: Similarity estimation techniques from rounding algorithms. In: STOC (2002)
12. Christiani, T.: A framework for similarity search with space-time tradeoffs using locality-sensitive filtering. In: SODA. SIAM, pp. 31–46 (2017)
13. Christiani, T.: Fast locality-sensitive hashing frameworks for approximate near neighbor search. In: SISAP (2019)
14. Christiani, T., Pagh, R., Thorup, M.: Confirmation sampling for exact nearest neighbor search. In: SISAP (2020)
15. Chum, O., Philbin, J., Sivic, J., Isard, M., Zisserman, A.: Total recall: automatic query expansion with a generative feature model for object retrieval. In: ICCV (2007)
16. Dasgupta, S., Sinha, K.: Randomized partition trees for exact nearest neighbor search. In: COLT (2013)
17. Datar, M., Immorlica, N., Indyk, P., Mirrokni, V.S.: Locality-sensitive hashing scheme based on p-stable distributions. In: SCG. SCG '04 (2004)
18. Dijkstra, E.W.: A Discipline of Programming. Prentice-Hall (1976)
19. Dong, W., Wang, Z., Josephson, W., Charikar, M., Li, K.: Modeling LSH for performance tuning. In: CIKM. ACM, pp. 669–678 (2008)
20. Douze, M., Guzhva, A., Deng, C., Johnson, J., Szilvasy, G., Mazaré, P., Lomeli, M., Hosseini, L., Jégou, H.: The faiss library. CoRR abs/2401.08281 (2024)
21. Efremenko, K., Kontorovich, A., Noivirt, M.: Fast and bayes-consistent nearest neighbors. In: AISTATS. Proceedings of Machine Learning Research, vol. 108. PMLR, pp. 1276–1286 (2020)
22. Houle, M.E.: Dimensionality, discriminability, density and distance distributions. In: ICDM Workshops. IEEE Computer Society, pp. 468–473 (2013)
23. Houle, M.E., Ma, X., Oria, V., Sun, J.: Query expansion for content-based similarity search using local and global features. ACM Trans. Multim. Comput. Commun. Appl. **13**(3) (2017)

24. Houle, M.E., Nett, M.: Rank cover trees for nearest neighbor search. In: SISAP (2013)
25. Hyvönen, V., Jääsaari, E., Roos, T.: A multilabel classification framework for approximate nearest neighbor search. J. Mach. Learn. Res. **25**(46), 1–51 (2024)
26. Hyvönen, V., Jääsaari, E., Roos, T.: A multilabel classification framework for approximate nearest neighbor search. In: Koyejo, S., Mohamed, S., Agarwal, A., Belgrave, D., Cho, K., Oh, A. (eds.) NeurIPS (2022)
27. Hyvönen, V., Pitkanen, T., Tasoulis, S., Jääsaari, E., Tuomainen, R., Wang, L., Corander, J., Roos, T.: Fast nearest neighbor search through sparse random projections and voting. In: Big Data 2016 (2016)
28. Indyk, P., Motwani, R.: Approximate nearest neighbors: towards removing the curse of dimensionality. In: STOC. ACM, pp. 604–613 (1998)
29. Iwasaki, M., Miyazaki, D.: Optimization of indexing based on k-nearest neigh-bor graph for proximity search in high-dimensional data. CoRR abs/1810.07355 (2018)
30. Johnsen, M.H.: Helping out a neighbor: a study of partition-based algorithms for k-approximate nearest neighbor search. Master's thesis, ITU (2024)
31. Li, P., Hastie, T.J., Church, K.W.: Very sparse random projections. In: KDD (2006)
32. Lv, Q., Josephson, W., Wang, Z., Charikar, M., Li, K.: Multi-probe LSH: efficient indexing for high-dimensional similarity search. In: VLDB. ACM (2007)
33. Malkov, Y.A., Yashunin, D.A.: Efficient and robust approximate nearest neighbor search using hierarchical navigable small world graphs. IEEE Trans. Pattern Anal. Mach. Intell. **42**(4), 824–836 (2020)
34. Manohar, M.D., Shen, Z., Blelloch, G.E., Dhulipala, L., Gu, Y., Simhadri, H.V., Sun, Y.: Parlayann: Scalable and deterministic parallel graph-based approximate nearest neighbor search algorithms. In: PPoPP. ACM, pp. 270–285 (2024)
35. Paulevé, L., Jégou, H., Amsaleg, L.: Locality sensitive hashing: A comparison of hash function types and querying mechanisms. Pattern Recognit. Lett. **31**(11) (2010)
36. Simhadri, H.V., Williams, G., Aumüller, M., Douze, M., Babenko, A., Baranchuk, D., Chen, Q., Hosseini, L., Krishnaswamy, R., Srinivasa, G., Subramanya, S.J., Wang, J.: Results of the NeurIPS'21 challenge on billion-scale approximate nearest neighbor search. In: NeurIPS (Competition and Demos) (2021)
37. Slaney, M., Lifshits, Y., He, J.: Optimal parameters for locality-sensitive hashing. Proc. IEEE **100**(9), 2604–2623 (2012)
38. Subramanya, S.J., Devvrit, Simhadri, H.V., Krishnaswamy, R., Kadekodi, R.: Rand-nsg: Fast accurate billion-point nearest neighbor search on a single node. In: NeurIPS (2019)
39. Sun, P., Simcha, D., Dopson, D., Guo, R., Kumar, S.: SOAR: improved indexing for approximate nearest neighbor search. In: NeurIPS (2023)

On the Design of Scalable Outlier Detection Methods Using Approximate Nearest Neighbor Graphs

Camilla Birch Okkels[1](✉), Martin Aumüller[1], and Arthur Zimek[2]

[1] IT University of Copenhagen, Copenhagen, Denmark
cabi@itu.dk, maau@itu.dk
[2] University of Southern Denmark, Odense, Denmark
zimek@imada.sdu.dk

Abstract. Efficient and reliable methods for distinguishing outliers in data remain crucial for data analysis. Although supervised methods based on neural networks have gained recent traction, unsupervised methods such as the kNN outlier method and local outlier factor (LOF) remain state-of-the-art solutions according to different standardized benchmarks. Unfortunately, exact outlier detection through nearest neighbor search queries provides a scalability bottleneck for the high-dimensional, big datasets that are routinely analyzed in data science applications. This paper explores benefits and limitations of using approximate nearest neighbor search via Hierarchical Navigable Small World graphs (HNSW) to overcome this scalability barrier. We evaluate direct implementations that compute the kNN and LOF score from approximate neighborhoods and show the robustness of the outlier detection even in settings where the approximation is far away from the exact neighborhoods. Furthermore, we design white-box methods that compute the outlier scores directly from the underlying graph. These methods show much more variability in the quality of the outlier scores and open new ground for the development of task-aware tools based on approximate nearest neighbor search techniques.

1 Introduction

At the core of many outlier detection methods is some notion of density and the intuition that outliers are observations appearing in areas of relatively low density. Most observations are clustered, i.e., in areas of relatively high density around their location. In this fundamental intuition, several aspects are defined in different ways when implemented in different methods. Firstly there would be some particular way of estimating the density (rarely as a proper kernel density estimate, such as in KDEOS [25], but usually using some assessment of ε-range queries or of k-nearest neighbors and their distances, which can be seen as proxies for a proper density estimation [31]). Secondly some particular way of deciding if the density is low or high, and finally some particular way to compare local density estimates, i.e., the aspect of the density being *relatively* low or high. The last aspect is reflected in the notion of "locality" in outlier detection methods [26]. Since the notion of outlierness itself is (perhaps necessarily) vague and inexact [31], and commonly used techniques for density estimation and their comparison are inexact by nature, it has been argued that the exact neighborhood might also be dispensable, if we only have a good approximation [15,31]. We follow this line of thought here and consider a more recent alternative [17] to the approximate methods explored in related work. The benefit of using approximate nearest neighbor search is obvious in terms of efficiency, addressing a bottleneck in many data mining algorithms, and particularly outlier detection algorithms.

Given the importance of outlier detection in large data collections, the guiding research question is: *How can recent advances in approximate nearest neighbor search improve the scalability of classical outlier detection methods without sacrificing the quality of the result?* In this paper, we focus on designing scalable variants for the well-known unsupervised outlier detectors kNN and LOF (local outlier factors) [7]. The basis of this investigation will be HNSW (Hierarchical Navigable Small World Graph) by Malkov and Yashunin [17], which stood the test of time, for example testified by an invited talk at SISAP 2023. Despite its weak worst-case guarantees [12], it is the basis of most implementations in the ann-benchmarks [4] project, and remains one of the best performing implementations even for approximate nearest neighbor search on billion-scale datasets [18]. Since both the kNN and LOF method rely on information about the k nearest neighbors of each data point, it is no surprise that a method with close to exact results will provide reliable outlier scores. However, while fast search time is paramount for approximate nearest neighbor search, the time to build the index is part of the running time considerations for downstream tasks such as outlier detection. Thus, optimal hyper-parameters for fast approximate searches might render the approach noncompetitive due to the index building time, and a closer investigation of the quality-performance trade-off is warranted.

While ANN methods can often be used as a "black-box" by replacing the exact nearest neighbor search routines in data mining applications, not much focus has been put on investigating if one can directly use the indexing data structure. Since popular outlier detection methods such as ODIN work directly on the kNN graph by using the in-degree as outlier score, a direct use of the HNSW graph could potentially further improve the scalability. However, the nature of approximate nearest neighbor graphs poses many challenges: first, these approaches use undirected edges and impose a strict degree bound on each node; second, a key insight of approximate nearest neighbor graphs is a pruning step which makes sure that direct edges between nodes are avoided if they are deemed unnecessary, for example because a node is still reachable through another neighbor. Both techniques make traditional approaches seem difficult to apply: There is no diversity in in-degree, rendering approaches like ODIN meaningless, and connections to near neighors are explicitly pruned away, making the computation of kNN and LOF scores unreliable.

After discussing related work in Sect. 2 and providing a summary of the necessary preliminaries in Sect. 3, we propose the "black-box" and "white-box" outlier detection methods in Sect. 4. The white-box method will use the sum of distances to neighbors as outlier score (imitating the kNN method) and contrast these scores to the score of their neighbors (imitating LOF). The evaluation of these methods is presented in Sect. 5. Our main findings are that HNSW-based kNN and LOF score computation provides a reliable and scalable alternative to exact alternatives. Despite the remarks made above, the white-box method provides competitive alternatives which further improve scalability. Our results indicate that, for some datasets, the inexactness of the approximation *helps* in achieving better results, adding further evidence that inexactness can improve downstream applications [15].

2 Related Work

Various methods for outlier detection are based on assessing the k nearest neighbors and their distances, resulting in some (proxy of a) local density estimate (or local model), which can be compared to other local models [31]. Classic examples are the kNN outlier score (simply the distance to the kth nearest neighbor) [23] or the aggregation (sum, average) of the distances to the k nearest neighbors [3]. The concept of locality (i.e., the local model of density is not compared to all other models, but to models in the vicinity) has been introduced with LOF [7], which inspired numerous variations [26]. Despite these many variants, the classic methods (kNN and LOF) can still be considered state-of-the-art [8].

The information of k nearest neighbors for each point can be interpreted as a directed graph (with the distances as edge weight), and has been used as such for outlier detection explicitly [11] (ODIN) or implicitly [22], yet not for approximate neighbor search. Approximate methods for outlier detection typically are filter-refinement methods targeting the top-n outliers only, or methods based on approximate nearest neighbor search (these two categories are not strictly mutually exclusive). Targeting the top-n outliers only allows to use approximate solutions in a first step (filter), and keeping only those candidates that have a chance to become one of the top-n outliers, when their (approximate) score is refined to the exact score [2,5,14,16,20,21].

While an advantage of using approximation techniques for filtering is the efficiency w.r.t. the number of instances in a dataset, the use of approximate neighborhood search additionally considers challenges of high dimensionality [32]. Several methods based on approximate neighborhood search are employing techniques based on random projections, for example, using LSH [28], LSH in combination with isolation forests [29], or projection-indexed nearest neighbors [6,9]. Other studies explored the use of space-filling curves [15,24] as well as LSH and NN-descent in combination with an ensemble approach over the approximate results [15].

The works by Campos et al. [8] and Han et al. [10] provide the basis for evaluating outlier detection methods. The former provides a large set of well-curated datasets (with and without duplicates and with different outlier ratios) and detailed benchmarking results, while the later includes state-of-the-art deep learning-based methods and has an easy to integrate pipeline.

3 Preliminaries

3.1 Outlier Detection

Let $\mathcal{X}$ be a metric space equipped with distance function d. Given a dataset $S \subseteq \mathcal{X}$, a data point $x \in S$, and an integer $k \geq 1$, we let $k\text{NN}(x)$ be the set of the k closest points to x in S. The computational problem *Outlier Detection* is to give each point $x \in S$ a score $score(x) \in \mathbb{R}$ that reflects its "outlierness." Given a threshold $\tau \in \mathbb{R}$, all points with a score of at least τ are considered the outliers, all other points are considered inliers.

The two most classic and popular outlier detection methods are the kNN outlier detection method (or kNN for short) and the local outlier factor (LOF). kNN [23] uses the distance to the k-th nearest neighbor as outlier score (or other variants, such as the mean of distances). This can be seen as a local density estimate, and the most prominent outliers are those with smallest local density. The Local Outlier Factor (LOF) [7] uses the kNN sets to calculate a "local reachability density" (lrd) for a data point $x \in S$, a particular form of a local density estimate. The local outlier factor for $x \in S$ is determined by comparing the lrd of a point x to the lrds of its k nearest neighbors, thus considering local differences whereas kNN gives a global perspective [26]. If $\text{LOF}_k(x)$ produces a high value (compared to the other data points in S), then x is an area of lower density compared to its neighbors and is therefore considered to be an outlier [7]. The time complexity of LOF is $O(k)$ operations per data point for a total of $O(kN)$. It requires that for each point $x \in S$, the set $k\text{NN}(x)$ is known, which requires $O(N^2)$ distances computations in the worst-case in high-dimensional spaces.

3.2 Graph-Based Approximate Nearest Neighbor Search

Given a dataset $S \subseteq \mathbb{R}^d$ and parameters M, ef, the goal is to build a graph $G = (V, E)$, where each point is represented by a vertex and edges exist between a point and a "diverse" set of at most M close points. Let us assume that such a graph G is given. To find the nearest neighbors of a query point q, HNSW [17] uses a hierarchy of graphs to find a good entry point into the bottom-layer graph that indexes all points. Given such a start point, carry out a greedy hill climbing. In each round, consider the currently closest point to the query not considered before. Inspect the neighborhood and compute the distances to the query point. After each round, trim the list of current closest points (inspected and non-inspected) to ef, which is usually called the beam width. Terminate if all points in the list have been considered. (Note that this is not a bound on the number of distance computations, since considered points might be trimmed.) To build the graph, order all the points and insert them one-by-one using the

Algorithm 1: BLACKBOX-HNSW($S, k, M, \textit{efC}, \textit{efS}$)

1 $k\text{NN} \leftarrow \{\}$, $\text{lrd}_k \leftarrow \{\}$, $\text{lof}_k \leftarrow \{\}$ $k\text{NN\_score}_k \leftarrow \{\}$ // empty dictionary to keep track of kNN score
2 $\mathcal{I} \leftarrow$ HNSW-BUILD($S, M, \textit{efC}$) // build an HNSW-based index $\mathcal{I}$
3 **foreach** $p \in S$ **do**
4 $\quad \lfloor$ $k\text{NN}[p] \leftarrow \mathcal{I}.\text{search}(p, k, \textit{efS})$
5 **foreach** $p \in S$ **do**
6 $\quad \lfloor$ Compute $\text{lrd}_k[p]$
7 **foreach** $p \in S$ **do**
8 $\quad |$ Compute $\text{lof}_k[p]$
9 $\quad \lfloor$ $k\text{NN\_score}_k[p] \leftarrow$ distance to k-th NN in $k\text{NN}[p]$
10 **return** ($k\text{NN\_score}_k, \text{lof}_k$)

Algorithm 2: WHITEBOX-HNSW($S, M, \textit{efC}$)

1 score $\leftarrow \{\}$ // empty dictionary to keep track of outlier score.
2 contrast_score $\leftarrow \{\}$ // empty dictionary to keep track of outlier score contrasted with neighborhood.
3 $(V, E) \leftarrow$ HNSW-BUILD($S, M, \textit{efC}$) // store groundlayer graph constructed by HNSW. Associate node $v \in V$ with its data point $p \in S$.
4 **foreach** $v \in V$ **do**
5 $\quad \lfloor$ score$[v] \leftarrow \frac{1}{\deg(v)} \sum_{(v,w) \in E} \text{dist}(v, w)$.
6 **foreach** $v \in V$ **do**
7 $\quad \lfloor$ contrast_score$[v] \leftarrow \frac{1}{\text{score}[v]\deg(v)} \sum_{(v,w) \in E} \text{score}[w]$.
8 **return** (score, contrast_score)

---

search algorithm, often with a smaller beam width than used for the queries. From the points inspected in this search, a pruned set of M points is chosen as neighbors of the inserted point (pruning might be necessary for its neighbors if the degree bound M is not met). There exist many other graph-based indexes that change details of this construction [13,27].

4 k-NN and LOF Computation Using HNSW

4.1 Computing kNN and LOF Using HNSW as Blackbox

Algorithm 1 presents pseudocode for our approach to compute kNN and LOF using HNSW as a black-box. In this construction, the user has to pick the k parameter for kNN and LOF. To build the HNSW index, two additional parameters M (the degree bound) and $\textit{efC}$ (the beam width during the construction search) are required. To find the (approximate) nearest neighbors, the user has to specify the beam width $\textit{efS}$ used during search. The running time of Algorithm 1 consists of building the index, search for each point for its k nearest neighbors, and—in the case of LOF—to compute the LOF scores from the kNN sets.

4.2 Computing kNN and LOF Using HNSW as Whitebox

Algorithm 2 presents our approach to directly compute kNN and LOF-like scores from the node neighborhoods in the approximate nearest neighbor graph. GANNOD (Graph ANN-based outlier detection) first builds a standard HNSW index, but only requires access to the ground layer graph. From the groundlayer graph (V, E) of the HNSW index, we compute an outlier score as follows: For each node $v \in V$, first the average distance to its neighbors is computed. This is the first outlier score that we assign to each point, similar to kNN. The *contrasted outlier score* of v is the ratio of the average score of its neighbors to its own score, similar to LOF.

As in Algorithm 1, the first step is to build an HNSW index for parameter choices M and efC. From this graph, we compute outlier scores directly and look at at most $O(M|V|)$ edges, which renders this approach much more efficient than Algorithm 1.

As motivated in the introduction, the design goals of approximate nearest neighbor graphs for high-dimensional data are typically not aligned with keeping the exact neighborhood information that the exact computation of kNN and LOF scores requires.

5 Evaluation

Baselines. We compare the HNSW-based outlier detection methods (Algorithm 1, "blackbox (BB)", and Algorithm 2, "whitebox (WB)") to the PyOD baseline [30] for computing kNN and LOF. As another baseline, we implemented ODIN [11] which builds the (exact) k-NN graph and uses the in-degree as outlier score. All methods use the popular implementation of NearestNeighbor provided with sklearn to speed up the neighborhood computation, in particular for low-dimensional datasets using tree-based indexes.[1]

Implementation notes. Our implementation of the white-box method builds upon hnswlib[2] and is implemented in C++ with a Python wrapper to run through the experimental pipeline. We use the same Python-based numpy code as sklearn's LOF implementation[3] to compute LOF scores for the blackbox. The implementation is available at: https://github.com/CamillaOkkels/ANN-outlier-detection.git.

Evaluation setup. All experiments are carried out on an Intel Xeon E5-2690v4 (2.60 GHz) with 512GB RAM using Ubuntu 20.04.6 LTS. The experimental

---
[1] https://scikit-learn.org/stable/modules/generated/sklearn.neighbors.NearestNeighbors.html.
[2] https://github.com/nmslib/hnswlib.
[3] https://github.com/scikit-learn/scikit-learn/blob/1.5.1/sklearn/neighbors/_lof.py.

pipeline is written in Python and uses some functionality provided by the ADBench outlier detection framework [10].

In the construction of the HNSW for the blackbox methods, *efC* was fixed at 100. We let *efS* take values [8, 16, 32, 48, 64, 128], and the value of M was set to [8, 16, 32, 48]. All combinations of values of M and *efS* were inspected in the evaluation. For the whitebox method we let *efC* take values [10, 50, 100, 200] and M the same values as in the blackbox evaluation. Again all combinations of *efC* and M were inspected. Each experiment was run 10 times, and we report on the mean over these runs because some parameter choices on some datasets had a high variance in the AUCROC scores produced.

Table 1. Overview of the datasets.

Dataset	#points (n)	dim. (d)	#Outliers
ALOI	49 534	27	1 508
BACKDOOR	95 329	196	2 329
CAMPAIGN	41 188	62	4 640
CELEBA	202 599	39	4 547
CENSUS	299 285	500	18 568
COVER	286 048	10	2 747
FRAUD	284 807	29	492
HTTP	567 498	3	2 211
INTERNETADS	1 966	1 555	368
MNIST	7 603	100	700
MUSK	3 062	166	97
SPEECH	3 686	400	61
KITSUNE SSDP FLOOD	4 077 263	115	1 439 603
KITSUNE MIRAI	764 136	116	642 516
KITSUNE ACTIVE WIRETAP	2 278 688	115	923 215

Datasets. Table 1 details the datasets used for the experimental evaluation. Several of the datasets (ALOI through SPEECH) are from the ADBench repository [10], which we refer to as small- to medium-sized datasets. The datasets selected had the highest value of $n \cdot d$ with n being the number of points and d being the dimension. Additionally we have some million-scale datasets from the Kitsune project [6,19]. Each dataset comes with ground truth labels of inliers/outliers.

Performance and Quality Metrics. As quality metric, we use the Area Under the ROC Curve (AUCROC) against the ground truth labels. The ROC curve plots the performance of the outlier detector at all thresholds with True Positive Rate on the y-axis and False Positive Rate on the x-axis. The AUCROC being the area under the ROC curve will range from 0 (for a model ranking outliers consistently after inliers) to 1 (for a model correctly ranking outliers before inliers), with 0.5

Table 2. Quality of Approximate Neighborhood of HNSW graph.

Dataset	M	recall@10 whitebox	recall@100 whitebox	recall100@100 blackbox
ALOI	4	0.38	0.86	0.99
	8	0.34	0.80	1.00
	20	0.34	0.79	1.00
	40	0.36	0.78	1.00
BACKDOOR	4	0.40	0.71	0.96
	8	0.37	0.68	0.99
	20	0.37	0.67	0.99
	40	0.39	0.72	0.99
MUSK	4	0.47	0.84	0.97
	8	0.42	0.81	1.00
	20	0.42	0.78	1.00
	40	0.34	0.66	1.00

being the expected value for a random ranking. As performance metric, we record the time it takes to compute the outlier score for a given dataset and parameter configuration. For the HNSW-based approaches, this includes building the index and either carrying out batched searches (Algorithm 1) or traversing the graph (Algorithm 2).

5.1 Results on Small to Medium-Sized Datasets

Quality of approximate neighborhood. Table 2 reports on the quality of the approximate neighborhood. For each dataset and M value, we report on the highest recall achieved by the neighborhoods that form the basis of Algorithm 1 and Algorithm 2. For the whitebox method, we measure the fraction of the 10-NN and 100-NN which have a direct connection to the point in the graph (are connected by an edge), normalized by the degree of the node. For the blackbox method, we report the fraction of true nearest neighbors for $k = 100$. We can observe from the results that the blackbox variant achieves close to perfect recall no matter which M parameter is used. The results are very different for the whitebox method. Over the datasets, we notice that each node is only connected to very few of its 10 nearest neighbors. Most of the connections are to nodes that are among the 100 nearest neighbors, but there is a large diversity between the different datasets. Thus, the approximate neighborhood of a neighbor includes a diverse set of points, showing the effectiveness of the pruning rules.

Global comparison between methods. Figure 1 presents a ranking of the proposed methods across all datasets. We rank the achieved quality of each method averaged over all datasets mentioned in Table 1. For each method we first took the

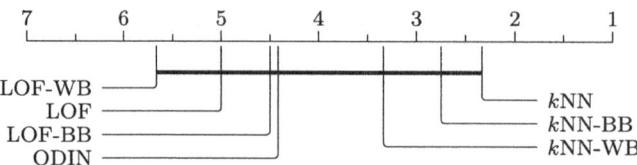

Fig. 1. Critical difference plot: the methods are shown with their average rank over all datasets. Methods within the critical difference (i.e., connected by a line) do not show quality differences that are statistically significant.

average AUCROC score over the 10 runs for each parameter setting. We then took the average AUCROC score achieved out of all parameter choices on a dataset and compared it to the score of the other methods to get the rankings. The plot indicates that global outlier detection methods rank best on average, indicating that the datasets mostly contain global outliers. However, we observe that the ranking differences are statistically insignificant, supporting the claim that the quality of the proposed methods is comparable to their baseline counterparts, kNN and LOF.

Quality evaluation of the individual methods. Figure 2 illustrates the difference in AUCROC scores of the kNN-BB, LOF-BB, kNN-WB and LOF-WB implementations respectively. The methods are evaluated against comparable methods, i.e. kNN-BB, kNN-WB against kNN and LOF-BB, LOF-WB against LOF. A score above 0 indicates that the proposed method provides better AUCROC than its baseline. The results shown are averages over the 10 runs.[4]

We observe that the blackbox methods (top left and right in Fig. 2) maintain a comparable AUCROC score to the baselines. For kNN-BB only MUSK seemed to stand out. Even then, the lowest value achieved with musk was about -0.25 whereas the highest value achieved for any other dataset was around 0.03 - a maximum difference of only 0.28. For LOF there is slightly more variability, only HTTP had too far spread of individual AUCROC values. This indicates that for the black-box methods, we can use small values of M and efS while still obtaining accurate outlier detection results.

The picture looks very different for the whitebox method (bottom row). For the kNN method (bottom left), we see improved results over the whole parameter range for datasets such as DONORS, ALOI, and SPEECH. However, we only see worse scores for datasets such as MNIST, FRAUD, and MUSK. For these, we usually require larger M and ef values to obtain comparable results. The LOF variant (bottom right) has again a larger result diversity. Except for HTTP, it often resembles or improves on the LOF baseline. A larger value of M is not as important as for the kNN variant, but ef should usually be set to a large value. This results indicate an interesting tradeoff: While the whitebox method

---

[4] Building the HNSW index using `hnswlib` in a multicore environment lead to slightly different graphs due to different pruning orders. This non-determinism resulted in different quality scores for the same hyperparameter settings.

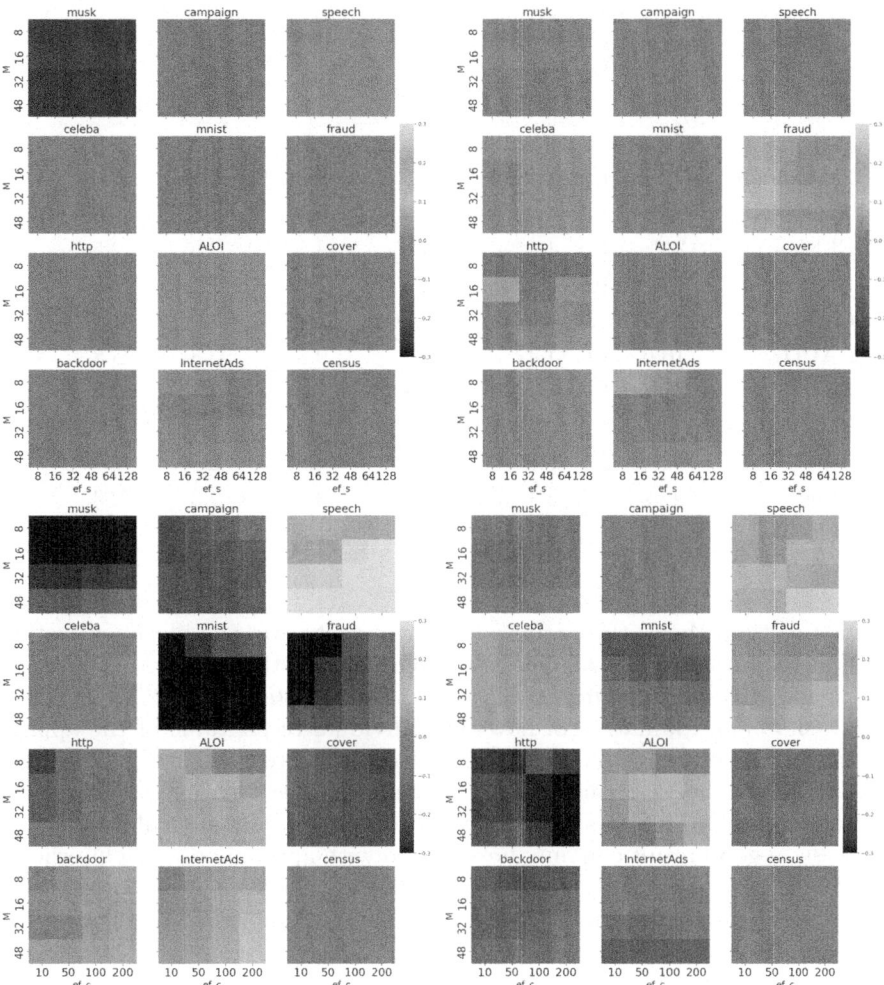

Fig. 2. Heatmap of difference in AUCROC scores compared to baseline of the blackbox kNN (top left), blackbox LOF (top right), whitebox kNN (bottom left), and whitebox LOF (bottom right) method as a function of parameters efS (blackbox) / ef (whitebox) and M (y-axis).

has faster running time, it seems that the blackbox method can work with much smaller parameters.

Scalability considerations. Table 3 summarizes the running time of the proposed methods on the small- to medium-sized datasets. We provide the time of the kNN method ($k = 100$), and the speedup over the kNN method to compare performance. Except for INTERNETADS and SPEECH, both the black-box and the white-box methods speed up the outlier detection significantly. In particular, the

Table 3. Speedup over PyOD's kNN method on small to medium-sized datasets. Values above 1 indicate a speedup, values below 1 a slowdown.

Dataset	Time	kNN	LOF	ODIN	kNN-WB	LOF-WB	kNN-BB	LOF-BB
ALOI	3.4 s	1.0	0.9	0.7	13.2	13.2	7.0	7.0
BACKDOOR	7.7 s	1.0	1.2	0.8	9.4	9.5	6.6	6.6
CAMPAIGN	2.9 s	1.0	1.0	0.7	9.9	9.9	5.6	5.6
CELEBA	3.7 s	1.0	1.1	0.5	4.6	4.5	2.5	2.5
CENSUS	107.1 s	1.0	1.1	1.1	16.2	16.1	10.0	10.0
COVER	18.1 s	1.0	1.0	0.6	14.6	14.5	8.0	8.0
FRAUD	16.6 s	1.0	1.1	0.7	9.6	9.5	5.8	5.8
HTTP	7.7 s	1.0	0.9	0.5	9.3	9.2	4.6	4.6
INTERNETADS	0.3 s	1.0	2.3	1.1	1.1	1.1	1.1	1.1
MNIST	0.3 s	1.0	0.8	0.5	4.8	4.8	2.8	2.8
MUSK	0.1 s	1.0	1.1	0.3	3.8	3.8	2.3	2.3
SPEECH	0.3 s	1.0	2.4	0.6	1.4	1.4	1.3	1.3

whitebox method shows the promised improvements over the blackbox method, usually being around twice as fast as the blackbox method.

Is better recall of the nearest neighbours always better for outlier detection? Figure 3 shows an evaluation of how the individual recall value of the approximate neighborhood ($k = 100$) influences the outlier detection quality. The figure shows a plot of the Pearson correlation coefficient between average recall and average AUCROC. The behavior is very dataset depend: In particular for FRAUD and MUSK, a good result can only be achieved with approximate neighborhoods. For other datasets, such as ALOI, CENSUS, and DONORS, the correlation is positive and points towards accurate neighborhood representations positively influencing the quality.

5.2 Results on the Large KITSUNE datasets

Now consider the bottom three datasets in Table 1. On these datasets the PyOD implementation of kNN and LOF could not finish the task in several hours.

Quality. Figure 4 presents an overview of the detection quality of the kNN-based methods on the three KITSUNE datasets. Outliers in the SSDP Flood and Active Wiretap datasets are well identified using the kNN method, with both the whitebox and black-box method achieving similar quality. In contrast the outliers in the MIRAI dataset are not identified through global methods as the AUCROC score achieved is comparable to what would be expected from random predictions. Figure 5 shows that the Mirai dataset is indeed identified by local methods, such as LOF-BB and LOF-WB. For the largest values of M and efC, the local methods outperform the kNN methods (0.69 vs. 0.5 AUCROC for blackbox and 0.77 vs. 0.48 for whitebox). In contrast, the outliers of the SSDP Flood and Active Wiretap datasets do not seem to be identified by the local methods.

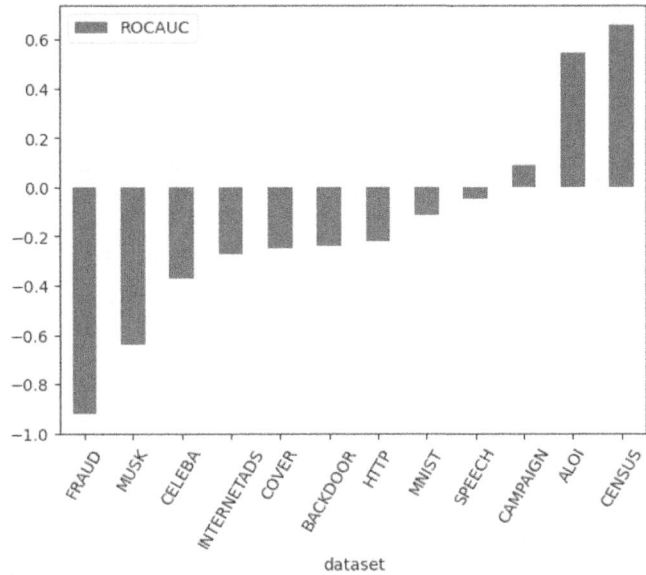

Fig. 3. The correlation coefficient between recall and AUCROC for kNN-BB.

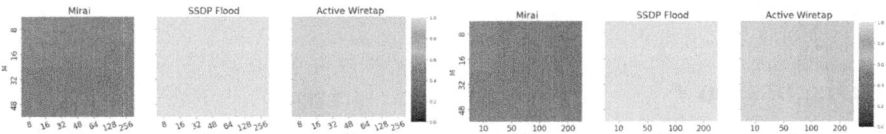

Fig. 4. Heatmap of AUCROC scores of the blackbox (left)/whitebox (right) kNN method for the Kitsune datasets as a function of parameters efS and efC (x-axis) and M (y-axis). Note that the scale is absolute due to the unavailability of baseline AUCROC values. For the blackbox method $k = 15$ and $efC = 200$ were chosen.

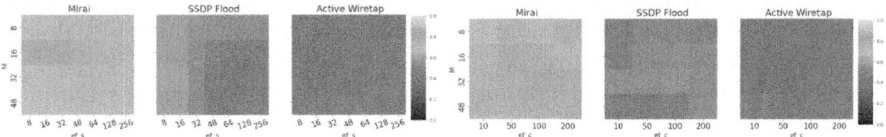

Fig. 5. Heatmap of AUCROC scores of the blackbox (left)/whitebox (right) LOF method for the Kitsune datasets as a function of parameters efS/efC (x-axis) and M (y-axis). For the blackbox method $k = 15$ and $efC = 200$ were chosen.

Scalability analysis. Figure 6 presents running time results on the three largest datasets for the whitebox and the blackbox method. The plots present the *minimum speed-up* of the white-box method vs. the black-box method, i.e., it is the smallest difference in performance we have observed in the tested parameter range. The plot shows that the majority of the running time is spent in building

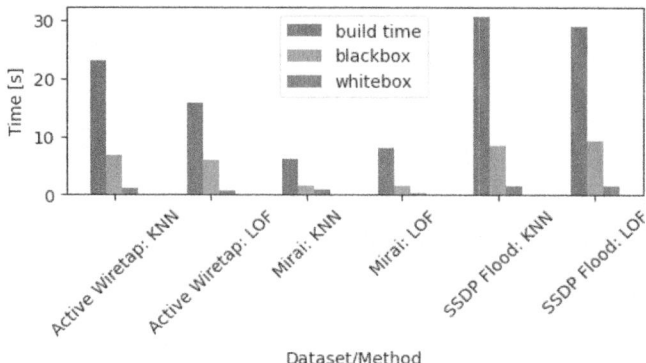

Fig. 6. Running time whitebox/blackbox on KITSUNE datasets.

the index. The black-box method is roughly a factor of 4 faster than the time it took to build the index; the white-box method can carry out the detection step 10—30x faster. As seen in the previous paragraph, this comes at *no cost in quality*. We remark once more that while the detection time is below one minute for the proposed methods, the PyOD-based baseline cannot carry out the task within several hours.

6 Discussion

In this paper, we considered the scalable computation of kNN and LOF-based outlier scores. We suggested two variants, one based on using HNSW as a black-box to compute approximate neighborhoods through k-NN searches, the other based on traversing the HNSW graph directly. Our experiments suggest that both variants provide a scalable alternative to the methods available in standard libraries such as `sklearn` and `PyOD`. In some cases, we noticed that the approximation helps the outlier detection result and provides better quality than an exact neighborhood computation.

In our evaluation we fixed certain values – namely k and efC for the black-box. While the results presented here do hold true more generally, it would be interesting for future work to study more closely the impact of these parameters. Especially k could be of interest as the optimal k for a NN-based outlier detection method can vary greatly from dataset to dataset. The parameter efC impacts the build time of the HNSW data structure. Seeing as build time had the greatest contribution to the running time it would be of interest to see if this could be improved while still maintaining a certain level of quality. In a more general sense it would be interesting to do a more in depth study of how to choose good parameters given certain dataset attributes.

In our study, we did not make use of the hierarchy that is provided with HNSW: Can this hierarchy be used to improve the whitebox approach? Moreover, `ODIN` used the kNN graph in a novel way by checking the in-degree, can the outlier detection goal be integrated into the graph building process? This could both allow better scalability, but could also give rise to novel estimators.

With this study, we hope to initiate a line of research that treats approximate methods as tools for algorithm design that go beyond replacing exact neighborhoods with their approximate counterparts. This idea can readily be applied in other applications: For example, in future work we would like to study how to use the HNSW graph to build a (hierarchical) density-based clustering. Many other data mining tasks can be studied under this research lens.

Acknowledgments. The initial research question was investigated in a Master thesis by Davídsson et al. [1]. We thank the anonymous reviewers for the careful comments on the paper. This project received funding from the Innovation Fund Denmark for the project DIREC (9142-00001B).

References

1. Davídsson, ÓA., Henriksen, S.B., Davídsson, T.B.: Improving the Efficiency of Outlier Detection Using Approximate Nearest Neighbor Search, Master thesis, IT University of Copenhagen (2023)
2. Angiulli, F., Fassetti, F.: DOLPHIN: an efficient algorithm for mining distance-based outliers in very large datasets. ACM TKDD **3**(1), 4, 1–57 (2009)
3. Angiulli, F., Pizzuti, C.: Outlier mining in large high-dimensional data sets. IEEE TKDE **17**(2), 203–215 (2005)
4. Aumüller, M., Bernhardsson, E., Faithfull, A.J.: Ann-benchmarks: a benchmarking tool for approximate nearest neighbor algorithms. Inf. Syst. **87** (2020)
5. Bay, S.D., Schwabacher, M.: Mining distance-based outliers in near linear time with randomization and a simple pruning rule. In: Proceedings of KDD, pp. 29–38 (2003)
6. Bhattacharya, A., Varambally, S., Bagchi, A., Bedathur, S.: Fast one-class classification using class boundary-preserving random projections. In: KDD (2021)
7. Breunig, M.M., Kriegel, H., Ng, R.T., Sander, J.: LOF: identifying density-based local outliers. In: SIGMOD Conference, pp. 93–104. ACM (2000)
8. Campos, G.O., et al.: On the evaluation of unsupervised outlier detection: measures, datasets, and an empirical study. Data Min. Knowl. Disc. **30**(4), 891–927 (2016)
9. de Vries, T., Chawla, S., Houle, M.E.: Density-preserving projections for large-scale local anomaly detection. KAIS **32**(1), 25–52 (2012)
10. Han, S., Hu, X., Huang, H., Jiang, M., Zhao, Y.: Adbench: anomaly detection benchmark. In: NeurIPS (2022)
11. Hautamäki, V., Kärkkäinen, I., Fränti, P.: Outlier detection using k-nearest neighbour graph. ICPR **3**, 430–433 (2004)
12. Indyk, P., Xu, H.: Worst-case performance of popular approximate nearest neighbor search implementations: guarantees and limitations. In: NeurIPS (2023)
13. Iwasaki, M., Miyazaki, D.: Optimization of Indexing Based on k-Nearest Neighbor Graph for Proximity Search in High-Dimensional Data (2018)
14. Jin, W., Tung, A.K., Han, J.: Mining top-n local outliers in large databases. In: Proceedings of KDD, pp. 293–298 (2001)

15. Kirner, E., Schubert, E., Zimek, A.: Good and bad neighborhood approximations for outlier detection ensembles. In: SISAP. Lecture Notes in Computer Science, vol. 10609, pp. 173–187. Springer (2017)
16. Kollios, G., Gunopulos, D., Koudas, N., Berchthold, S.: Efficient biased sampling for approximate clustering and outlier detection in large datasets. IEEE TKDE **15**(5) (2003)
17. Malkov, Y.A., Yashunin, D.A.: Efficient and robust approximate nearest neighbor search using hierarchical navigable small world graphs. IEEE TPAMI **42**(4) (2020)
18. Manohar, M.D., et al.: Parlayann: scalable and deterministic parallel graph-based approximate nearest neighbor search algorithms. In: PPoPP, pp. 270–285. ACM (2024)
19. Mirsky, Y., Doitshman, T., Elovici, Y., Shabtai, A.: Kitsune: an ensemble of autoencoders for online network intrusion detection. In: The Network and Distributed System Security Symposium (NDSS) (2018)
20. Nguyen, H.V., Gopalkrishnan, V.: Efficient pruning schemes for distance-based outlier detection. In: Proceedings of ECML PKDD, pp. 160–175 (2009)
21. Orair, G.H., Teixeira, C., Wang, Y., Meira, W., Jr., Parthasarathy, S.: Distance-based outlier detection: consolidation and renewed bearing. PVLDB **3**(2), 1469–1480 (2010)
22. Radovanovic, M., Nanopoulos, A., Ivanovic, M.: Reverse nearest neighbors in unsupervised distance-based outlier detection. IEEE Trans. Knowl. Data Eng. **27**(5) (2015)
23. Ramaswamy, S., Rastogi, R., Shim, K.: Efficient algorithms for mining outliers from large data sets. In: Proceedings of SIGMOD (2000)
24. Schubert, E., Zimek, A., Kriegel, H.P.: Fast and scalable outlier detection with approximate nearest neighbor ensembles. In: Proceedings of DASFAA (2015)
25. Schubert, E., Zimek, A., Kriegel, H.: Generalized outlier detection with flexible kernel density estimates. In: SDM, pp. 542–550. SIAM (2014)
26. Schubert, E., Zimek, A., Kriegel, H.P.: Local outlier detection reconsidered: a generalized view on locality with applications to spatial, video, and network outlier detection. Data Min. Knowl. Disc. **28**, 190–237 (2014)
27. Subramanya, S.J., Devvrit, F., Simhadri, H.V., Krishnaswamy, R., Kadekodi, R.: DiskANN: fast accurate billion-point nearest neighbor search on a single node. In: NeurIPS, pp. 13748–13758 (2019)
28. Wang, Y., Parthasarathy, S., Tatikonda, S.: Locality sensitive outlier detection: a ranking driven approach. In: Proceedings of ICDE (2011)
29. Zhang, X., et al.: LSHiForest: a generic framework for fast tree isolation based ensemble anomaly analysis. In: Proceedings of ICDE (2017)
30. Zhao, Y., Nasrullah, Z., Li, Z.: Pyod: a python toolbox for scalable outlier detection. J. Mach. Learn. Res. **20**(96) (2019)
31. Zimek, A., Filzmoser, P.: There and back again: outlier detection between statistical reasoning and data mining algorithms. Data Mining Knowl. Discov. **8**(6) (2018)
32. Zimek, A., Schubert, E., Kriegel, H.: A survey on unsupervised outlier detection in high-dimensional numerical data. Stat. Anal. Data Min. **5**(5), 363–387 (2012)

A Topological Evaluation Model for Manifold Learning and Embedding Techniques

Victor Reyes, Margarita Liarou, and Stephane Marchand-Maillet[✉]

Viper Group, University of Geneva, Geneva, Switzerland
stephane.marchand-maillet@unige.ch

Abstract. Data arising from sensors is generally high-dimensional and manifold learning or, more generally, embedding techniques are applied for dimension reduction, possibly with the hope of circumventing the curse of dimensionality. Low-dimensional data may then be the basis for further processing, such as clustering or learning. It is therefore critical that the reduced data representation is faithful to the information contained in the original data. Manifold learning methods are generally evaluated either by visual inspection or by quantifying globally the preservation of neighborhood structures over known dataset. In this paper, we argue for measures that behave smoothly along increasing unfaithfulness to the original data. To build such measures, we return to the manifold assumption and exploit topological information. We further and principally argue for the utility of local measurements of unfaithfulness of representation, as distortions may not be distributed uniformly over the data. The aim here is less to compare manifold techniques than to assess a given technique for its faithfulness. Experiments demonstrate the value of our proposals.

Keywords: Evaluation measure · Dimension reduction · Information embedding · Manifold learning · Local scale estimation

1 Introduction

Data analysis techniques require the initial step on representing the data into a d-dimensional space. Embedding techniques map the information into a d-dimensional space where each information content is represented by a d-dimensional feature vector. A typical class of embedding techniques applies to vocabulary words where a co-occurrence in sentences is known (e.g., [7]).

Similarly, dimension reduction techniques map the initial (e.g., sensor) D-dimensional information onto d-dimensional data ($d \leq D$) while preserving information structure (e.g., [5,6,9]). Dimension reduction techniques generally

---

This work is partly funded by the Swiss National Science Foundation under grant number 207509 "Structural Intrinsic Dimensionality".

© The Author(s), under exclusive license to Springer Nature Switzerland AG 2024
E. Chávez et al. (Eds.): SISAP 2024, LNCS 15268, pp. 185–192, 2024.
https://doi.org/10.1007/978-3-031-75823-2_15

aim at discovering "essential" components from within the information in view of compacting its representation, as a preprocessing step to dimension-sensitive operations (e.g., clustering). Dimension reduction is also used to generate representations one can visualize, as visualization is limited to 2D or 3D (or few more dimensions via metaphors). Be it for visualization or as a preprocessing for further operations, it is critical to study in detail whether the projected information corresponds (in some sense) to the original information.

In this paper, we are interested in quantifying *locally* the "faithfulness" of the reduced representation to the original representation, as a way to evaluate dimension reduction techniques.

2 Models for Dimension Reduction and Its Evaluation

Given a N-sized D-dimensional dataset $\mathcal{X} = \{\boldsymbol{x}_i\}_{i=1}^{N}$ where $\boldsymbol{x}_i \in \mathbb{X} \subset \mathbb{R}^D \; \forall i \in [\![N]\!]$, dimension reduction seeks to embed $\mathcal{X}$ as set $\mathcal{X}' = \{\boldsymbol{x}'_i\}_{i=1}^{N}$ into d-dimensional space $\mathbb{X}' \subset \mathbb{R}^d$, where $d \leq D$ is either given or subject to heuristics or prior knowledge. We denote the embedding operator[1] as

$$\begin{cases} \boldsymbol{\phi} : \mathbb{X} \to \mathbb{X}' \\ \boldsymbol{x}_i \mapsto \boldsymbol{\phi}(\boldsymbol{x}_i) = \boldsymbol{x}'_i \end{cases}$$

Note that although this notation applies point-wise, the frequent transductive nature of embedding operations does not explicit such an operator in practice. In that case, we abuse the notation and also write $\mathcal{X}' = \boldsymbol{\phi}(\mathcal{X})$.

There is no true unified view of dimension reduction techniques $\boldsymbol{\phi}$ [10]. However, a common feature that most of the modern strategies share for their implementation is that they can be cast in a physical framework of attraction-repulsion [1]. Here, internal attractive forces keep (generally local) structures together while a general repulsion force spreads the data.

A similar commonly made assumption is that the data not only lies on a manifold $\mathcal{M} \subset \mathbb{X}$ (so-called *manifold assumption*) but may also be composed of clusters i.e., separable high density regions. Some techniques (such as UMAP versus t-SNE) even explicitly base their evaluation on their ability to distinguish such structures in the data. Here, our goal is to refine this aspect and to quantify how much such assumptions encourage the creation of substructures (such as clusters) that do not actually exist in the original data. Such a quantification is often lacking while it is critical since some data analysis use dimension reduction as a preprocessing step to their data analysis [8]. Evaluating dimension reduction methods is generally proposed as a global measure over the complete dataset. Our main proposal is to investigate the creation of *local* evaluation measures. Dimension reduction models generally minimize a global loss function $\mathcal{L}(\mathcal{X}, \mathcal{X}')$, often taken as a sum of local losses (i.e., of the form $\mathcal{L}(\mathcal{X}, \mathcal{X}') = \sum_i \mathcal{L}_i(\mathcal{X}, \mathcal{X}')$) incurred at every point $\boldsymbol{x}_i \in \mathcal{X}$. Because the loss $\mathcal{L}_i$ or faithfulness is likely not to be distributed uniformly across $\mathcal{X}$, we advocate for ways of detailing global measures into local measures.

---
[1] For notation brevity, we will note with a $\lrcorner'$ images through $\boldsymbol{\phi}$, e.g., $\boldsymbol{x}' = \boldsymbol{\phi}(\boldsymbol{x})$.

The Earth map design is a classical argument to illustrate the above: one wishes to create a 2D map out of a 3D structure while concentrating the distortion (un-faithfulness) at maximum. The Mercator projection chooses to concentrate that distortion over the most unpopulated area along a line across the Pacific Ocean. Elsewhere, despite a smooth transform, the mapping is faithful to the spherical reality. Only lines across the Pacific Ocean get disrupted, or, equivalently, shortest paths are stretched across the map. A local visualization of faithfulness would therefore attribute low distortion values all over the Earth map but at border points.

2.1 Constructing Evaluation Measures

Basing dimension reduction assessment over individual inter-data distance preservation, as in measures classically proposed (e.g., NPR [4]), may be too drastic or too sensitive and finally little informative. Not only should the measure assess the preservation of structures by ϕ at a larger scale than just local but should also degrade smoothly with distortions in the preservation of structures.

Considering the dataset $\mathcal{X}$ as a joint sample of N i.i.d D-dimensional random variables distributed along some distribution in $\mathbb{X}$ whose support is (approximately) a manifold $\mathcal{M} \subset \mathbb{X}$, we start by studying the mapping ϕ in the continuous domain $\mathbb{X}$. We state the general principle that the assessment of the faithfulness of ϕ over $\mathcal{M}$ is a measure (eval) of the preservation of structures built over $\mathcal{M}$ and $\mathcal{M}' = \phi(\mathcal{M})$.

Definition 1. (Evaluation measure eval) *Given a manifold $\mathcal{M}$ subset of $\mathbb{X} \subset \mathbb{R}^D$ and the embedding operator ϕ, an evaluation measure for ϕ over $\mathcal{M}$ is a measure of divergence D of any structure $\mathbf{G}$ build over $\mathbb{X}'$, $\mathbf{G}(\mathcal{M}')$ from that built over $\mathbb{X}$, $\mathbf{G}(\mathcal{M})$.*

$$\mathsf{eval}(\phi; \mathcal{M}) \stackrel{def}{=} \mathsf{D}(\mathbf{G}(\mathcal{M}) \| \mathbf{G}(\mathcal{M}'))$$

Definition 1 requires that structure $\mathbf{G}$ can be built independently over $\mathcal{M}$ and $\mathcal{M}'$. The divergence D then compares their intrinsic characteristics, for example degree, length or curvature. Here, and in contrast to [3], we measure how much ϕ is a *local* isometry. We therefore focus on length, which we identify to the length of geodesic curves on the respective manifolds (assumed to bear a metric). We use geodesic curves of limited length to evaluate the preservation of local intrinsic structures of limited size through ϕ.

Since we want focus on local structures $\mathbf{G}_x$ at every point x, we translate Definition 1 locally by defining a local evaluation measure at $x \in \mathcal{M}$ by

$$\mathsf{eval}_x(\phi; \mathcal{M}) \stackrel{def}{=} \mathsf{D}(\mathbf{G}_x(\mathcal{M}) \| \mathbf{G}_{x'}(\mathcal{M}')) \quad \forall x \in \mathcal{M}$$

where local structures $\mathbf{G}_x(\mathcal{M})$ are the subset of the structures $\mathbf{G}(\mathcal{M})$ relevant to (involving) x. Conversely, from given local structures generating a local evaluation of the embedding, a global evaluation measure is easily defined (e.g., summing over all points).

For example, consider, given $r > 0$, that $\mathbf{G}_x$ is the geodesic ball $\mathcal{S}_r(x)$ of radius r in centered at $x \in \mathcal{M}$. In this case, a possible $\text{eval}_x(\phi; \mathcal{M})$ would measure the overlap between $\phi(\mathcal{S}_r(x))$ and $\mathcal{S}_{r'}(x')$ for an appropriate value of r' (see discussion below) as a measure of neighborhood preservation (akin to a local continuous NPR measure).

This example further points to the fact that ϕ cannot be restricted to be considered as a isometry or local isometry (in which case we have $\frac{r}{r'} = 1$) but that the local scaling that ϕ induces ($\frac{r}{r'} \neq 1$) should be suppressed by accounting for a local scale.

Definition 2 (Local scale). *Given $\varepsilon > 0$ a local scale on $\mathcal{M}$, the local scale on $\mathcal{M}'$ set by ϕ is:*

$$\varepsilon' \stackrel{def}{=} \mathbb{E}_{x,y \in \mathcal{M} \mid d(x,y)=\varepsilon}[d(x', y')]$$

Clearly, if ϕ is a local isometry, $\varepsilon' = \varepsilon$, if ϕ is a scaling then $\varepsilon' = \sigma\varepsilon$ for some constant scaling factor σ. Parameter ε and its consequence ε' therefore become our *unit scales* over $\mathcal{M}$ and $\mathcal{M}'$, respectively. From there, we can define (local) faithfulness as follows.

Definition 3. (L-faithfulness) *Given a scale $\varepsilon > 0$ on $\mathcal{M}$ and a size $L > 0$, we say that ϕ is L-faithful to $\mathcal{M}$ if for any geodesic $\gamma : [0,1] \to \mathcal{M}$ in $\mathcal{M}$ of length $\lambda(\gamma) = \varepsilon L$ it holds*

$$|\lambda(\gamma) - \sigma\lambda(\gamma')| < \varepsilon \qquad \sigma = \frac{\varepsilon}{\varepsilon'}$$

where γ' is the geodesic curve over $\mathcal{M}'$ between $\phi(\gamma(0))$ and $\phi(\gamma(1))$ and ε' is given by Definition 2.

Definition 3 states that an L-faithful mapping ϕ maps a structure (geodesic) of size L onto another structure whose length is indiscernible from L at the scale ε. From there, an obvious evaluation measure for ϕ is given by measuring the distortion of L-faithfulness. Following Definitions 1 and 3:

Definition 4. (L-faithfulness measure) *Given a scale $\varepsilon > 0$ on $\mathcal{M}$ and a size $L > 0$, we define for every embedding ϕ:*

$$L\text{-eval}(\phi; \mathcal{M}) \stackrel{def}{=} \mathbb{E}_{\substack{\gamma \subset \mathcal{M} \\ \lambda(\gamma) \leq \varepsilon L}} \left[\frac{|\lambda(\gamma) - \sigma\lambda(\gamma')|}{\lambda(\gamma)} \right] \qquad (1)$$

where $\sigma = \frac{\varepsilon}{\varepsilon'}$ and ε' is given by Definition 2.

$L\text{-eval}(\phi; \mathcal{M})$ simply integrates L-faithfulness over all structures up to size L. This is an intermediate between assessing ϕ as a global isometry [3] and as a neighborhood-preserving operator [1,4]. We expect this relaxation to make the measure smoother and more robust along the degradation of ϕ. Further, this measure naturally fits our goal for a local evaluation measure:

Definition 5. (Local L-faithfulness measure) *Given a scale $\varepsilon > 0$ on $\mathcal{M}$, a size $L > 0$ and an embedding ϕ, we define at every point $x \in \mathcal{M}$:*

$$L\text{-eval}_x(\phi; \mathcal{M}) \stackrel{def}{=} \mathbb{E}_{\substack{\gamma \subset \mathcal{M} \\ \lambda(\gamma) \leq \varepsilon L \\ \gamma(0)=x}} \left[\frac{|\lambda(\gamma) - \sigma\lambda(\gamma')|}{\lambda(\gamma)} \right] \qquad (2)$$

where $\sigma = \frac{\varepsilon}{\varepsilon'}$ and ε' is given by Definition 2.

The expectation in Eq (2) can be replaced by any operator such as min, max, med. We can also read $\frac{|\lambda(\gamma) - \sigma\lambda(\gamma')|}{\lambda(\gamma)}$ in Eq (2) as $\frac{f(\lambda(\gamma) - \sigma\lambda(\gamma'))}{f(\lambda(\gamma))}$ where $f(\cdot)$ plays the role of a (semi-linear, $f(u) = |u|$ here) penalty per geodesic. This provides an opportunity to generalize $f(\cdot)$ to e.g., quadratic or logarithmic. Further, Eq (2) provides a *recall-like* evaluation, seeking to recover in $\mathcal{M}'$ what was in $\mathcal{M}$. A symmetric *precision-like* evaluation, would simply base the expectation on $\gamma' \in \mathcal{M}'$ rather than on $\gamma \in \mathcal{M}$ to evaluate whether structures present in $\mathcal{M}'$ actually exist in $\mathcal{M}$.

Discrete Evaluation Model We map the above model onto the discrete setting where we are given the dataset $\mathcal{X} \subset \mathbb{X}$. We first note that most of the existing measures can easily be cast in our setup. We can use the index equivariance to instantiate a evaluation measure $\text{eval}(\phi, \mathcal{X})$ in the discrete space by pushing $\mathbf{G}(\mathcal{X})$ onto $\mathcal{X}'$ as $(\mathbf{G}(\mathcal{X}))'$ and compare it to $\mathbf{G}(\mathcal{X}')$. For example, the NPR [4] is defined using K-nearest neighborhoods as structure $\mathbf{G}$ and size of intersection ($\mathsf{D}(\cdot \| \cdot) = |\cdot \cap \cdot|$) as divergence measure. Similarly, applying the MDS loss as a measure would use the complete graph as structure $\mathbf{G}$ and D derived from the classical MDS loss.

Now, we define the discrete-equivalent $L\text{-eval}_x(\phi; \mathcal{X})$ to the above $L\text{-eval}_x(\phi; \mathcal{M})$ using the following steps:

- The manifold $\mathcal{M}$ is tracked via $\mathcal{X}$ using a proximity graph $\mathbf{G}$ (e.g., symmetrized KNN graph preserving connectivity) allowing to highlight intrinsic local interactions only.
- The definition of K in $\mathcal{X}$ is equivalent to providing a local scale ε on $\mathcal{M}$. This scale is transferred as is on $\mathcal{X}' = \phi(\mathcal{X})$
- Geodesic curves $\gamma \subset \mathcal{M}$ find their canonical equivalent in shortest paths in graph $\mathbf{G}$.
- Checking path integrity through ϕ is reduced to comparing their lengths in both $\mathcal{X}$ and $\mathcal{X}'$. To avoid the need to compute the scaling parameter σ, we directly measure the length L of paths over $\mathbf{G}$ by counting proximity steps they cover (as opposed for example to defining a graph distance by weighting edges with their length).

In practice, this allows to use the simple Breadth First Search procedure (rather than Dijkstra's algorithm) to build a single-source shortest path tree from every node within limited range L. This can be seen by rewriting Eq (2) into an empirical estimate:

$$\widehat{L\text{-eval}}_{x_i}(\phi; \mathcal{X}) = \frac{1}{N_{x_i}} \sum_{\substack{\gamma(0) = x \\ \lambda(\gamma) \leq L}} \left[\frac{|\lambda(\gamma) - \lambda(\gamma')|}{\lambda(\gamma)} \right] \qquad (3)$$

where N_{x_i} is the number of BFS-paths of length at most L from x_i in $\mathcal{X}$.

As a result, we obtain the following base procedure (Algorithm 1) to evaluate the local L-faithfulness of $\mathcal{X}' = \phi(\mathcal{X})$ to $\mathcal{X}$. Algorithm 1 returns an array mapping local L-faithfulness onto every data $x_i \in \mathcal{X}$.

Algorithm 1 Measuring the local L-faithfulness of $\mathcal{X}' = \phi(\mathcal{X})$ to $\mathcal{X}$

1: **procedure** L-FAITHFUL($\mathcal{X}, \mathcal{X}', K$)
2:     $\mathbf{G}_\mathcal{X}$ is the symmetrized KNN graph of $\mathcal{X}$
3:     $\mathbf{G}_{\mathcal{X}'}$ is the symmetrized KNN graph of $\mathcal{X}'$
4:     $\boldsymbol{\lambda}$ and $\boldsymbol{\lambda}'$ are $N \times N$ dimensional arrays
5:     **for** every index $i \in [\![N]\!]$ **do**
6:         Compute $\mathbf{T}_\mathcal{X}(i)$ as BFS tree in $\mathbf{G}_\mathcal{X}$ rooted at $\boldsymbol{x}_i$ with max depth L
7:         Compute $\mathbf{T}_{\mathcal{X}'}(i)$ as BFS tree in $\mathbf{G}_{\mathcal{X}'}$ rooted at $\boldsymbol{x}'_i$ with max depth L
8:         **for** every index $j \in [\![N]\!]$ **do**
9:             $\boldsymbol{\lambda}(i,j) \leftarrow$ depth of $\boldsymbol{x}_j$ in $\mathbf{T}_\mathcal{X}(i)$
10:            $\boldsymbol{\lambda}'(i,j) \leftarrow$ depth of $\boldsymbol{x}'_j$ in $\mathbf{T}_{\mathcal{X}'}(i)$
11:     **return** $L\text{-eval}_x(\phi;\mathcal{X}) = \left[\frac{1}{N_{x_i}}\sum_j\left(\frac{|\lambda(i,j)-\lambda'(i,j)|}{\lambda(i,j)}\right)\right]_{i=1}^N$

3 Results and Discussion

We now apply our measure $L\text{-eval}_x(\phi;\mathcal{X})$ for known dataset. The values for K and L should be investigated and discussed. However, due to space limitations, we report results with $K = 20$ and $L = 10$ as reasonable values considering our dataset. We also did not notice significant variations in the results with different values. We use UMAP (Python `umap-learn` package) as a most commonly used manifold learning technique.

We start with a basic setup where we use the classical 3D SwissRoll dataset which has an obvious unfolding as a 2D manifold (sheet) with zero structural distortion. Although this setup does not correspond to the most critical high-to-low-dimensional setup, it is useful to pinpoint the interest for a local measure.

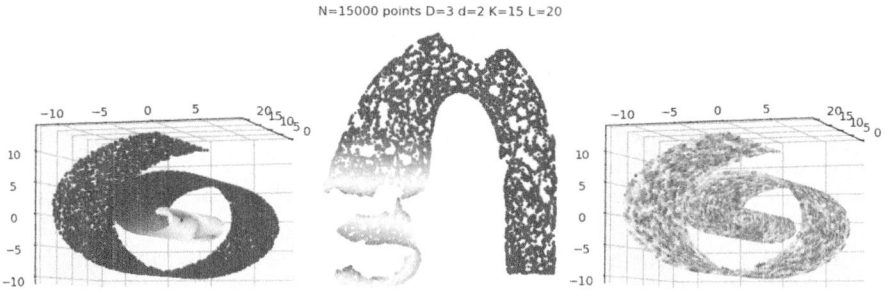

Fig. 1. 3D-to-2D SwissRoll using UMAP (center plot). Most of the data manifold incurs low (green) distortion whereas the high (red) distortion concentrates in some parts. Left and center: $L\text{-eval}_x(\phi;\mathcal{X})$, right: NPR measure ($K = 20$)

Figure 1(center) shows how our $L\text{-eval}_x$ measure highlights the fact that UMAP spuriously rips apart the 2D data manifold and illustrates how faithfulness is unevenly distributed as a consequence. Arguably, this mapping is due to repulsive forces pushing away borders of low density parts of the data but a detailed assessment of UMAP action is out of the scope of this paper. The fact that the initial data is 3-dimensional and the equivariance of indices allow us to also illustrate this in a 3D view (left).

Figure 1(right) shows the local mapping of the NPR evaluation measure. Clearly, the sensitivity of this neighborhood-based measure makes it non-smooth and therefore locally less informative than our proposal. More, this simple example pinpoints the fact that the global NPR value that would be reported here integrates non-informative local values of the measure.

We further study the case of the 3D sphere as an extension of the Earth mapping example. Clearly, the mapping is not made via a clear cut along a line between the poles but rather as a partitioning along lower density regions, for the same reason as above. Again our faithfulness measure is displayed locally (Fig. 2(left and center)) whereas the NPR measure (right) is less informative.

Fig. 2. 3D-to-2D sphere. Left: $L\text{-eval}_x(\phi; \mathcal{X})$ over the original data. Center: $L\text{-eval}_x(\phi; \mathcal{X})$ over the mapped data. Right: NPR over the original data ($K = 20$)

As a practical, high-dimensional, real-world example, Fig. 3 finally illustrates the mapping of the MNIST Test dataset ($D = 250$ first PCA components) mapped onto $d = 100$D, where the evaluation are shown on the 2D mapping.

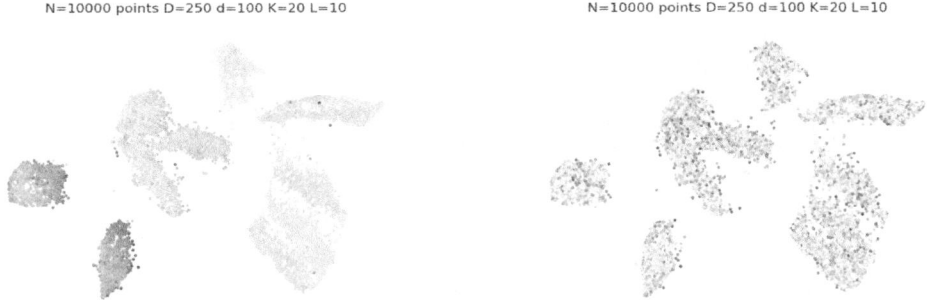

Fig. 3. MNIST Test dataset ($D = 250$, $d = 100$, displayed on the 2D mapping). Left: $L\text{-eval}_x(\phi; \mathcal{X})$ measure. Right: NPR measure ($K = 20$)

We note the disparity of faithfulness values $L\text{-eval}_x$ (left) for various clusters. This can be interpreted by the fact that some clusters (green) are submanifolds that can be unfolded within the target dimension without any critical distortion, wheres the ones colored in red cannot. This is to be thought of in relation to Local Intrinsic Dimensionality (lID). Here again, the NPR measure focusing on point-wise scale fails to surface this information.

4 Conclusion

In this preliminary study, we have demonstrated the usefulness of the definition of a local measure for the faithfulness of an embedding to the original data. We have proposed a topological model based on geodesics rather than on the pointwise neighborhood preservation criterion generally used. The goal is to obtain a consistent degradation of the measure along decreasing faithfulness. While our topological model is reliable and relevant for the assessment of the local faithfulness of a given mapping ϕ, the use of geodesics is at the cost of a non-obvious normalization, making it difficult to use this measure to compare techniques with each other. We will continue developing our model in this direction.

This study poses the more general question of characterization of relevant structures in high-dimensional data. This question is yet to be fully answered and impacts both the manifold learning models and their evaluation.

References

1. Böhm, J.N., Berens, P., Kobak, D.: Attraction-repulsion spectrum in neighbor embeddings (2020). arXiv:abs/2007.08902
2. Gracia, A., González, S., Robles, V., Menasalvas, E.: A methodology to compare dimensionality reduction algorithms in terms of loss of quality. Inf. Sci. **270**, 1–27 (2014)
3. Lara-Cabrera, R., González-Prieto, Á., Pérez-López, D., Trujillo, D., Ortega, F.: An evaluation framework for dimensionality reduction through sectional curvature (2023). arXiv:abs/2303.09909
4. Li, Y., Lu, R.: Applying Ricci flow to high dimensional manifold learning (2017). arXiv:abs/1703.10675
5. van der Maaten, L., Hinton, G.: Visualizing data using t-SNE. J. Mach. Learn. Res. **9**, 2579–2605 (2008)
6. McInnes, L., Healy, J., Melville, J.: UMAP: uniform manifold approximation and projection for dimension reduction (2018). arXiv:abs/1802.03342
7. Pennington, J., Socher, R., Manning, C.D.: GloVe: global vectors for word representation. In: Empirical Methods in Natural Language Processing (EMNLP), pp. 1532–1543 (2014)
8. Quintelier, K., Couckuyt, A., Emmaneel, A., Aerts, J., Saeys, Y., Van Gassen, S.: Analyzing high-dimensional cytometry data using FlowSOM. Nature Protocols **16**, 1–27 (2021)
9. Tenenbaum, J.B., de Silva, V., Langford, J.C.: A global geometric framework for nonlinear dimensionality reduction. Science **290**(5500), 2319–2323 (2000)
10. Wang, Y., Huang, H., Rudin, C., Shaposhnik, Y.: Understanding how dimension reduction tools work: an empirical approach to deciphering t-SNE, UMAP, TriMAP, and PaCMAP for data visualization. J. Mach. Learn. Res.. 1–73 (2021)

Local Intrinsic Dimensionality and the Convergence Order of Fixed-Point Iteration

Michael E. Houle[1,2], Vincent Oria[1], and Hamideh Sabaei[1](✉)

[1] New Jersey Institute of Technology, Ying Wu College of Computing, Newark, NJ 07102, USA
{michael.houle,oria,hs833}@njit.edu
[2] The University of Melbourne, School of Computing and Information Systems, Melbourne, VIC 3010, Australia

Abstract. Fixed-point iteration (FPI) is a crucially important technique at the foundation of many scientific and engineering fields, such as numerical analysis, dynamical systems, optimization, and machine learning. In these domains, algorithmic efficiency and stability is often assessed using the notion of convergence order, a quantity whose estimation has typically involved line fitting in log-log space, or finding the limit of an associated function on differences of sequence values. In this paper, we establish a theoretical equivalence between the convergence order of fixed-point iteration and the local intrinsic dimensionality (LID) of the update function as measured from its fixed-point limit. We then show how an existing MLE estimator of LID can be adapted for the context of FPI to produce novel estimators of convergence order, even for those cases where the update function and the limit point of the iteration are unknown. Although most estimators of LID assume that the data samples are drawn from some distribution of distances to a reference point, we show how this assumption can be relaxed using the LID representation theorem. Experiments are provided for a variety of functions that show competitive performance against traditional estimators of convergence order.

Keywords: Fixed-Point Iteration · Convergence Order · Local Intrinsic Dimensionality · LID · Estimation

1 Introduction

The convergence of iterative processes is of fundamental interest in many scientific and engineering disciplines, including numerical analysis, dynamical systems, optimization, and machine learning. In addition to questions as to whether or where an iteration may converge, the overall performance can greatly depend on how rapidly it approaches its limit point. When the process can be described as a continuously differentiable function, convergence is usually characterized by the polynomial degree of the difference between a function and its limit. More generally, a point sequence is said to *converge with order* m to its limit L if

$$\lim_{j\to\infty} \frac{|x_{j+1} - L|}{|x_j - L|^m} = \lambda \qquad (1)$$

for some positive *convergence rate* $\lambda > 0$ [23,28]. It should be noted that some sources use the term 'convergence rate' as a synonym for the convergence order.

In numerical analysis and the analysis of dynamical systems, the order of convergence determines the efficiency and stability of algorithms solving differential equations whose closed forms are not available [4]. In signal processing, convergence order estimation supports adaptive filtering, which is crucial for applications such as echo cancellation and noise reduction [29]. Financial mathematics utilizes convergence insights to enhance the accuracy of models used in risk management and derivatives pricing, essential for financial forecasting and decision-making [30]. In optimization areas such as machine learning, convergence order is used to categorize the performance of iterative search algorithms such as gradient descent, the Newton-Raphson method, the secant method, and the conjugate gradient method [23]. The speed at which these algorithms converge to a minimum or maximum can significantly affect their computational efficiency and reliability in practical applications.

These applications are examples of fixed-point iteration (FPI), in which the goal is to find a point L in the vicinity of an initial point x_0 for which the update function U produces no change at L; that is, a point for which $U(L) = L$. The iteration generates a sequence of points defined by $x_{n+1} = U(x_n)$, for integers $n \geq 0$. For example, with the Newton-Raphson method for finding the zeros of a continuously differentiable function f, the update function

$$U(x) = x - \frac{f(x)}{f'(x)}$$

has the property that $U(L) = L$ whenever $f(L) = 0$ and $f'(L) \neq 0$. The Newton-Raphson method is known to have a convergence order of $m = 1$ ('linear') when $U'(L) \neq 0$; however, when $U'(L) = 0$, the convergence order is at least $m \geq 2$ ('quadratic' or higher) [28].

This paper explores the relationship between the convergence properties of fixed-point iteration and the local intrinsic dimensionality (LID) [12–14] of its update function. The LID model was originally developed to characterize local complexity in the distribution of data, in terms of the number of latent features required to explain the growth rate in probability measure in an expanding neighborhood centered at a location of interest [12]. In this distributional setting, the probability measure is described by the cumulative distribution function (CDF) of distances from the neighborhood center, and the intrinsic dimensionality is derived from the growth order of the CDF. However, the LID model has a more general interpretation in terms of the growth properties of 'smooth' functions F having a fixed point at the origin (that is, with $F(0) = 0$).

Many practical estimators of LID have been developed based on interpoint distances within neighborhood samples [2,11]. Recently, theoretical connections have been established between LID and the asymptotic properties of entropies and divergences when restricted to local distributions, allowing for estimators for LID to be repurposed for estimating entropy or divergence (and vice versa) [6,7].

LID and its estimators have had many recent applications in databases, data mining, and machine learning. These include the design and analysis of similarity search heuristics [5,8,16], dependency analysis [25], feature selection and ranking [17], and outlier detection [3,18]. In deep learning, LID has been used to characterize and detect adversarial examples [1,21,32] and other anomalies [31], and has appeared in loss functions and regularization terms for classification [19,22]. LID estimation has also been employed in the study of deformation in materials [33], and in the interpretation of sports data [27].

Existing estimators of LID generally operate under the assumption that the data has been sampled from the lower tail of some (unknown) distribution of distances. Taking advantage of the strong relationship between LID and the statistical theory of extreme values (EVT) [13], LID estimators typically make use only of the distribution of the locations of the samples within the lower tail, without any knowledge of the CDF probabilities themselves. As such, they cannot be directly applied to the estimation of convergence order in general settings, where the sequence values need not conform to distributional assumptions.

The main original contributions of this paper are:

1. A formal analysis of the LID properties of update functions for fixed-point iteration, including a proof of equivalence between the convergence order of an update function U and the theoretical LID value of U after translation of its fixed point to the origin.
2. A general framework by which existing estimators of LID can be adapted for the context of FPI to produce novel estimators of convergence order, even for those cases where the update function and the limit point of the iteration are unknown. Although estimators of LID generally assume that the data samples are drawn from some distribution of distances, we show how this assumption can be relaxed using a key result from the theory of LID.
3. Using our proposed framework, we derive an estimator of convergence order for FPI from the MLE (Hill) estimator of LID, and provide an experimental analysis that shows the effectiveness of the estimator in comparison with traditional estimators. We also provide experiments that demonstrate how the LID can be effectively estimated using samples chosen at arbitrary locations, rather than drawn according to a distance distribution or as the sequence produced by a fixed-point iteration process.

2 Estimation of Convergence Order

Here, we review established methods for estimating the convergence order of a sequence, which can be divided into two broad types: sequence-based, and curve fitting. We focus only on estimators that do not rely on explicit knowledge of the true limit of the sequence.

2.1 Sequence-Based Estimation

Estimators of convergence have been developed by identifying a function that takes several elements of the sequence as input, and whose output is known to tend to the convergence order. The estimate is obtained by iterating this sequence of function values until convergence.

The Iterative Ratio (IR) estimator determines the convergence order from two sets of log-ratios of differences of consecutive sequence elements. These ratios have been shown to tend to the exponent m shown in Eq. 1 [28]:

$$m \approx \log\left|\frac{x_{j+1} - x_j}{x_j - x_{j-1}}\right| \bigg/ \log\left|\frac{x_j - x_{j-1}}{x_{j-1} - x_{j-2}}\right|.$$

The Half-Step (HS) estimator focuses on the effect on convergence when the step size h in a fixed-point iteration is varied. For a given choice of h, an estimate of the limit L is calculated, which we denote by $\hat{L}_h$. The estimator of convergence order uses successive approximations of $\hat{L}_h$, halving h each time, to obtain an estimate of m [26]:

$$\frac{\hat{L}_h - \hat{L}_{h/2}}{\hat{L}_{h/2} - \hat{L}_{h/4}} = 2^m + \mathrm{O}(h).$$

The expression is derived by considering the relationship between the error $|\hat{L}_h - L|$ and the step size h, which has been shown to satisfy $|\hat{L}_h - L| \leq ch^m$ for some constant $c > 0$. It should be noted that the accuracy of the method depends greatly on the choice of step size h, and may not be effective for problems where the required value h is impractically small.

2.2 Curve Fitting

In contrast to sequence-based approaches, the convergence order can often be estimated more directly by fitting a polynomial curve to a larger sequence of sample values. Curve fitting requires that each element in the sequence, x_j, be associated with a measurement r_j along some scale with variable r. The goal is to determine the convergence order of the unknown function $x = U(r)$ using ordered pairs (r_j, x_j) as data samples.

The Log-Log (LL) estimation technique proposed by Grassberger and Procaccia [10] assumes that the function is of the form $U(r) = cr^m + b$ for some real constants b and $c > 0$. The convergence order m is identified by fitting a linear regression model to the data under a log-log transformation, characterized by the linear equation

$$\log(U(r) - b) = m \log r + \log c, \qquad (2)$$

where the log-transformed variables are $\log r$ and $\log(U(r) - b)$, the slope is m, and the intercept is $\log c$.

When U is an update function for FPI, b is an offset value that establishes a fixed point for $U(r) - b$ at the origin (since $U(b) - b = 0$). The log-log technique can therefore be regarded as an estimation of the convergence order of FPI. However, for effective estimation of m, the offset b must either be known or be determined in some way.

Log-Log estimation was originally proposed by Grassberger and Procaccia for the estimation of the correlation dimension (CD), a measure of the global intrinsic dimensionality (GID) of datasets useful in the characterization of complexity in dynamical systems and chaos theory [24]. CD quantifies the fractal scaling properties of data, capturing the rate at which the number of close pairs of points increases as the distance threshold increases. CD can also be interpreted as a form of expectation of LID values across a global distribution [14]. In the next section, we give an overview of the theory of LID.

3 Local Intrinsic Dimensionality

The Local Intrinsic Dimensionality (LID) model [12,13] is a continuous extension of the discrete expansion dimension (ED) of Karger and Ruhl [20], wherein dimensionality is inferred from the relationship between the volume and radius of an expanding ball. Given two measurements of radii (r_1 and r_2) and volume (V_1 and V_2), the dimension m can be obtained from the ratios of the measurements:

$$\frac{V_2}{V_1} = \left(\frac{r_2}{r_1}\right)^m \implies m = \frac{\ln(V_2/V_1)}{\ln(r_2/r_1)}.$$

Early expansion models are discrete, in that they estimate volume by the number of data points captured by the ball. The LID model differs from ED in that it allows volume to be viewed as probability measure, and the data as samples drawn from an underlying distribution. For balls sharing a common center, the probability measure is denoted by a function $F(r)$ of the radius r. F can be regarded as the CDF of the distribution of distances induced by samples from the global data distribution. However, the LID model has been formulated in more general terms: it is not necessary for the distances to be Euclidean, nor does the function F need to satisfy the conditions of a CDF.

Definition 1 ([13]). *Let F be a real-valued function that is non-zero over some open interval containing $r \in \mathbb{R}$, $r \neq 0$. The intrinsic dimensionality of F at r is defined as follows, whenever the limit exists:*

$$\mathrm{IntrDim}_F(r) \triangleq \lim_{\epsilon \to 0} \frac{\ln\left(F((1+\epsilon)r)/F(r)\right)}{\ln(1+\epsilon)}.$$

When F is 'smooth' (continuously differentiable) at r, its intrinsic dimensionality can be determined by applying l'Hôpital's rule to the limit:

Theorem 1 ([13]). *Let F be a real-valued function that is non-zero over some open interval containing $r \in \mathbb{R}$, $r \neq 0$. If F is continuously differentiable at r,*

$$\mathrm{ID}_F(r) \triangleq \frac{r \cdot F'(r)}{F(r)} = \mathrm{IntrDim}_F(r).$$

The local intrinsic dimensionality at the center of expansion is modeled as the limit of $\mathrm{ID}_F(r)$ as the distance r tends to 0:

$$\mathrm{ID}_F^* \triangleq \lim_{r \to 0} \mathrm{ID}_F(r).$$

For the local intrinsic dimensionality of a function F, or of a query from which the CDF of its induced distance distribution is F, we will sometimes use the terms 'LID' or 'LID value' when referring to the quantity ID_F^*.

The intrinsic dimensionality function ID_F has been shown to fully characterize F. The following LID Representation Theorem [13] is analogous to a fundamental result from the statistical theory of extreme values (EVT), the Karamata Representation Theorem for regularly varying functions [9,13].

Theorem 2 (LID Representation [13]**)** *Let $F : \mathbb{R} \to \mathbb{R}$ be a real-valued function, and assume that ID_F^* exists. Let r and w be values for which r/w and $F(r)/F(w)$ are both positive. If F is non-zero and continuously differentiable everywhere in the interval $[\min\{r,w\}, \max\{r,w\}]$, then*

$$\frac{F(r)}{F(w)} = \left(\frac{r}{w}\right)^{\mathrm{ID}_F^*} \cdot A_F(r,w), \text{ where } A_F(r,w) \triangleq \exp\left(\int_r^w \frac{\mathrm{ID}_F^* - \mathrm{ID}_F(t)}{t} \, dt\right),$$

whenever the integral exists.

In distributional settings, w corresponds to the lower tail boundary and r to a value of the random variable. The auxiliary factor $A_F(r,w)$ is related to the slowly-varying functions studied in EVT [9]. In [13], $A_F(r,w)$ is shown to tend to 1 as $r, w \to 0$, under the condition that the log-ratio $\ln(r/w)$ remains bounded. Under these conditions, the auxiliary factor vanishes asymptotically as the distances tend to 0.

4 LID and Fixed-Point-Iteration

We now establish an equivalence relationship between the convergence order of an FPI update function U with fixed point ϕ and the LID of its translate $F(r) = U(r+\phi) - U(\phi)$ mapping ϕ to the origin. In this version of the paper, several of the proofs are omitted due to space constraints.

Before proving the main result, we establish a property of the auxiliary function of Theorem 2 for FPI sequences tending to zero, that implies that the auxiliary grows slower than any monomial function.

Theorem 3. *Let $U : \mathbb{R} \to \mathbb{R}$ be an update function for FPI with fixed point at the origin. Consider the auxiliary function $A_U(U(r), r)$ parameterized by r. If U is non-zero and continuously differentiable in some interval containing the origin, and if ID_U^* exists, then*

$$\lim_{r \to 0} \frac{\ln A_U(U(r), r)}{\ln r} = 0.$$

The following result shows that a function F is asymptotically proportional to a monomial with degree equal to ID_F^*. It can be considered as a stronger version of a special case of the statement of Lemma 1 from [15].

Lemma 1. *Let $F : \mathbb{R} \to \mathbb{R}$ be an analytic function with a fixed point at the origin. If ID_F^* exists and is positive, then there exists a constant $\lambda \neq 0$ such that*

$$\lim_{r \to 0} \frac{F(r)}{r^m} = \lambda$$

if and only if $m = \text{ID}_F^*$.

Proof. (*sketch*) The proof proceeds by expressing $F(r)$ as a product $G(r)H(r)$, where $H(r) = r^{\text{ID}_F^* - n}$ with n being the largest integer less than or equal to ID_F^*. We then show that $\text{ID}_G^* = n$. Concentrating then on G, we demonstrate that its first $n-1$ derivatives satisfy $G^{(s)}(0) = 0$. An argument based on iterated application of l'Hôpital's rule then shows that the result holds for G and r^n, from which we deduce that it also holds for F and r^m. □

We are now in position to state and prove the equivalence between LID and the convergence order of FPI.

Theorem 4 (Equivalence of LID & Convergence Order). *Let $U : \mathbb{R} \to \mathbb{R}$ be a function that is analytic in the vicinity of a fixed point $U(\phi) = \phi$. Define $F(r) = U(r + \phi) - U(\phi)$ to be the translate of U with the origin as its fixed point. Assume that ID_F^* exists and is positive. Then fixed-point iteration on U has convergence order m if and only if $m = \text{ID}_F^*$.*

Proof. (*sketch*) Consider the fixed-point iteration generated by U, with $x_{j+1} = U(x_j)$. If U has convergence order m, then there exists $\lambda > 0$ such that

$$\lambda = \lim_{j \to \infty} \frac{|U(x_j) - \phi|}{|x_j - \phi|^m} = \lim_{x \to \phi} \frac{|U(x) - \phi|}{|x - \phi|^m} = \lim_{r \to 0} \frac{|U(r + \phi) - U(\phi)|}{|r|^m} = \lim_{r \to 0} \frac{|F(r)|}{|r|^m},$$

under the substitution $x = r + \phi$. From Lemma 1, $m = \text{ID}_F^*$. □

5 LID-Based Estimation of FPI Convergence Order

We now turn our attention to the problem of adapting existing estimators of LID for the purpose of estimating the convergence order of FPI. We begin by considering a general framework for estimation, and then provide details for the case of the MLE estimator of LID. Heuristics are then proposed to handle the case where the fixed point of convergence is unknown.

5.1 A General Framework

Given an update function U with fixed point at $\phi = U(\phi)$, we consider its translate $F(r) = U(r + \phi) - \phi$ that maps ϕ to the origin. With the translate F

having fixed point $0 = F(0)$, we now consider its restriction to an interval $[0, w]$ within which it is strictly monotone, through the following normalization:

$$F_w(r) \triangleq \frac{F(r)}{F(w)} = \frac{U(r+\phi) - \phi}{U(w+\phi) - \phi} \qquad (3)$$

After the translation and normalization, $F_w(0) = 0$ and $F_w(w) = 1$, satisfying the conditions of a CDF. From Theorem 1, or alternatively from Lemma 1, it is easy to see that $\mathrm{ID}^*_{F_w} = \mathrm{ID}^*_F$.

Consider a set of k consecutive sequence values produced by the FPI with update function U, finishing with the t-th sequence element $\phi_t = U^{(t)}(\phi_0)$, where ϕ_0 is the initial sequence element. For simplicity, we assume that these values are greater than ϕ, and are generated in descending order. After the translation, and reversing the ordering, these values form the set $R = \{r_i | 1 \leq i \leq k\}$, where

$$r_i = \phi_{t-k+i} - \phi = U^{(t-k+i)}(\phi_0) - \phi. \qquad (4)$$

From Theorem 4, estimating the LID of F using the set R as samples is equivalent to using the set $\Phi = \{\phi_{t-k+i} | 1 \leq i \leq k\}$ to estimate the convergence order of U. However, the elements of R do not necessarily satisfy the distributional assumptions of known estimators.

To overcome this difficulty, we use the LID representation theorem to adjust the observations R, in the following way. Let $\mathcal{E}$ be an estimator of ID^*_F that operates on random samples to produce an estimate

$$\widehat{\mathrm{ID}^*}_{\mathcal{E}} = \mathcal{E}(X),$$

for samples $X = \{x_1, \ldots, x_k\}$ drawn from a distributional process with F_w as its CDF, where $0 < x_i < x_j \leq w$ when $1 \leq i < j \leq k$. Theorem 2 tells us that

$$\frac{F_w(x_i)}{F_w(r_i)} = \frac{F(x_i)}{F(r_i)} = \left(\frac{x_i}{r_i}\right)^{\mathrm{ID}^*_F} \cdot A_F(x_i, r_i) \implies x_i = r_i \left(\frac{F(x_i)}{F(r_i) A_F(x_i, r_i)}\right)^{\frac{1}{\mathrm{ID}^*_F}}.$$

As x_i and r_i together tend to 0, A_F can be ignored, which suggests the following substitution into the LID estimator:

$$\widehat{\mathrm{ID}^*}_{\mathcal{E}} = \mathcal{E}(X) \quad \text{where} \quad x_i = r_i \left(\frac{F(x_i)}{F(r_i)}\right)^{\frac{1}{\mathrm{ID}^*_F}}. \qquad (5)$$

5.2 Special Case: the MLE Estimator

We now derive an estimator for the LID of the translate F of U, by substituting Eq. 5 into the MLE (Hill) estimator [2,11]:

$$\widehat{\mathrm{ID}^*}_{\mathrm{MLE}} \triangleq -\left(\frac{1}{k}\sum_{i=1}^{k} \ln \frac{x_i}{w}\right)^{-1} = -\left(\frac{1}{k}\sum_{i=1}^{k} \ln \left[\frac{r_i}{w}\left(\frac{F(x_i)}{F(r_i)}\right)^{\frac{1}{\mathrm{ID}^*_F}}\right]\right)^{-1}$$

$$= -\left(\frac{1}{k}\sum_{i=1}^{k} \ln \frac{r_i}{w} + \frac{1}{k}\sum_{i=1}^{k} \ln \left(\frac{x_i}{w}\right) - \frac{1}{\mathrm{ID}^*_F} \cdot \frac{1}{k}\sum_{i=1}^{k} \ln \frac{F(r_i)}{F(w)}\right)^{-1}$$

from Theorem 2, ignoring the auxiliary factors $A_F(x_i, w)$, which all tend to 1 during the convergence.

The three summations can be expressed in terms of functions of the same form as the Hill estimator. For a set of values V, we introduce the notation

$$\mathcal{H}(V, w) \triangleq -\left(\frac{1}{|V|} \sum_{v \in V} \ln \frac{v}{w}\right)^{-1}.$$

Continuing the derivation, and taking the reciprocals of both sides yields

$$\frac{1}{\mathcal{H}(X,w)} = \frac{1}{\mathcal{H}(R,w)} + \frac{1}{\mathcal{H}(X,w)} - \frac{1}{\text{ID}_F^*} \cdot \frac{1}{\mathcal{H}(F(R), F(w))},$$

where $F(R)$ is the set produced by applying F to each element of R.

At this point, we solve for the only remaining ID_F^* term and declare the resulting formulation to be our proposed MLE-based estimator for the convergence order of U, which we refer to as COE (Convergence Order Estimation).

$$\widehat{\text{ID}^*}_{\text{COE}} \triangleq \frac{\mathcal{H}(R,w)}{\mathcal{H}(F(R), F(w))} = \frac{\mathcal{H}(\Phi - \phi, w)}{\mathcal{H}(U(\Phi) - U(\phi), U(w+\phi) - U(\phi))}. \quad (6)$$

5.3 Implementing the FPI Estimator

In FPI, the limit point ϕ is often unknown, or inconvenient to compute in advance. In such cases, we adopt the convention of approximating ϕ by ϕ_{t+2}, the element in the sequence appearing immediately after the samples in $U(\Phi)$. Similarly, we set the tail boundary w such that $w + \phi$ equals ϕ_{t-k}, the sequence element appearing immediately before the first sample in Φ. We refer to this FPI-specific variant of the estimator as FIE (Fixed-point Iteration Estimation).

$$\widehat{\text{ID}^*}_{\text{FIE}} \triangleq \sum_{i=1}^{k} \ln \frac{\phi_{t+1-k+i} - \phi_{t+2}}{\phi_{t+1-k} - \phi_{t+2}} \bigg/ \sum_{i=1}^{k} \ln \frac{\phi_{t-k+i} - \phi_{t+2}}{\phi_{t-k} - \phi_{t+2}}. \quad (7)$$

As long as the samples in Φ and their updates $U(\Phi)$ are all strictly less than the approximation for w and greater than the approximation for ϕ, the formulation of Eq. 7 is valid. In practice, for the sake of robustness when the estimator is applied to a non-monotone sample, we sort the sample window beforehand.

We conclude this section by noting that the estimator can be applied to functions that do not satisfy the FPI update property. For a non-FPI function G, we would require an explicit list of ordered pairs $(\phi_t, G(\phi_t))$ for each of the sequence samples. Using the ordered pairs, we employ the estimator as expressed in Eq. 6, with $(\phi, G(\phi))$ approximated by $(\phi_{t+1}, G(\phi_{t+1}))$, and $(w+\phi, G(w+\phi))$ set to $(\phi_{t-k}, G(\phi_{t-k}))$. We refer to this more general variant as GIE (General Iteration Estimation).

6 Experiments

This section describes the experimental framework and performance results of our estimators of convergence order, FIE and GIE, against the traditional estimators Log-Log (LL) and Iterative Ratio (IR). Half-Step (HS) was also considered, but excluded from the analysis due to numerical issues arising from the need to generate sequences at extremely fine resolution (small step sizes).

6.1 Experimental Framework

For testing, we considered two tasks: estimating the convergence order of root finding, and estimating the LID of general sequences.

Tasks. For the root finding task, we considered the three functions listed in Table 1, using either Newton-Raphson (NR) or the Secant Method (SM). For each of the three, the ground-truth convergence order of the update function is known; typically, for NR the order is 2 (or sometimes 1), and for functions with simple roots, SM converges with order equal to the golden ratio (roughly 1.618).

For the general iteration task, three functions were considered, with the goal of estimating the LID of each as it grows away from a specified limit point—an ordered pair consisting of a sequence value and its associated function value.

For both tasks, knowledge of the true limit point (ϕ for the FPI tasks, and $(\phi, G(\phi))$ for the general iteration tasks) was withheld from all estimators tested.

Sequence Data Generation. For the root finding tasks, sequence data was generated from an initial point drawn uniformly at random from within a range centered at a large value. This central value was chosen large enough so that at least 60 iterations would be produced before convergence, and that 10 sequence elements would fall within a relatively narrow band before the convergence limit (referred to as the 'convergence tail' in Table 1). However, we excluded any sequence elements falling within an extremely small tolerance threshold from the convergence limit itself. Thereafter, the $k \in \{10, 20\}$ most recently-generated elements (whether in the convergence tail or not) were retained as the sequence values presented to FIE, LL, and IR. This process was repeated until 10 convergent sequences were generated, with the estimates of convergence order reported in the table being the averages of the estimates produced for each run.

For each general iteration task, a sequence was generated as the endpoints of k uniform intervals within the convergence tail listed in Table 1, for the choices of $k \in \{32, 64, 128, 256\}$. The convergence point itself was excluded from the data, leaving k values available to the estimators GIE, LL, and IR. The function was then applied to these values, producing a sequence of ordered pairs.

The FIE, GIE and LL estimators all make use of a parameter representing the limit value, and in addition, FIE and GIE require the specification of a value for w, the tail boundary. These values were taken from the k samples (or sample pairs) available to them, without reuse in other parts of the estimator formulation. It should also be noted that despite having a large number of sequence samples available, the IR estimation uses only the last four elements generated. For the general iteration tasks, IR uses only function values, since it is not able to take advantage of data in the form of ordered pairs.

Log-Log Estimator. The LL estimator requires that linear regression be performed on the sequence data after transformation to log-log space. This was implemented using PyTorch's linear regression from nn.Module using the Stochastic Gradient Descent (SGD) optimizer torch.optim.

Due to the relatively high computational cost of linear regression, we implemented LL in two ways: prioritizing accuracy (LL:A), and prioritizing execution speed (LL:S). For better accuracy, we set the learning rate at 0.01 and the number of iterations at 2000, and for faster execution, a learning rate of 0.1 over 100 iterations. For the offset b in LL estimator we supplied the same values as was used by FIE/GIE for the limit parameters, namely ϕ (for FPI tasks) and $(\phi, G(\phi))$ (for non-FPI tasks).

6.2 Experimental Results

The results of our experiments are presented in Table 1. For the FPI tasks, we observe generally good performance from the FIE estimator relative to the ground truth value, with slight degradation in accuracy when the number of samples is larger ($k = 20$). Both LL versions performed less consistently over the data, which indicates that they may be overweighting sequence elements that are farther from the convergence limit. The tradeoff between the speed of LL:S and accuracy of LL:A is not favorable relative to the performance of FIE.

On the other hand, IR, with its focus on only the four elements closest to the convergence point, performs consistently best for the FPI tasks once it is sufficiently close to the limit. For the general iteration tasks, although IR performs well for the trigonometric function $2 - \cos(r)$, its performance degrades sharply for the higher-degree tasks, showing instability due to the small number of samples employed, and its inability to make use of ordered-pair information.

Both GIE and the LL variants performed well on the trigonometric function $2 - \cos(r)$ and the monomial function $4r^3$, with the LL:A variant achieving slightly better accuracy than GIE, and the fast LL:S variant having lower accuracy. However, all methods had difficulty with the higher-degree polynomial. Over the range of samples $[0, 0, 2.5]$, the r^3 term dominates for values less than 0.5, whereas the $4r^5$ term dominates for greater values. Perhaps unsurprisingly, GIE and the LL variants reported a blend of the dominant influences over the interval. Here again, even the fast LL:S variant required execution times that were orders of magnitude greater than those of GIE.

Table 1. Estimation of convergence order of 6 functions, for FPI tasks (Newton-Raphson and Secant Method) and general iteration (GI) tasks. Execution times are reported in seconds.

Function		Limit	k	CO $\equiv$ LID	FIE GIE	LL:A	LL:S	IR	Time FIE GIE	Time LL:A	Time LL:S	Time IR	Conv Tail	Initial Point
NR	$(r-2)^2$	2	10	1	1.008	1.497	2.800	1.009	8.2e−5	1.1e+0	1.3e−2	1.9e−5	[2.0, 4.0]	1.0e+10 ± 1.0e+8
NR	$(r-2)^2$	2	20	1	1.001	1.466	2.754	0.998	1.5e−4	1.1e+0	1.4e−2	2.2e−5	[2.0, 4.0]	1.0e+10 ± 1.0e+8
SM	$(r-2)(r^2+1)$	2	10	1.618	1.525	1.384	0.605	1.628	8.6e−5	1.0e+0	1.7e−2	2.2e−5	[2.0, 4.0]	3.0e+06 ± 3.0e+4
SM	$(r-2)(r^2+1)$	2	20	1.618	1.347	1.611	0.303	1.619	1.4e−4	1.1e+0	1.5e−2	2.2e−5	[2.0, 4.0]	3.0e+06 ± 3.0e+4
SM	$e^{r-1}-1$	2	10	1.618	1.562	2.001	1.906	1.641	9.6e−5	1.1e+0	1.5e−2	2.9e−5	[2.0, 4.0]	4.0e+01 ± 2.0e+0
SM	$e^{r-1}-1$	2	20	1.618	1.462	2.190	2.380	1.623	1.4e−4	9.6e−1	1.4e−2	2.3e−5	[2.0, 4.0]	4.0e+01 ± 2.0e+0
NR	$(r-2)(r^2+1)$	2	10	2	1.764	2.071	0.656	2.001	8.7e−5	1.9e+0	1.4e−2	2.1e−5	[2.0, 4.0]	8.0e+09 ± 8.0e+7
NR	$(r-2)(r^2+1)$	2	20	2	1.443	1.319	0.887	2.001	1.2e−4	1.9e+0	1.3e−2	2.1e−5	[2.0, 4.0]	8.0e+09 ± 8.0e+7
NR	$e^{r-1}-1$	2	10	2	1.867	1.667	0.891	2.050	8.4e−5	1.9e+0	1.4e−2	2.8e−5	[2.0, 4.0]	5.5e+01 ± 2.0e+0
NR	$e^{r-1}-1$	2	20	2	1.688	1.779	1.251	2.015	1.1e−4	2.0e+0	1.4e−2	2.1e−5	[2.0, 4.0]	5.5e+01 ± 2.0e+0
GI	$2-\cos(r)$	(0,1)	32	2	1.863	1.943	1.836	2.155	2.2e−4	2.1e+0	1.0e−2	1.9e−5	[0.0, 1.5]	—
GI	$2-\cos(r)$	(0,1)	64	2	1.867	1.949	2.033	2.151	2.0e−4	1.9e+0	1.0e−2	1.2e−5	[0.0, 1.5]	—
GI	$2-\cos(r)$	(0,1)	128	2	1.869	1.953	1.873	2.151	4.2e−4	2.2e+0	1.2e−2	1.2e−5	[0.0, 1.5]	—
GI	$2-\cos(r)$	(0,1)	256	2	1.870	1.954	1.744	2.151	1.7e−3	2.9e+0	1.4e−2	1.5e−5	[0.0, 1.5]	—
GI	$4r^3$	(0,0)	32	3	2.999	2.999	2.919	1.948	1.1e−4	1.9e+0	2.5e−2	1.3e−5	[0.0, 2.5]	—
GI	$4r^3$	(0,0)	64	3	2.999	2.999	2.978	1.948	2.2e−4	2.9e+0	1.0e−2	1.3e−5	[0.0, 2.5]	—
GI	$4r^3$	(0,0)	128	3	2.999	2.999	2.945	1.948	7.9e−4	4.0e+0	2.4e−2	1.9e−5	[0.0, 2.5]	—
GI	$4r^3$	(0,0)	256	3	2.999	2.999	2.967	1.948	1.6e−3	2.3e+0	1.5e−2	2.0e−5	[0.0, 2.5]	—
GI	$4r^5+r^3+1$	(0,1)	32	3	4.528	4.141	4.141	1.774	2.0e−4	1.9e+0	1.4e−2	1.6e−5	[0.0, 2.5]	—
GI	$4r^5+r^3+1$	(0,1)	64	3	4.496	4.055	4.055	1.893	2.2e−4	1.9e+0	1.0e−2	1.3e−5	[0.0, 2.5]	—
GI	$4r^5+r^3+1$	(0,1)	128	3	4.476	4.001	4.001	1.933	4.2e−4	2.0e+0	1.4e−2	1.3e−5	[0.0, 2.5]	—
GI	$4r^5+r^3+1$	(0,1)	256	3	4.465	3.966	3.966	1.944	9.0e−4	2.1e+0	1.1e−2	1.4e−5	[0.0, 2.5]	—

7 Conclusion

In this paper, we established a theoretical equivalence between the convergence order of fixed-point iteration and the local intrinsic dimensionality of the update function as measured from its limit. With this equivalence, we showed how existing distributional estimators of LID can be leveraged to create novel estimators of the convergence order of FPI, and for the LID of the growth of functions away from a fixed point that is not necessarily the origin. Starting with the well-established MLE estimator of LID, we implemented practical and efficient estimators for FPI and for general iteration that are competitive in accuracy and execution time with the most popular estimators of convergence order, the Iterative Ratio estimator and the Log-Log technique, even when the convergence point itself is unknown.

These new LID-based convergence order estimation techniques are sufficiently fast and general to be applied in situations such as machine learning, where efficient estimation of LID or convergence order have the potential to identify and control issues with the convergence of learning processes. As future work, we plan to expand our approach to include other estimators of LID as the basis of new estimators of convergence order, and to explore their uses in machine learning and deep learning.

References

1. Amsaleg, L., Bailey, J., Barbe, A., Erfani, S.M., Furon, T., Houle, M.E., Radovanović, M., Vinh Nguyen, X.: High intrinsic dimensionality facilitates adversarial attack: theoretical evidence. IEEE TIFS **16**, 854–865 (2021)
2. Amsaleg, L., Chelly, O., Furon, T., Girard, S., Houle, M.E., Kawarabayashi, K., Nett, M.: Extreme-value-theoretic estimation of local intrinsic dimensionality. Data Min. Knowl. Discov. **32**(6), 1768–1805 (2018)
3. Anderberg, A., Bailey, J., Campello, R.J.G.B, Houle, M.E., Marques, H.O., Radovanović, M., Zimek, A.: Dimensionality-aware outlier detection. In: Proceedings of SDM, pp. 652–650 (2024)
4. Atkinson, K.: An Introduction to Numerical Analysis. Wiley (1991)
5. Aumüller, M., Ceccarello, M.: The role of local intrinsic dimensionality in benchmarking nearest neighbor search. In: Proceedings of SISAP, pp. 113–127 (2019)
6. Bailey, J., Houle, M.E., Ma, X.: Local intrinsic dimensionality, entropy and statistical divergences. Entropy **24**(9), 1220 (2022)
7. Bailey, J., Houle, M.E., Ma, X.: Relationships between tail entropies and local intrinsic dimensionality and their use for estimation and feature representation. Inf. Syst. **118**, 102245 (2023)
8. Casanova, G., Englmeier, E., Houle, M.E., Kröger, P., Nett, M., Schubert, E., Zimek, A.: Dimensional testing for reverse k-nearest neighbor search. PVLDB **10**(7), 769–780 (2017)
9. Coles, S.: An Introduction to Statistical Modeling of Extreme Values. Springer (2001)
10. Grassberger, P., Procaccia, I.: Characterization of strange attractors. Phys. Rev. Lett. **50**, 346–349 (1983). Jan
11. Hill, B.M.: A simple general approach to inference about the tail of a distribution. Annals Stat. **3**(5), 1163–1174 (1975)
12. Houle, M.E.: Dimensionality, discriminability, density and distance distributions. In: Proceedings of ICDM Workshops, pp. 468–473 (2013)
13. Houle, M.E.: Local intrinsic dimensionality I: an extreme-value-theoretic foundation for similarity applications. In: Proceedings of SISAP, pp. 64–79 (2017)
14. Houle, M.E.: Local intrinsic dimensionality II: multivariate analysis and distributional support. In: Proceedings of SISAP, pp. 80–95 (2017)
15. Houle, M.E.: Local intrinsic dimensionality III: density and similarity. In: Proceedings of SISAP, pp. 248–260 (2020)
16. Houle, M.E., Ma, X., Oria, V.: Effective and efficient algorithms for flexible aggregate similarity search in high dimensional spaces. IEEE TKDE **27**(12), 3258–3273 (2015)
17. Houle, M.E., Oria, V., Wali, A.M.: Improving k-NN graph accuracy using local intrinsic dimensionality. In: Proceedings of SISAP, pp. 110–124 (2017)
18. Houle, M.E., Schubert, E., Zimek, A.: On the correlation between local intrinsic dimensionality and outlierness. In: Proceedings of SISAP, pp. 177–191 (2018)
19. Huang, H., Campello, R.J.G.B., Erfani, S.M., Ma, X., Houle, M.E., Bailey, J.: LDReg: local dimensionality regularized self-supervised learning. In: Proceedings of ICLR, pp. 1–26 (2024)

20. Karger, D.R., Ruhl, M.: Finding nearest neighbors in growth-restricted metrics. In: Proceedings of STOC, pp. 741–750 (2002)
21. Ma, X., Li, B., Wang, Y., Erfani, S.M., Wijewickrema, S.N.R., Schoenebeck, G., Song, D., Houle, M.E., Bailey, J.: Characterizing adversarial subspaces using local intrinsic dimensionality. In: Proceedings of ICLR, pp. 1–15 (2018)
22. Ma, X., Wang, Y., Houle, M.E., Zhou, S., Erfani, S.M., Xia, S., Wijewickrema, S.N.R., Bailey, J.: Dimensionality-driven learning with noisy labels. In: Proceedings of ICML, pp. 3361–3370 (2018)
23. Nocedal, J., Wright, S.J.: Numerical Optimization, 2nd edn. Springer (2006)
24. Robinson, R.C.: An Introduction to Dynamical Systems: Continuous and Discrete, 2nd edn. AMS (2012)
25. Romano, S., Chelly, O., Nguyen, V., Bailey, J., Houle, M.E.: Measuring dependency via intrinsic dimensionality. In: Proceedings of ICPR, pp. 1207–1212 (2016)
26. Runborg, O.: Verifying Numerical Convergence Rates. Technical report, KTH Computer Science and Communication (2012)
27. Santos-Fernandez, E., Denti, F., Mengersen, K., Mira, A.: The role of intrinsic dimension in high-resolution player tracking data—insights in basketball. Ann. Appl. Stati. **16**(1), 326–348 (2022)
28. Senning, J.R.: Computing and Estimating the Rate of Convergence. Technical report, Gordon College, Wenham, MA, USA (2007)
29. Sergio, D.P., Diniz, R.: Adaptive Filtering: Algorithms and Practical Implementation (2002)
30. Soleymani, F., Barfeie, M.: Pricing options under stochastic volatility jump model: a stable adaptive scheme. Appl. Num. Math. **145**, 69–89 (2019)
31. Wang, Q., Erfani, S.M., Leckie, C., Houle, M.E.: A dimensionality-driven approach for unsupervised out-of-distribution detection. In: Proceedings of SDM, pp. 118–126 (2021)
32. Weerasinghe, W., Abraham, T., Alpcan, T., Erfani, S.M., Leckie, C., Rubinstein, B.I.P.: Closing the big-lid: an effective local intrinsic dimensionality defense for nonlinear regression poisoning. In: Proceedings of IJCAI, pp. 3176–3184 (2021)
33. Zhou, S., Tordesillas, A., Pouragha, M., Bailey, J., Bondell, H.: On local intrinsic dimensionality of deformation in complex materials. Sci. Rep. **11**(1), 10216 (2021)

Identifying Propagating Signals with Spatio-Temporal Clustering in Multivariate Time Series

Jan David Hüwel[✉][iD], Georg Stefan Schlake[iD], Kevin Albrechts, and Christian Beecks[iD]

University in Hagen, Hagen, Germany
{jan.huewel,georg.schlake,christian.beecks}@fernuni-hagen.de

Abstract. Recordings via multiple sensors positioned on a grid result in multivariate time series, whose subsequences can be compared in aspects of time, space and shape. Existing methods for tracking propagating signals in this high-content data format require a starting position, which has to be determined by a domain expert. In this paper, we propose a fully unsupervised method to discover propagating signals via density-based clusters with respect to the three aspects mentioned above. For this purpose, we adapt the DBSCAN algorithm to our specific setting and present an exemplary application of this method on pharmacological data.

Keywords: Clustering · Multivariate Time Series · Signal Propagation

1 Introduction

Time series data is among the most prevalent data types for analysis. A time series can be any sequence of data points or values collected over time, thus offering numerous potential applications, like meteorological measurements [11], developments of stock prices [9] or recordings of biological or chemical processes [7]. When multiple features are recorded simultaneously, the data is referred to as a multivariate time series. In instances where multivariate time series originate from multiple physical sensor recordings in parallel, the dimensions of the data are spatially correlated additionally to the temporal dimension of the recordings. Examples for this are temperature or sound volume measurements in different rooms of a house or traffic flow data throughout a city. Such spatio-temporal data is the focus of this work.

A potential task within these settings is to identify the propagation of a signal, i.e. a distinct one-dimensional subsequence, through the dataset [7]. Knowledge of such signals can inform users about unknown properties of the observed system. While there exists a method that can solve this problem [7], it requires a starting subsequence to be selected by a domain expert and can then determine the propagation of that specific signal. Here, we propose a fully unsupervised method to find all propagating signals within a spatio-temporal multivariate time series. To do so, we utilize the similarity of adjacent occurrences of a propagating signal as well as the spatial and temporal closeness to define a multi-dimensional similarity measure, that, when used in a density-based clustering approach, automatically encapsulates a signal propagation as a single cluster.

The clustering of time series subsequences is prone to errors caused by the similarity of directly neighboring subsequences [8]. To prevent flawed results, we expand the clustering process to automatically filter out trivial matches of subsequences. This will be explained in more detail in Sect. 3.

2 Related Work

Clustering is a classical task in unsupervised learning, where the goal is to split a dataset into different groups of objects based on their properties. *Density-based clustering* algorithms like *DBSCAN* [13] have the emphasis on identifying clusters which are defined by a constant minimal density. This allows algorithms to find clusters of any shape as long as the minimal density is maintained throughout the entire area of any cluster. Objects in sparse areas will be considered as noise. As the signal propagations, that we seek, may form arbitrary shapes, a density-based approach is most suitable for our solution.

While DBSCAN is a widely known algorithm, a few algorithms have been proposed to adapt DBSCAN for different scenarios like clustering spatio-temporal subsequences. *E-DBSCAN* [3] is an approach to cluster trajectories of positioning data. However, these trajectories are not comparable to our use case as they use a widely different feature set. Another extension is *CoExDBSCAN* [4], which performs DBSCAN in a user-defined subspace and using further user-defined constraints for the clusterings. These constraints change the neighbours considered to be in a dense region for an object. This might help to mitigate problems in higher dimensional spaces and regarding complex relationships between objects. There exist multiple adaptions of DBSCAN for spatio-temporal time-series called *ST-DBSCAN*, so we will suffix these algorithms with abbreviations of their authors for clarity. *ST-DBSCAN-BK* [2] and *ST-DBSCAN-WWL* use different neighbourhoods, with two separate density thresholds for the spatial and for the non-spatial distance between points. An object is in the neighborhood of another, if both the spatial distance and the formal distance are smaller than their respective thresholds. An additional third view on the data is used for temporal dimensions. If these exist, only objects measured consecutively will be considered neighbours.

Neither of these methods are directly applicable to identify propagating signals as clusters, as one has to take into account specific limitations for what constitutes as a spreading signal. Therefore, we will define our own method in the next section.

3 Method

In this section, we will introduce the necessary notation to apply density-based clustering to spatially connected time series data. The definitions will focus on the metrics involved and the neighborhood of a subsequence, which can then be used in the DBSCAN algorithm [5].

To start, we first quantify the data, that we are working with.

Definition 1 (Time series). *A univariate time series* $(x_1, ..., x_n) \in \mathbb{T} = \mathbb{R}^n$ *is a sequence of n values representing measurements over time. A multivariate time series* $T = (t_1, ..., t_n) \in \mathbb{T}^d$ *with* $d > 1$ *is a sequence of n real-valued d-dimensional vectors. We denote* $T_i = (t_{1,i}, ..., t_{n,i})$ *as the ith dimension of T.*

In our application in Sect. 4, the time series data are recorded by sensors located on a grid. Therefore, we denote a single time series as $T_{(i,j)}$, using a two-dimensional index for dimensions to represent row and column of the corresponding sensor. The spatial distance between time series thus depends on the indices in question.

While the clustering of entire time series is an interesting problem (cf. Section 2), our work focuses on subsequence clustering.

Definition 2 (Subsequence). *A subsequence of a time series T is a continuous section of values in T and is denoted as* $T[s : e] = (t_s, t_{s+1}, ..., t_{e-1}, t_e)$.

Propagating signals exhibit closeness to their predecessor and successors in three dimensions: space, time and form. To determine these neighbors for a given subsequence, we thus require distance measures for these dimensions. As described above, the spatial distance can be determined via the time series index.

Definition 3 (Spatial distance). *The spatial distance between two subsequences* $T_{(i,j)}[s : e]$ *and* $T_{(i',j')}[s' : e']$ *uses a maximum distance on the location given by the index and is defined as*

$$d_s\left(T_{(i,j)}[s:e], T_{(i',j')}[s':e']\right) = max\left(|i-i'|, |j-j'|\right). \quad (1)$$

The distance between two subsequences is thus the amount of direct "neighbor-to-neighbor" jumps required to reach one from the other, including the possibility of diagonal jumps. This is motivated by the assumption that signals can spread to directly neighboring sensors.

If a signal is propagated to a neighboring time series, the temporal difference between the two occurrences is specified as the difference in their starting time.

Definition 4 (Temporal distance). *The temporal distance between two subsequences* $T_{(i,j)}[s : e]$ *and* $T_{(i',j')}[s' : e']$ *determines the difference between the starting points and is defined as*

$$d_t\left(T_{(i,j)}[s:e], T_{(i',j')}[s':e']\right) = |s-s'|. \quad (2)$$

Lastly, the formal resemblance of two subsequences is vital factor in determining whether or not they belong to the same signal propagation. As activity levels and sensor sensibility can vary in *in vivo* and *in vitro* data, we want to normalize the sequences in question and then use a classic distance measure, like the Euclidean distance. This is the only dimension that requires equal lengths of the subsequences.

Definition 5 (Formal distance). *The formal distance between two subsequences $T_{(i,j)}[s:s+l]$ and $T_{(i',j')}[s':s'+l]$ is given by the Z-normalized Euclidean distance*

$$d_f\left(T_{(i,j)}[s:s+l], T_{(i',j')}[s':s'+l]\right) = \sqrt{\sum_{k=0}^{e-s} \left(\bar{t}_{(i,j)s+k} - \bar{t}_{(i',j')s'+k}\right)^2}. \quad (3)$$

Here, $\bar{t}$ denotes the normalized values in the sequence.

When determining the amount of "similar" matches within a subsequence's neighborhood, we do not want to consider subsequences that are just slightly shifted versions of the original. These trivial matches can have a large impact on the quality of the resulting clusters [8]. Therefore, given a labelled subsequence S, we aim to exclude every similar subsequence S', where every subsequence in between the two is also similar to S. These subsequences form the set of trivial matches.

Definition 6 (Trivial matches). *Given a similarity threshold ε, the set of trivial matches of a subsequence is defined as follows.*

$$TM_\varepsilon(T_{(i,j)}[s:s+l]) = \{T_{(i,j)}[s':s'+l] \mid \forall s \leq x \leq s' \vee s' \leq x \leq s : \\ d_f\left(T_{(i,j)}[s:s+l], T_{(i,j)}[x:x+l]\right) \leq \varepsilon\} \quad (4)$$

If needed, the threshold for trivial matches can be treated separately from the one for the formal similarity in a cluster. In some settings, a broader definition of trivial matches is desirable to guarantee unique motifs in clusters.

A fundamental requirement for DBSCAN and its variants is a notion of the neighborhood of an element [5,13]. The neighborhood consists of all elements, that have a lower distance to the element, than the predetermined threshold. With the three distance measures, a threshold is required for each one. Here, we further deviate from approaches like ST-DBSAN by introducing a *minimum* temporal distance as well as the maximum thresholds for each dimension. The reason is, that signal propagation in most settings does not happen instantly. For example, cell activity might take a few hundred milliseconds to be able to spread to surrounding cells [7]. Potential matches, that lie even closer together, are not likely to be genuine propagations.

Definition 7 (Neighborhood). *Given thresholds $\varepsilon_s, \varepsilon_{t,min}, \varepsilon_{t,max}, \varepsilon_f$ for spatial, temporal and formal similarity respectively, the neighborhood of a subsequence S is defined as the set of all unique subsequences, that are spatially, temporally and formally similar to S as defined in Definitions 3, 4 and 5.*

$$\mathcal{N}(S) = \{S' \mid 1 \leq d_s(S,S') \leq \varepsilon_s \wedge \varepsilon_{t,min} \leq d_t(S,S') \leq \varepsilon_{t,max} \wedge \\ d_f(S,S') \leq \varepsilon_f \wedge \nexists S'' \in TM_{\varepsilon_f}(S') : d_f(S,S'') < d_f(S,S')\} \quad (5)$$

This definition replaces the definition for the ε-neighborhood in the regular DBSCAN algorithm (cf. [5]).

Determining this neighborhood for every subsequence in a dataset is a very costly approach. Hence we employ the *MASS* algorithm [10] to accelerate computations. A more detailed explanation on this method is beyond the scope of this short paper.

4 Experiments

In this section, we present an exemplary application of our method on data from the pharmacological domain. We visually show an example of an identified cluster and discuss its validity. Since there are no methods with the same goal that we are aware of, a comparison to existing methods is omitted.

4.1 Data

The pancreas can be divided into the endocrine and exocrine glands, which fulfill different tasks in the body. The endocrine part contains pancreatic islets, also called islets of Langerhans, which in turn consist of α-, β-, γ-, δ- and ε-cells. Of these, the β-cells are the most important in the production of insulin. [1,6,7,12]

Our data were generated by placing an islet of Langerhans unto a multielectrode array (MEA), i.e. a grid of sensors, stimulating the cells externally with a 10mM glucose solution, and then measuring their electrical activity with a frequency of 1000 Hz for five minutes. The stimulation happens before the recording begins and includes an activation time, so the measured activity is not spiking in the beginning. The MEA is composed of 65×65 sensors with a distance of 16 μm between them. Given an average distance of 15 μm between β-cells [12], it's reasonable to assume one cell per sensor. Not all sensors are covered by cells, so the inactive border areas are cut after measurement. This results in 20×19 time series with 300,000 datapoints each.

It is currently not known, to what degree β-cells are influenced by the activity of adjacent cells [1], however, earlier experiments indicate at least some propagation of signals [7]. Here, we aim to reconfirm that notion.

4.2 Results

We extract subsequences from the presented dataset via a sliding window of stepsize 1 and length 12,000, corresponding to 12 s of activity. This was deemed a realistic timeframe to include a full activation cycle by domain experts [7]. For the same reason, we use temporal limits of $\varepsilon_{t,min} = 100$ and $\varepsilon_{t,max} = 1000$, corresponding to 0.1 s and 1 s respectively. This is broad enough to catch cases, where an intermediate occurrence of a signal is not detected. Spatially, we use a threshold of $\varepsilon_s = 1$, treating subsequences as neighboring if they belong to directly adjacent sensors. The value for the formal threshold is based on previous

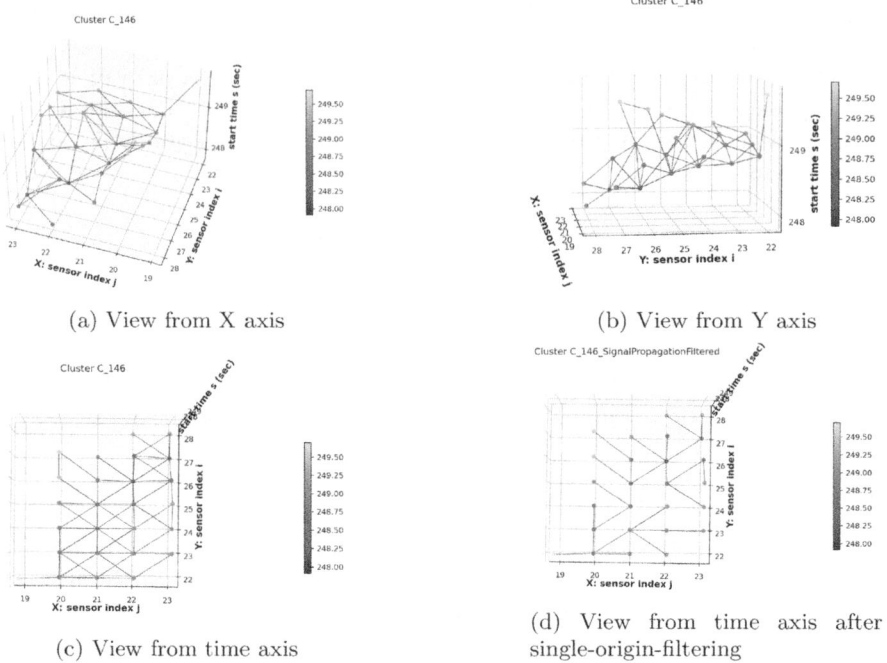

Fig. 1. Exemplary cluster identified in the dataset, visualized from three sides. The time of occurrence is additionally marked with a color scale. When restricting each subsequence to one predecessor, the graph depicts a realistic propagation of the identified signal.

research with the same distance function [7] and is set to $\varepsilon_f = 110$. Finally, we define core-subsequences as having at least two neighbors: $minPoints = 2$.

Figure 1 shows one of 184 identified clusters in the data. Each point in the three-dimensional space consists of the row- and column index and the starting time (in seconds) of the corresponding subsequence. The cluster is presented from three different viewpoints and a colorscale is employed to highlight the temporal development. A connection between two points indicates, that they are neighbors after Definition 7.

For the final image, we limit each subsequence to a maximum of one predecessor. In the context of our dataset, it is reasonable to assume a singular point of origin instead of multiple sources accumulating in a signal. If multiple predecessors are found, the one with the strongest formal similarity is treated as the point of origin. The resulting graph depicts a potential propagation of the signal in question. For domain experts, these results can offer a valuable step towards the analysis of cell activity.

In this application, only 0.003% of all subsequences were appointed to a cluster, while the rest is identified as noise or excluded. Consequently, for a second experiment, we take the average of all subsequences in our data (with a step-

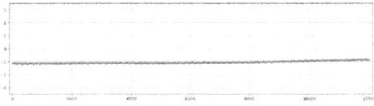
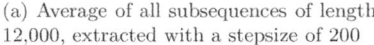
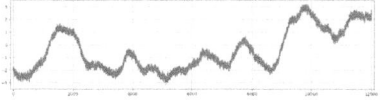

(a) Average of all subsequences of length 12,000, extracted with a stepsize of 200

(b) Average of all subsequences of length 12,000, that were assigned a cluster

Fig. 2. Comparison between the general average of all subsequences and the average of all subsequences, that were assigned a cluster.

size of 200, due to technical limitations) and compare it to the average of all subsequences, that belong to a cluster. The results are shown in Fig. 2. The overall average follows the expectation, that the mean of a large amount of subsequences is a constant [8]. The cluster average on the other hand contains clear patterns of activity. We can infer, that there is higher inter-cluster and intra-cluster similarity, than there is similarity between clusters and noise. Within the pharmacological context, subsequences that contain meaningful activity appear to be assigned to clusters, facilitating further analysis on the extracted data.

5 Conclusion

In this paper, we have used a spatio-temporal density-based clustering approach to identify patterns of similar behavior in spatially connected multivariate time series data. We adjusted common variants of the DBSCAN algorithm to this specific problem statement and improved efficiency by using the MASS algorithm for formal similarity between subsequences. Afterwards, we applied the resulting method to pharmacological data depicting the electrical activity of pancreatic β-cells. The results show clusters of varying sizes, which can be interpreted as propagations of cell activity. The experiments also confirm, that the exclusion of subsequences, that are identified as noise, leads to a more distinct average pattern between subsequences, indicating similarities between clusters.

It is yet undetermined how well this approach generalizes to other domains. Moreover, the manual examination of clusters still needs to be assessed, which ones depict genuine signal propagations within the context of the data.

Acknowledgments. This work is funded by DFG grant No. 454630593: EPIX - Efficient Ptolemaic Indexing. Additionally, this research was supported by the research training group "Dataninja" (Trustworthy AI for Seamless Problem Solving: Next Generation Intelligence Joins Robust Data Analysis) funded by the German federal state of North Rhine-Westphalia.

References

1. Benninger, R.K., Kravets, V.: The physiological role of β-cell heterogeneity in pancreatic islet function. Nat. Rev. Endocrinol. **18**(1), 9–22 (2022)
2. Birant, D., Kut, A.: ST-DBSCAN: An algorithm for clustering spatial-temporal data. Data Knowl. Eng. **60**(1), 208–221 (2007)
3. Cheng, D., Yue, G., Pei, T., Wu, M.: Clustering indoor positioning data using E-DBSCAN. ISPRS Int. J. Geo Inf. **10**(10), 669 (2021)
4. Ertl, B., Meyer, J., Schneider, M., Streit, A.: CoExDBSCAN: density-based clustering with constrained expansion. In: KDIR, pp. 104–115 (2020)
5. Ester, M., Kriegel, H.P., Sander, J., Xu, X., et al.: A density-based algorithm for discovering clusters in large spatial databases with noise. In: KDD, vol. 96, pp. 226–231 (1996)
6. Hüwel, J.D., Gresch, A., Berger, T., Düfer, M., Beecks, C.: Analysis of extracellular potential recordings by high-density micro-electrode arrays of pancreatic islets. In: International Conference on Database and Expert Systems Applications, pp. 270–276. Springer (2022)
7. Hüwel, J.D., Gresch, A., Berns, F., Koch, R., Düfer, M., Beecks, C.: Tracing patterns in electrophysiological time series data. In: DSAA, pp. 1–10. IEEE (2022)
8. Keogh, E., Lin, J.: Clustering of time-series subsequences is meaningless: implications for previous and future research. Knowl. Inf. Syst. **8**, 154–177 (2005)
9. Mondal, P., Shit, L., Goswami, S.: Study of effectiveness of time series modeling (ARIMA) in forecasting stock prices. Int. J. Comput. Sci. Eng. Appl. **4**(2), 13 (2014)
10. Mueen, A., et al.: The fastest similarity search algorithm for time series subsequences under Euclidean distance (2017). http://www.cs.unm.edu/~mueen/FastestSimilaritySearch.html
11. Murat, M., Malinowska, I., Gos, M., Krzyszczak, J.: Forecasting daily meteorological time series using ARIMA and regression models. Int. Agrophys. **32**(2) (2018)
12. Rorsman, P., Ashcroft, F.M.: Pancreatic β-cell electrical activity and insulin secretion: of mice and men. Physiol. Rev. **98**(1), 117–214 (2018)
13. Schubert, E., Sander, J., Ester, M., Kriegel, H.P., Xu, X.: DBSCAN revisited, revisited: why and how you should (still) use DBSCAN. TODS **42**(3), 1–21 (2017)

Robust Statistical Scaling of Outlier Scores: Improving the Quality of Outlier Probabilities for Outliers

Philipp Röchner[1](✉), Henrique O. Marques[2](✉),
Ricardo J. G. B. Campello[2], Arthur Zimek[2], and Franz Rothlauf[1]

[1] Johannes Gutenberg University Mainz, Mainz, Germany
{roechner,rothlauf}@uni-mainz.de
[2] University of Southern Denmark, Odense, Denmark
oli@sdu.dk, {campello,zimek}@imada.sdu.dk

Abstract. Outlier detection algorithms typically assign an outlier score to each observation in a dataset, indicating the degree to which an observation is an outlier. However, these scores are often not comparable across algorithms and can be difficult for humans to interpret. Statistical scaling addresses this problem by transforming outlier scores into outlier probabilities without using ground-truth labels, thereby improving interpretability and comparability across algorithms. However, the quality of this transformation can be different for outliers and inliers. Missing outliers in scenarios where they are of particular interest—such as healthcare, finance, or engineering—can be costly or dangerous. Thus, ensuring good probabilities for outliers is essential. This paper argues that statistical scaling, as commonly used in the literature, does not produce equally good probabilities for outliers as for inliers. Therefore, we propose robust statistical scaling, which uses robust estimators to improve the probabilities for outliers. We evaluate several variants of our method against other outlier score transformations for real-world datasets and outlier detection algorithms, where it can improve the probabilities for outliers.

Keywords: Outlier detection · Anomaly detection · Outlier probabilities · Calibration · Probability estimates · Unsupervised learning · Robust statistics

1 Introduction

Outlier detection algorithms compute real-valued outlier scores to identify outliers, which are significantly different from the other observations in the dataset, the so-called inliers [5]. The real-valued outlier scores are often difficult for humans to interpret and are not comparable across algorithms. Therefore, several transformations have been proposed to convert outlier scores into *outlier*

The extended version of this article contains additional content, including examples and results: https://arxiv.org/abs/2408.15874 [13].

© The Author(s), under exclusive license to Springer Nature Switzerland AG 2024
E. Chávez et al. (Eds.): SISAP 2024, LNCS 15268, pp. 215–222, 2024.
https://doi.org/10.1007/978-3-031-75823-2_18

probabilities [1,2,4,6,10], which quantifies the probability that an observation is an outlier [12].

Good outlier probabilities are concentrated around zero and one, called sharpness, separate outliers from inliers, called refinement, and reflect the frequency of outliers for observations with a similar outlier probability, called calibration [12]. It is usually important to discuss the quality of the probabilities for outliers and inliers separately: Users are often particularly interested in outliers, where it is necessary to determine good probabilities for outliers; otherwise, the probabilities of the outliers can be misleading to users and subsequent methods.

To our knowledge, there is a lack of research on the differences between the quality of probabilities of outliers and inliers. For supervised classification on imbalanced datasets, the quality of probabilities for observations from different classes can be significantly different [15]. In general, we expect the quality of probabilities to be different for outliers and inliers because outliers are rare compared to inliers and significantly different from inliers.

We study outlier score transformations that use only the outlier scores and no external information, such as which observation is an inlier or an outlier. We argue and empirically show that statistical scaling [6], a commonly used outlier score transformation [8,11], computes inferior probabilities for outliers than for inliers. Therefore, we propose robust statistical scaling, which uses robust estimators to compute outlier probabilities. We evaluate several variants of our method against other outlier score transformations for real-world datasets and outlier detection algorithms, where it can improve the probabilities for outliers.

2 Problem Statement

We study an n-dimensional real-valued dataset $\boldsymbol{X} = \{\boldsymbol{x}_i\}_{i=1}^{N}$ with N observations, where $\boldsymbol{x}_i \in \mathbb{R}^n$, and an outlier detection algorithm $\mathrm{D}_{\boldsymbol{X}} : \mathbb{R}^n \to \mathbb{R}$ with outlier scores $\boldsymbol{S} := \{s_i\}_{i=1}^{N}$, where $s_i := \mathrm{D}_{\boldsymbol{X}}(\boldsymbol{x}_i)$.

We seek a transformation $\mathrm{T}_{\boldsymbol{S}} : \mathbb{R} \to [0,1]$ of outlier scores $\boldsymbol{S}$ into outlier probabilities $\boldsymbol{p} = \{p_i\}_{i=1}^{N}$, where

$$p_i := \mathrm{T}_{\boldsymbol{S}}(s_i) = \mathrm{T}_{\boldsymbol{S}}(\mathrm{D}_{\boldsymbol{X}}(\boldsymbol{x}_i)) \in [0,1],$$

so that the probabilities $\boldsymbol{p}$ are sharp, refined, and calibrated for outliers and inliers: Sharp outlier probabilities are concentrated around zero and one, refined probabilities have pure ground-truth labels for observations with similar outlier probabilities, and calibrated outlier probabilities match the fraction of outliers for observations with similar outlier probabilities [12].

The outlier score transformation $\mathrm{T}_{\boldsymbol{S}}$ is unsupervised; that is, it depends only on the outlier scores $\boldsymbol{S}$ and has no ground truth information about whether an observation is an outlier or an inlier.

3 Background: Non-robust Statistical Scaling of Outlier Scores

Statistical scaling first fits a parametric distribution to the frequency distribution of outlier scores computed by a given outlier detection algorithm on a dataset [6]. Therefore, the parameters of the parametric distributions are determined from the data using, for example, the method of moments (MoM) or maximum likelihood estimation (MLE). Outlier probabilities are then calculated using a modified cumulative distribution function of the approximated frequency distribution [6].

In the following, we refer to statistical scaling as non-robust statistical scaling [6]. We also limit the following discussion to non-robust Gaussian scaling, a variant of non-robust statistical scaling that uses Gaussian distributions, but similar arguments apply to other distributions.

Definition 1 (Non-robust Gaussian Scaling). *For an outlier score distribution $S \subset \mathbb{R}$ and an outlier score $s \in S$, its outlier probability using non-robust Gaussian scaling GS_S is*

$$GS_S(s) := \max\left(0, erf\left(\frac{s - \mu_S}{\sigma_S \sqrt{2}}\right)\right), \quad (1)$$

with the mean μ_S and the standard deviation (SD) σ_S of the Gaussian distribution fitted to the outlier scores S; erf is the error function.

The error function is a scaled and translated variant of the cumulative distribution function of the Gaussian distribution: it is sigmoid-shaped, monotonically increasing, and maps the real numbers to the interval $]-1, 1[$. According to Eq. (1) and because the Gaussian error function is negative for negative arguments, all observations with outlier scores less than or equal to the mean μ_S have an outlier probability of zero.

In the following, we assume that a sufficiently well-functioning outlier detection algorithm has calculated the outlier scores such that high outlier scores correspond to outliers and low outlier scores correspond to inliers. We also assume that we have determined the parameters of the Gaussian distribution in Definition 1 using an approach sensitive to long-tailed distributions, such as the MoM or, equivalently, MLE.

Since there are typically fewer outliers than inliers, the outlier score distribution of a proper outlier detection algorithm will have a long upper tail. This asymmetry of the outlier score distribution shifts the mean to the right because the mean is sensitive to extreme values. As a result, Gaussian scaling maps many outlier scores to an outlier probability of zero (see Eq. (1)), which would be correct for inliers but incorrect for outliers. Because the SD is also sensitive to extreme values, the scale of the Gaussian distribution that approximates the outlier score distribution is large. As a result, the outlier probabilities of observations with outlier scores larger than the outlier score mean increase slowly (see Eq. (1)): outliers with scores larger than the score mean may have probabilities that are too low, and inliers may have outlier probabilities that are too high.

4 Robust Statistical Scaling of Outlier Scores

To improve the probabilities of outliers, we propose *robust statistical scaling*, which uses robust estimators to fit a distribution to the outlier scores. We discuss *robust Gaussian scaling* as an example of robust statistical scaling, but similar arguments apply to robust statistical scaling using other distributions.

Definition 2 (Robust Gaussian Scaling). *For an outlier score distribution $S \subset \mathbb{R}$ and an outlier score $s \in S$, its outlier probability using robust Gaussian scaling rGS_S is*

$$rGS_S(s) := \max\left(0, erf\left(\frac{s - \mu_S^{robust}}{\sigma_S^{robust}\sqrt{2}}\right)\right), \tag{2}$$

with a center μ_S^{robust} and a scale σ_S^{robust} of the Gaussian distribution robustly fitted to the outlier scores S; erf is the error function.

Because a robustly fitted Gaussian distribution is less sensitive to extreme values, its center and scale are smaller than the sample mean and sample SD. As a result, robust Gaussian scaling maps fewer outlier scores to outlier probabilities of 0. Moreover, for outlier scores greater than the center of the outlier score distribution, the outlier probabilities increase faster. As a result, probabilities of outliers computed with robust Gaussian scaling are larger than those computed with non-robust Gaussian scaling.

There are several robust estimators for outlier score distributions. For example, we can replace the sample mean μ_S and SD σ_S in Equation (1) with robust estimates for the center and scale of the outlier score distribution.

Examples of robust estimators of a distribution's center are the median or the *asymmetric trimmed mean* [7]. For the asymmetric trimmed mean, we remove a certain percentage of the largest scores from the outlier score distribution, leaving the lower tail unchanged, and calculate the mean of the remaining outlier scores.

The *normalized median absolute deviation from the median* (nMAD), *normalized interquartile range* (nIQR), and *trimmed standard deviation* are examples of robust estimators of a distribution's scale [7].

Instead of computing a robust center and a robust scale separately, *M-estimators* can robustly estimate the center and scale of distributions simultaneously [7]: They iteratively estimate moments of distributions starting from initial values, such as the median for the center and the nMAD for the scale. The center and scale estimates are then updated based on the residuals between the current center and scale estimates and the observed data, with larger residuals weighted less than smaller residuals [7], which can be thought of as soft trimming [14].

We discuss a detailed example comparing non-robust and robust statistical scaling in the extended version of this article: https://arxiv.org/abs/2408.15874 [13].

5 Experiments

We first compute outlier scores for real-world datasets using outlier detection algorithms, then transform the outlier scores into outlier probabilities using outlier score transformations, and finally evaluate their outlier probabilities.

We compute outlier scores on 21 real-world datasets [3], excluding the KDD and ALOI datasets for computational reasons, using 11 outlier detection algorithms. Combining the real-world datasets and outlier detection algorithms results in 231 sets of outlier scores.

We investigate linear scaling [6], non-robust Gaussian scaling [6], and 10 variants of robust Gaussian scaling (Definition 2) to convert the 231 outlier score distributions mentioned above into outlier probabilities.

We evaluate outlier probabilities using the Brier score, sharpness, refinement, and calibration errors and their stratified variants [12]. All of the above measures have values between zero and one, with lower values corresponding to better outlier probabilities [12].

To compare an outlier probability distribution p with reference outlier probabilities p_{ref} for an outlier probability measure M, we introduce a *skill score* [9]; this also makes the quality of outlier probabilities computed from different outlier score distributions better comparable.

Definition 3 (Skill Score). *For an outlier probability measure M using ground-truth labels y, the skill score MSS of outlier probabilities p with respect to reference outlier probabilities $p_{ref} \subset \mathbb{R}$ is*

$$MSS(p, p_{ref}) := MSS(p, p_{ref}, y) := -\log_2\left(\frac{M(p, y)}{M(p_{ref}, y)}\right). \tag{3}$$

The Brier score, sharpness, refinement, and calibration errors are candidates for the outlier probability measure M in Definition 3 (skill score). For these measures, a positive skill score indicates better, a skill score of zero indicates equal, and a negative skill score indicates inferior outlier probabilities p compared to the reference outlier probabilities p_{ref}.

We describe our experiments in more detail in the extended version of this article: https://arxiv.org/abs/2408.15874 [13].

6 Results

6.1 Are the Outlier Probabilities Computed by Non-robust Gaussian Scaling Similarly Good for Outliers and Inliers?

First, we evaluate whether non-robust Gaussian scaling has equally good probabilities for outliers and inliers. Figure 1 shows the skill score MSS (Definition 3) of non-robust Gaussian scaling for the stratified Brier score, calibration, refinement, and separation errors for outliers compared to the corresponding stratified measure for inliers.

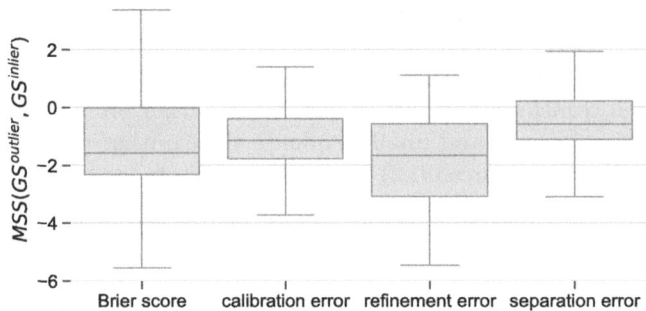

Fig. 1. Skill scores MSS($GS^{outlier}, GS^{inlier}$) of the probabilities computed by non-robust Gaussian scaling for outliers $GS^{outlier}$ compared to the probabilities for inliers GS^{inlier}.

For all four outlier probability measures examined, the distributions of the skill scores for non-robust Gaussian scaling are predominantly negative; that is, for non-robust Gaussian scaling, the stratified measures for outliers are inferior to those for inliers, consistent with our discussion in Sect. 3.

6.2 Does Robust Gaussian Scaling Improve the Probabilities of Outliers Compared to Non-robust Gaussian Scaling?

We examine whether robust Gaussian scaling improves the probabilities of outliers compared to non-robust Gaussian scaling. Figure 2 shows skill scores MSS (Definition 3) comparing the stratified Brier score, sharpness, refinement, and calibration errors for outliers of linear scaling and variants of robust Gaussian scaling $T^{outlier}$ with non-robust Gaussian scaling $GS^{outlier}$.

Compared to non-robust Gaussian scaling, all variants of robust Gaussian scaling shift the distributions of the stratified Brier skill scores for outliers above zero; this means that overall, the probabilities for outliers computed by robust Gaussian scaling have better Brier scores than those computed by non-robust Gaussian scaling. Robust Gaussian scaling with trimmed mean as center and nMAD as scale improves the Brier score for outliers the most. Most variants of robust Gaussian scaling improve the sharpness and refinement of probabilities for outliers compared to non-robust Gaussian scaling. The stratified sharpness error for outliers is best for the mean as the center and nMAD as the scale for robust Gaussian scaling. Linear scaling improves the stratified refinement and calibration errors for outliers the most compared to non-robust Gaussian scaling, followed by robust Gaussian scaling using the median and nMAD for the refinement and Huber's T for the calibration error.

As expected, robust Gaussian scaling can improve the Brier score, sharpness, and refinement of outliers compared to non-robust Gaussian scaling; only its calibration is inferior.

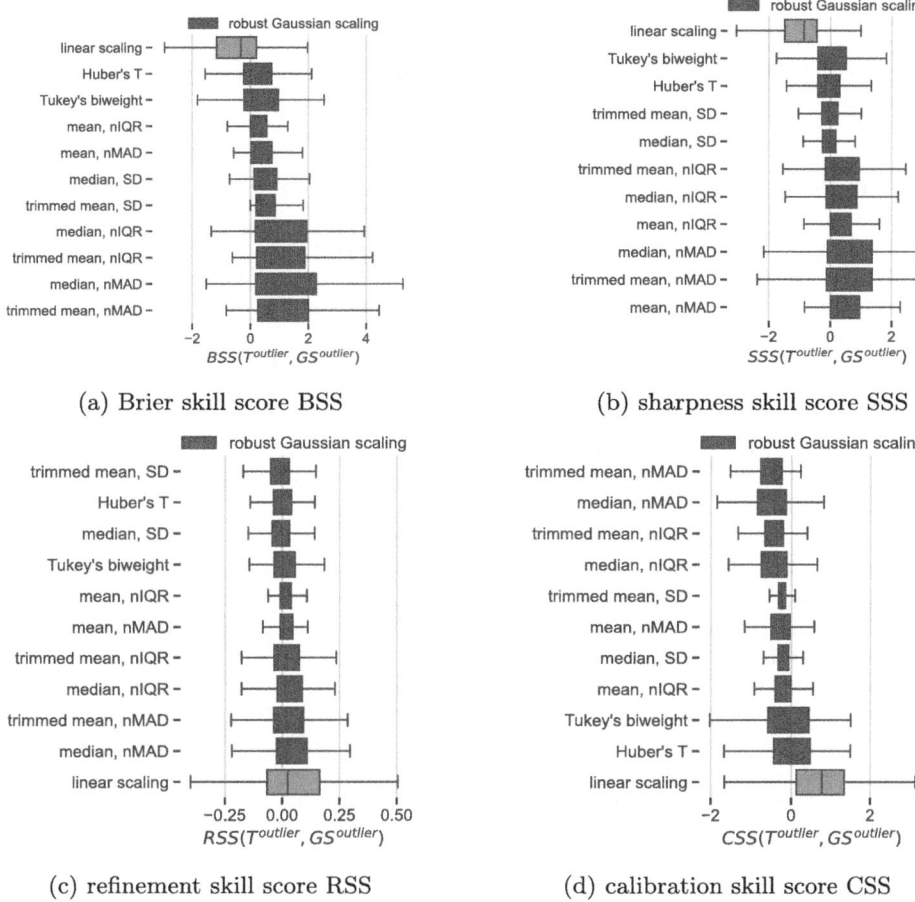

Fig. 2. Skill scores MSS($T^{outlier}$, $GS^{outlier}$) of probabilities for outliers computed by outlier score transformations $T^{outlier}$, which are linear scaling and variants of robust Gaussian scaling, compared to probabilities for outliers computed by non-robust Gaussian scaling $GS^{outlier}$. For clarity, we do not display skill scores less (greater) than 1.5 times the first (third) quartile.

7 Conclusions

We argue and empirically demonstrate that non-robust statistical scaling, a commonly used outlier score transformation, computes inferior probabilities for outliers than for inliers. Therefore, we propose robust statistical scaling using robust estimators. We empirically evaluate several variants of our method and show that it can improve the probabilities of outliers. The extended version of this article discusses additional results: https://arxiv.org/abs/2408.15874 [13].

Acknowledgments. The Independent Research Fund Denmark partly funded this study in the project Reliable Outlier Detection.

Disclosure of Interests. The authors have no competing interests to declare that are relevant to the content of this article.

References

1. Bauder, R.A., Khoshgoftaar, T.M.: Estimating outlier score probabilities. In: IEEE IRI, pp. 559–568. IEEE (2017)
2. Bouguessa, M.: Modeling outlier score distributions. In: ADMA, pp. 713–725. Springer (2012)
3. Campos, G.O., Zimek, A., Sander, J., Campello, R.J.G.B., Micenková, B., Schubert, E., Assent, I., Houle, M.E.: On the evaluation of unsupervised outlier detection: measures, datasets, and an empirical study. Data Min. Knowl. Disc. **30**, 891–927 (2016)
4. Gao, J., Tan, P.N.: Converting output scores from outlier detection algorithms into probability estimates. In: ICDM, pp. 212–221 (2006)
5. Hawkins, D.M.: Identification of Outliers. Springer (1980)
6. Kriegel, H.P., Kroger, P., Schubert, E., Zimek, A.: Interpreting and unifying outlier scores. In: SDM, pp. 13–24. SIAM (2011)
7. Maronna, R.A., Martin, R.D., Yohai, V.J., Salibián-Barrera, M.: Robust Statistics: Theory and Methods (with R). Wiley (2019)
8. Marques, H.O., Campello, R.J., Sander, J., Zimek, A.: Internal evaluation of unsupervised outlier detection. ACM TKDD **14**(4), 1–42 (2020)
9. Murphy, A.H.: Hedging and skill scores for probability forecasts. J. Appl. Meteorol. Climatol. **12**(1), 215–223 (1973)
10. Perini, L., Vercruyssen, V., Davis, J.: Quantifying the confidence of anomaly detectors in their example-wise predictions. In: ECML PKDD, pp. 227–243 (2020)
11. Rayana, S., Zhong, W., Akoglu, L.: Sequential ensemble learning for outlier detection: A bias-variance perspective. In: ICDM, pp. 1167–1172 (2016)
12. Röchner, P., Marques, H.O., Campello, R.J.G.B., Zimek, A.: Evaluating outlier probabilities: assessing sharpness, refinement, and calibration using stratified and weighted measures. Data Min. Knowl. Disc. (2024)
13. Röchner, P., Marques, H.O., Campello, R.J.G.B., Zimek, A., Rothlauf, F.: Robust statistical scaling of outlier scores: improving the quality of outlier probabilities for outliers (extended version) (2024), https://arxiv.org/abs/2408.15874
14. Venables, W., Ripley, B., Venables, W., Ripley, B.: Robust statistics. Modern Applied Statistics with S-PLUS, pp. 247–266 (1997)
15. Wallace, B.C., Dahabreh, I.J.: Improving class probability estimates for imbalanced data. Knowl. Inf. Syst. **41**(1), 33–52 (2014)

Advancing the PAM Algorithm to Semi-supervised *k*-Medoids Clustering

Miriama Jánošová[1]([✉]), Andreas Lang[2], Petra Budikova[1],
Erich Schubert[2], and Vlastislav Dohnal[1]

[1] Faculty of Informatics, Masaryk University, Brno, Czech Republic
{janosova,budikova,dohnal}@fi.muni.cz
[2] TU Dortmund University, Dortmund, Germany
{andreas.lang,erich.schubert}@tu-dortmund.de

Abstract. The analysis of complex, weakly labeled data is increasingly popular, presenting unique challenges. Traditional unsupervised clustering aims to uncover interrelated sets of objects using feature-based similarity of the objects, but this approach often hits its limits for complex multimedia data. Thus, semi-supervised clustering that exploits small amounts of labeled training data has gained traction recently. In this paper, we propose LabeledPAM, a semi-supervised extension of FasterPAM, a state-of-the-art k-medoids clustering algorithm. Our approach is applicable in semi-supervised classification tasks, where labels are assigned to clusters with minimal labeled data, as well as in semi-supervised clustering scenarios, identifying new clusters with unknown labels. We evaluate our proposal against other semi-supervised clustering techniques suitable for arbitrary distances, demonstrating its efficacy and versatility.

Keywords: Semi-supervised clustering · k-medoids · Partitioning around medoids · FasterPAM · Semi-supervised classification

1 Introduction

Clustering is a core technique in data analysis and machine learning. It involves grouping unlabeled objects based on their similarity, ensuring that objects within the same cluster are more similar to each other than to those in different clusters. This method is widely used across various fields to uncover patterns, organize information, and facilitate further analysis. Various types of clustering algorithms exist, including density-based, hierarchical, and partitioning methods, each tailored to address specific data characteristics and analytical requirements [6].

Partitioning clustering algorithms, such as k-means and k-medoids, are popular for their simplicity and efficiency. These algorithms aim to split the data into a predefined number of clusters by minimizing the distance between data points and their respective cluster center. Unlike k-means, which computes centers as arithmetic means of data points, k-medoids selects objects from the dataset as cluster centers. This characteristic makes any k-medoids algorithm particularly advantageous for handling various distance measures, including non-metric measures [5]. However, identifying optimal medoids (centers of clusters that minimize the distances from the data points within those clusters) is an NP-hard problem, necessitating the development of several heuristics for approximation [2].

A recent advancement in this area is the FasterPAM algorithm [18], which improves the k-medoids state-of-the-art by carefully identifying computational steps to skip. So, it accelerates the previous algorithm more than $\mathcal{O}(k)$ times. Despite its effectiveness, FasterPAM, like other unsupervised algorithms, might struggle to produce clusters that align well with the underlying latent structure of the data, particularly in the presence of noise or overlapping clusters.

Unsupervised clustering is a crucial technique in data analysis, yet increasing data complexity makes it difficult to create semantically consistent clusters. At the same time, many domains have available only limited amounts of labeled data. Injecting labeled data into unlabeled datasets can guide the clustering process to yield more semantically meaningful and pure clusters. This approach falls within semi-supervised learning, which combines the principles of supervised and unsupervised learning by using both labeled and unlabeled data. In semi-supervised clustering, the goal is to enhance data organization by refining clusters so that data objects from different classes get better separated in dense areas. This approach should also more effectively distinguish clusters containing unlabeled data from unknown classes. Another valuable application is semi-supervised classification, where the identified clusters are used to predict labels for unlabeled objects based on their similarity to the known clusters.

1.1 Related Work

One of the earliest proposed heuristics of the k-medoids algorithm, the Partitioning Around Medoids (PAM) algorithm [14], iteratively *swaps* some current medoid with the non-medoid object that yields the biggest improvement. Unfortunately, PAM suffers from high computational times. Later work focused on improving its memory efficiency [9] and early stopping strategies [4]. Whereas, Schubert et al. [18] demonstrated that eager swaps provide clustering of similar quality with reduced runtime and parallel computation of which medoid should be swapped, resulting in the FasterPAM algorithm.

Recent surveys on semi-supervised learning present a wide range of methods for integrating labeled knowledge into the clustering process [20]. In this work, we limit our attention to techniques that are based on the k-medoid principle, because of its applicability to arbitrary distance functions.

A straightforward extension of the k-medoid idea are the seeded k-medoids [3], where the initialization phase picks initial cluster centers (seeds) exclusively from the labeled data. Conversely, the COP k-medoids algorithm [8], which is an extension of COP k-means [21] for non-Euclidean spaces, uses labels throughout the entire process. This method adapts the random initialization to ensure at least one medoid is selected for each label. Each cluster then adopts a label based on the label of medoid. In particular, COP k-medoids uses *must-link* and *cannot-link* constraints [11] to determine which objects must be clustered together and which must not. The algorithm not only clusters objects together if their labels are the same, but also when at least one of them is unlabeled.

However, a weakness of the COP k-medoids algorithm is that medoids only shift locally, potentially getting stuck in local minima [16]. The optimization strategy involves two alternating steps: (i) computing the cluster medoid, and (ii) reassigning each sample to its closest medoid. These steps are repeated until no medoid is shifted. Despite its fast execution, this approach has a notable limitation: it is unable to reassign objects located near cluster boundaries to neighboring clusters effectively and allows only minor shifts of the medoids.

Last but not least, Sparse Partitioning Around Medoids [13] is a variant of FasterPAM where not all distances are known, and the algorithm tries to cover the data set with as few cluster centers as possible given distance and capacity constraints. This shares some optimization strategies with our approach.

1.2 Contribution of This Paper

We propose an extension of the FasterPAM algorithm that integrates prior knowledge from labeled data with a global optimization strategy. Our method serves a dual purpose. It can function as (i) an independent clustering algorithm or (ii) a post-processing step to refine clusters generated by FasterPAM. The primary objective is to improve the purity and relevance of resulting clusters, thereby enhancing overall clustering performance. In our experimental evaluation, we demonstrate the effectiveness of our algorithm in both semi-supervised clustering and semi-supervised classification scenarios. We compare our approach with FasterPAM and semi-supervised variations of the k-medoids to show that our enhanced swap method identifies better data partitions.

2 Foundations

In this section, we recall the concepts of k-medoids and semi-supervised clustering. We review particularly the swap procedure of FasterPAM [18] and introduce the notation that will be used to describe the algorithms.

2.1 k-Medoids

The k-medoids is a clustering method that identifies k representative objects (medoids) within a dataset that serve as cluster centers, denoted as m_i. The objective is to minimize overall dissimilarity between the medoids and the remaining data, i.e., to reduce the Total Deviation (TD), see Eq. 1. Formally, given a dataset X, the k-medoids loss is defined as:

$$\text{TD} = \sum_{x \in X} \min_{i \in \{1,\ldots,k\}} d(x, m_i). \tag{1}$$

PAM is a particular algorithm that solves the k-medoids clustering problem. It starts by choosing k initial medoids randomly from X. In each iteration, it identifies and performs the most beneficial swap between any non-medoid and any medoid objects to minimize the TD, with the change denoted as ΔTD. The new total deviation is then TD = TD $-\Delta$TD. This process continues until the maximum number of iterations is reached or ΔTD becomes negligible.

FasterPAM is a state-of-the-art algorithm accelerating PAM. It eagerly attempts to swap each non-medoid object with all medoids in parallel. FasterPAM internally stores the distances from each object to its closest and second closest medoid to reduce the overhead of re-evaluating distances. We denote the functions returning the respective medoid indexes as $nearest(x_i)$ and $second(x_i)$.

To decide whether a swap between any current medoid m_i and a new non-medoid object x_c that acts as a candidate is favorable, the algorithm computes ΔTD from two components: *penalty* and *benefit*. Namely, a one-dimensional array ΔTD^{-m_i} contains the penalties for removing m_i, and ΔTD^{+x_c} stores a scalar value equal to the benefit of picking x_c as a medoid. In ΔTD^{-m_i}, the penalty for replacing m_i with x_c is accumulated as follows: it is computed as the penalty for reassigning all objects from m_i's cluster elsewhere, i.e., to their second nearest medoids. This can be computed at the beginning of the algorithm for all current medoids. Then the negative penalty is added for objects that stay in the cluster, i.e., will be assigned to x_c. As for ΔTD^{+x_c}, for all objects x_o where $d(x_c, x_o) < d_{nearest}(x_o)$ we gain a decrease of ΔTD by $d_{nearest}(x_o) - d(x_c, x_o)$. This gain is computed in the same manner regardless of which medoid is to be replaced by x_c, so a single variable ΔTD^{+x_c} is sufficient here.

The crucial aspect is that, at the end of each iteration, the sum of ΔTD^{-m_i} and ΔTD^{+x_c} is the loss of replacing m_i with x_c. Although only one of these options can actually occur, all the alternatives are computed in one pass. Let us emphasize that ΔTD^{-m_i} contains only the sum of penalties, i.e., increase of ΔTD, for objects that must be reassigned when x_c replaces m_i. Conversely, ΔTD^{+x_c} holds the sum of gains, i.e., a decrease of ΔTD, from all objects that are newly assigned to the cluster of x_c. When $\Delta\text{TD}^{-m_i} + \Delta\text{TD}^{+x_c} < 0$, the overall ΔTD is improved (decreased) by replacing m_i with x_c. If this inequality holds for multiple m_i, the swap with the lowest penalty is selected.

2.2 Semi-supervised Clustering

Assume a function $lbl(x)$ that returns a label for an object $x \in X$ as follows:

$$\forall (x \in X) \exists (i \in \{None\} \cup \{lbl_1, lbl_2, \ldots, lbl_{nl}\}) : lbl(x) = i, \tag{2}$$

where lbl_i are individual labels, *None* marks an unlabeled/*None*-labeled object, and nl is the total number of labels. In semi-supervised clustering, each cluster consists of objects that have the same label or are *None*-labeled. A cluster j has a label lbl_i if and only if it contains at least one object with lbl_i; otherwise, the cluster is unlabeled. The label of a cluster j is denoted as $lbl_{cl}(j)$.

Semi-supervised k-medoids modifies the original objective to respect labels. The algorithm produces k clusters, which can be both labeled and unlabeled. If $k > nl$, multiple clusters can share the same label; $k < nl$ is considered invalid.

An object x can be assigned to a cluster j with medoid m_j if and only if x has the same label as the cluster or x is unlabeled. The distance metric must be extended to account for labels, i.e., the *labeled distance* function $d_l(x, y)$:

$$d_l(x, m_j) = \begin{cases} d(x, y) & \text{if } lbl(x) = lbl_{cl}(j) \vee lbl(x) = None, \\ \infty & \text{otherwise.} \end{cases} \quad (3)$$

The objective of semi-supervised k-medoids is to minimize TD is defined analogously but uses d_l:

$$\text{TD} = \sum_{x \in X} \min_{j \in \{1, \ldots, k\}} d_l(x, m_j). \quad (4)$$

3 LabeledPAM

We propose a novel semi-supervised clustering algorithm, *LabeledPAM*, that extends FasterPAM by incorporating labels and cannot-link constraints. It benefits from the flexibility of PAM and the accelerated medoid selection introduced with FasterPAM, enabling it to adjust the partitioning according to the provided labels. Our algorithm is expected to be more robust than COP k-medoids due to the usage of PAM-style swaps compared to the more local alternating, k-means like, optimization of COP k-medoids. This was previously shown in [18] that COP k-medoids inclines to get stuck in inferior local optima.

3.1 Initialization Strategies

For LabeledPAM, we first need to find an initial state that does not violate cannot-link constraints. Our goal is twofold: (i) to ensure that each label is represented in the initial set of medoids, which assumes $k \geq nl$, and (ii) to distribute the medoids across the entire dataset. We propose and later compare three distinct initialization strategies:

The **Random** initialization samples random medoids from X. First, one medoid is sampled from each label. Then, the remaining medoids are chosen randomly from the remainder, independent of the labels.

The **Medoids** strategy calculates for each label a medoid from all objects having this label, and takes them as initial medoids. Additional $(k - nl)$ objects are sampled from X randomly.

The **FasterPAM** initialization acquires the initial medoids by clustering X with FasterPAM. The clusters are labeled based on the most common label in their cluster. If there still exists a label lbl_A that is not yet represented by a cluster, then a cluster that contains some objects labeled with lbl_A is relabeled. Such relabeling must not cause the removal of another label completely. If no such cluster exists, a random sample of the missing label replaces an unlabeled medoid. If all medoids are labeled, an existing medoid, whose label is already represented by more than one cluster, will be chosen instead.

3.2 Implementation Details

Similarly to FasterPAM, LabeledPAM iteratively searches for any beneficial swap of a non-medoid x_c and m_i that reduces TD. The best potential swap for x_c is determined in one pass for all medoids m_i. However, in LabeledPAM the computation of $\Delta\,\mathrm{TD}^{+x_c}$ must be decomposed to individual labels. The set of objects that can enter the cluster of x_c depends on the cluster's label. As explained earlier, the label of x_c's cluster is known immediately if $lbl(x_c) \neq None$, but an unlabeled x_c can create a cluster of an arbitrary label.

In detail, $\Delta\,\mathrm{TD}^{+x_c}$ becomes a 2-dimensional array of $k \times nl$ (the number of clusters by the number of labels). In summary, we will use these references:

$\Delta\,\mathrm{TD}^{-m}[i]$ – loss of removing m_i, which can be precomputed,
$\Delta\,\mathrm{TD}^{+x_c}[i][l]$ – benefit of introducing the new medoid x_c with the label l and undoing loss stored in $\Delta\,\mathrm{TD}^{-m}[i]$.

Following the implementation of FasterPAM, we begin by calculating the initial loss associated with removing any cluster i, denoted as $\Delta\,\mathrm{TD}^{-m}[i]$. At this stage, we consider only the removal of the cluster without replacement, so future cluster labels are not yet relevant. This loss can be precomputed and stored for efficiency until a swap occurs. After a medoid is swapped, the initial loss must be updated accordingly.

However, there is a special case to consider: there might be an object o_{sad} with the second nearest medoid undefined, i.e., $d_l(o_{sad}, second(o_{sad})) = \infty$. This occurs when there is no second cluster with the same label as o_{sad} and no unlabeled cluster. In that case, the medoid $nearest(o_{sad})$ cannot change the label, as this would make a cannot-link constraint unsatisfiable. Naively, one could set $\Delta\,\mathrm{TD}^{-m}[nearest(o_{sad})] = \infty$ or to some very high constant, but this leads to numerical issues. Instead, we remove the loss of o_{sad} completely, $\Delta\,\mathrm{TD}^{-m}[nearest(o_{sad})] \mathrel{-}= d_l(o_{sad}, nearest(o_{sad}))$. The satisfaction of the cannot-link constraint is ensured when calculating the reassignment loss for an actual candidate x_c based on its label. Here, the complete loss of the assignment of o_{sad} is added.

After computing $\Delta\,\mathrm{TD}^{-m}$, LabeledPAM decides whether each x_c can become a medoid and replace an existing one. There are two possible cases:

1: x_c has a label (e.g., lbl_A), then x_c can form a cluster with the label lbl_A and containing objects having lbl_A or $None$;
2: x_c is $None$-labeled, then x_c can form a cluster of *any label*, including $None$.

We describe these cases in greater detail, following the methodology of FasterPAM to be comprehensive and enable a direct comparison. To simplify the description of the algorithm, we add an additional loop over all medoids m_i, when describing how the losses are updated.

Case 1: $lbl(x_c) = lbl_A$. The only label the new cluster can adopt is lbl_A. Therefore, we invalidate all other potential labels of the new cluster besides lbl_A, i.e., setting $\Delta\,\mathrm{TD}^{+x_c}[i][l] = \infty$ for all labels $l \neq lbl_A$. Conversely, for lbl_A, the accumulator is set to $\Delta\,\mathrm{TD}^{+x_c}[i][lbl_A] = 0$.

Algorithm 1: $updateAssignmentLoss(x_o, x_c, l, i)$

1 $d_{oc} \leftarrow d_l(x_o, x_c)$
2 $idx_n \leftarrow nearest(x_o)$
3 **if** $d_{second}(x_o) = \infty \land lbl(x_o) \neq l$ **then**
      /* no valid assignment possible   */
4     $\Delta \text{TD}^{+x_c}[idx_n][l] \leftarrow \infty$
5 **else if** $d_{second}(x_o) = \infty \land lbl(x_o) = l$ **then**
      /* new cluster requests the same label as x_o */
6   **if** $i = idx_n$ **then** $\Delta \text{TD}^{+x_c}[i][l] \leftarrow \Delta \text{TD}^{+x_c}[i][l] + d_{oc}$ **else**
       $\Delta \text{TD}^{+x_c}[i][l] \leftarrow \Delta \text{TD}^{+x_c}[i][l] + d_{oc} - d_{nearest}(x_o)$
7 **else if** $d_{oc} < d_{nearest}(x_o)$ **then**
      /* new cluster is the best option   */
8   **if** $i = idx_n$ **then** $\Delta \text{TD}^{+x_c}[i][l] \leftarrow \Delta \text{TD}^{+x_c}[i][l] + d_{oc} - d_{second}(x_o)$ **else**
       $\Delta \text{TD}^{+x_c}[i][l] \leftarrow \Delta \text{TD}^{+x_c}[i][l] + d_{oc} - d_{nearest}(x_o)$
9 **else if** $d_{oc} < d_{second}(x_o)$ **then**
      /* the new cluster is the best option if idx_n is replaced   */
10   **if** $i = idx_n$ **then** $\Delta \text{TD}^{+x_c}[i][l] \leftarrow \Delta \text{TD}^{+x_c}[i][l] + d_{oc} - d_{second}(x_o)$ **else** pass
11 **return** $\Delta \text{TD}, \Delta \text{TD}^{+x_c}$

Next, we iterate through all non-medoid objects x_o to determine whether they can be included in the new lbl_A-labeled cluster. This depends on the label of x_o with the possible scenarios:

1. $lbl(x_o) = lbl_A$ or $lbl(x_o) = None$: x_o can be assigned to the new cluster. So, $d_l(x_o, x_c)$ is computed and all changes to the loss are performed by running the subroutine $updateAssignmentLoss(x_o, x_c, lbl(x_o), i)$;
2. otherwise: x_o cannot be part of the new cluster. Here, we need to check whether x_o is o_{sad}, by verifying that $d_l(x_o, second(x_o)) \neq \infty$. If there exists a second cluster for x_o, then the loss does not need to be updated, as x_o can stay with the second medoid. Otherwise, $\Delta \text{TD}^{+x_c}[nearest(x_o)][l] = \infty$ for *all labels* l, suggesting that the swap between the medoid $nearest(x_o)$ and x_c is not viable.

Case 2: $lbl(x_c) = None$. Provided that $lbl(x_c) = None$, the new cluster can adopt any label. Therefore, the loss of removing a medoid needs to be computed for *all labels*. For each label l and medoid m_i, $\Delta \text{TD}^{+x_c}[i][l] = 0$.

Next, we iterate through the entire dataset to determine under what conditions each non-medoid object x_o could join the new cluster of x_c. The situations are as follows:

1. $lbl(x_o) = None$: Regardless of the label of the new cluster, x_o can join it. We run the subroutine $updateAssignmentLoss(x_o, x_c, l, i)$ for all labels l and medoids i.
2. otherwise: x_o can join the new cluster only if the cluster will be assigned the label $lbl(x_o)$. We run the subroutine $updateAssignmentLoss(x_o, x_c, lbl(x_o), i)$.

Algorithm 1 specifies the subroutine *updateAssignmentLoss*. This highlights significant modifications to FasterPAM in the swap phase of x_c and m_i. For readability purposes, we condensed the subroutine to a single label l and medoid m_i. It can also be computed in one pass by combining the benefits and losses, like in FasterPAM and adding them to the corresponding arrays.

Our new update assignment loss function differs in a few significant ways from the FasterPAM version. First, we need to handle the clusters that cannot be replaced by a cluster of different label. If the label does not match the label of the new medoid, then the now medoid cannot satisfy the cannot-link criteria. If the labels do match, we need to remember that some data objects, i.e., o_{sad} are not included in the loss calculation and, therefore, we need to add the whole loss to the new medoid. The other two cases are nearly the same as in FasterPAM, but the change in loss is added to all the labels.

Like in FasterPAM, the best potential swap for replacing each medoid with the new medoid x_c and label l is calculated:

$$\underset{i \in 1,\ldots,k;\ l \in 1,\ldots,nl}{\arg\min} (\Delta \operatorname{TD}^{+x_c}[i][l] + \operatorname{TD}^{-m}[i]). \tag{5}$$

If this swap reduces TD, it is performed eagerly without searching for the optimal one. In between swaps, two additional steps need to be taken care of. Clusters can drift away from labeled data, at which point the label is removed from the cluster. Those clusters thus encompass areas that are not well explained by labeled data objects. At the same time, other clusters may change their medoid, which allows labeled objects to govern a previously unlabeled cluster. To recognize those cases, it is checked if the labeling of previously unlabeled medoids is beneficial.

The resulting runtime complexity is $O(N^2(nl+k))$ where N is the number of samples in the data set, nl the number of labels, and k the number of clusters.

4 Experimental Evaluation

To compare the proposed technique and evaluate it objectively, we aim to answer the following research questions:

1. **Classification**: How accurately can new data be assigned to existing labels when using partially labeled training data?
2. **Clustering**: How well does the clustering result align with the given labels?
3. **Overclustering**: How much does clustering/classification performance change when the number of clusters is increased for the same dataset?
4. **Missing labels**: How do the algorithms react to missing labels, and how does it impact clustering quality?

Table 1. Overview of the datasets used in the experiments: the dataset name; sizes of *Train* and *Test* partitions; the vector dimensionality D; the number of labels nl, and the contents' description of data object in *Domain*.

Name	Train	Test	D	nl	Domain
ALOI [19]	9000	1000	63	100	HSB histograms
MNIST [12]	55,800	9200	768	10	Black and white images
OPTIDIGITS [1]	4620	1000	64	10	Black and white images
CIFAR [10]	50,000	10,000	768	10	CLIP embeddings

4.1 Methodology

Algorithms: We compare LabeledPAM (*LPAM*) with the existing techniques, namely *Seeded k-medoids* [3], *COP k-medoids* [8], and the unsupervised *Faster-PAM* [18] as baseline. LabeledPAM is evaluated for each initialization strategy.

Implementation: To ensure a fair and comparable evaluation, we implemented all algorithms in the ELKI framework [17], using the original implementation of FasterPAM. First, all algorithms compute the full distance matrix.

Datasets: To assess the performance thoroughly, we use four real-world datasets of different domains, see Table 1. For CIFAR [10], we used the 10-class version and extracted CLIP embeddings [15] from pictures. A sample from MNIST [12] was taken to accommodate the maximum array size of Java. For the same reason, we restrict ALOI to the first 100 classes. The Optical Recognition of Handwritten Digits dataset (OPTIDIGITS) was used as is.

Semi-supervised setup: We split all the datasets into train and test subsets. For the train set, we used a Stratified Shuffle Split to generate 10 splits for varying amounts of labeled data (from 5% to 50%). We calculate the pairwise distance matrix for all datasets and apply the semi-supervised clustering algorithms to the training data.

Metrics: Clustering quality is assessed by Adjusted Rand Index (ARI) [7] directly on the training set. For the classification, we use the test set and based on the result of the particular algorithm, we assign for each test object the label of the closest medoid of the labeled cluster, i.e., we perform nearest-neighbor classification using the medoids of labeled clusters only. Medoids of clusters that are not labeled are not used. For the research question **Missing labels**, we discard the labels of 20% of classes.

4.2 Results

Classification and clustering: In Fig. 1, we study the quality of the basic case where the number of clusters is the same as the number of labels, focusing on the MNIST and ALOI datasets. In this setup, all label classes have at least one labeled instance, enabling us to answer our first two research questions.

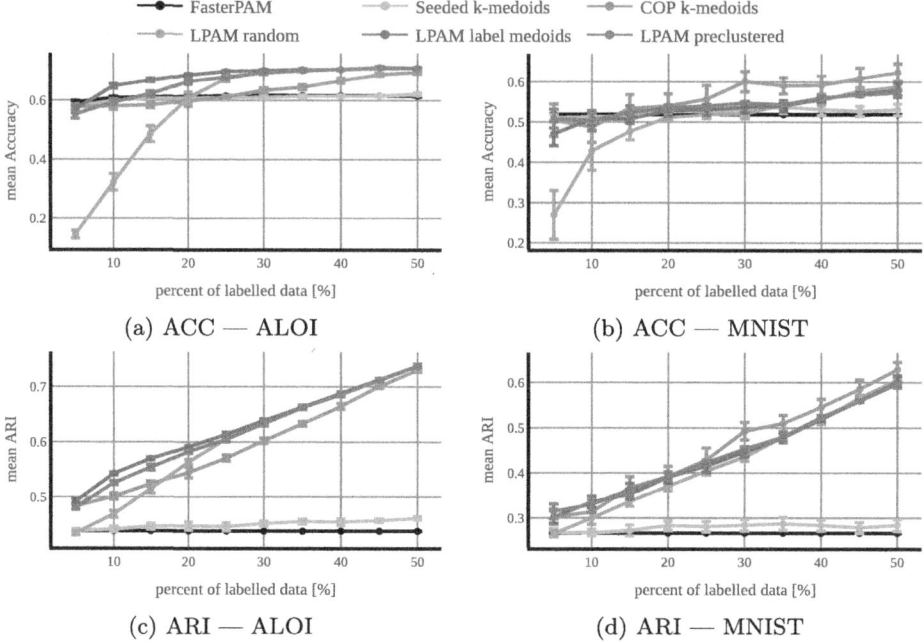

Fig. 1. Classification and clustering quality for k equal to the number of labels in the dataset for various percentages of labeled data. For each experiment, the graphs show mean accuracy/ARI as well as 95% confidence intervals.

The classification accuracy on the ALOI dataset, as shown in Fig. 1a, reveals that the unsupervised FasterPAM algorithm does not really benefit from an increase in labeled data, as can be expected. Similarly, the Seeded k-medoids algorithm, which only alters the initialization, shows no substantial improvement, indicating that the PAM algorithm is not getting stuck in a poor local minimum for this dataset.

In contrast, our LabeledPAM algorithm demonstrates different behavior. The randomly initialized version performs poorly, especially for a small amount of labeled data, due to falling in a local minimum. With the more sophisticated initializations, the highest accuracy can be achieved with both versions, particularly for 10–40% of labeled data points. The COP k-medoids algorithm can reach a similar accuracy but requires more labeled objects to do so.

The high dependence on the initialization for our method suggests that there are many local optima where the optimization procedure can get stuck. In particular, when the number of clusters is similar to the number of labels, many swaps are prevented by the cannot-link constraints, making the optimization more challenging than in regular k-medoids.

Figure 1b indicates the classification accuracy for the MNIST dataset. FasterPAM, Seeded k-medoids, and LabeledPAM with random initialization perform similarly to their results in the previous dataset. However, the COP k-

Table 2. Accuracy with 95% confidence intervals.

Algorithm Dataset		COP k-medoids	FasterPAM	LPAM lbl medoid	LPAM pre-cluster	LPAM random	Seeded k-medoids
ALOI	10%	0.581 ± 0.013	0.608 ± 0.004	0.649 ± 0.008	0.593 ± 0.009	0.322 ± 0.028	0.598 ± 0.005
	30%	0.634 ± 0.009	0.617 ± 0.002	0.702 ± 0.003	0.693 ± 0.005	0.695 ± 0.004	0.611 ± 0.005
CIFAR	10%	0.611 ± 0.022	0.630 ± 0.000	0.633 ± 0.016	0.632 ± 0.004	0.502 ± 0.079	0.630 ± 0.000
	30%	0.688 ± 0.012	0.630 ± 0.001	0.689 ± 0.010	0.683 ± 0.017	0.686 ± 0.013	0.630 ± 0.001
MNIST	10%	0.487 ± 0.038	0.519 ± 0.000	0.508 ± 0.015	0.504 ± 0.025	0.428 ± 0.048	0.514 ± 0.007
	30%	0.601 ± 0.025	0.519 ± 0.000	0.537 ± 0.014	0.547 ± 0.002	0.521 ± 0.013	0.531 ± 0.015
OPTI	10%	0.809 ± 0.020	0.794 ± 0.000	0.846 ± 0.020	0.809 ± 0.001	0.487 ± 0.099	0.792 ± 0.003
	30%	0.840 ± 0.011	0.794 ± 0.000	0.856 ± 0.009	0.861 ± 0.000	0.861 ± 0.000	0.800 ± 0.012

Table 3. ARI with 95% confidence intervals.

Algorithm Dataset		COP k-medoids	FasterPAM	LPAM lbl medoid	LPAM pre-cluster	LPAM random	Seeded k-medoids
ALOI	10%	0.501 ± 0.005	0.438 ± 0.000	0.543 ± 0.002	0.526 ± 0.004	0.468 ± 0.007	0.441 ± 0.005
	30%	0.602 ± 0.005	0.438 ± 0.000	0.639 ± 0.003	0.634 ± 0.005	0.635 ± 0.004	0.451 ± 0.004
CIFAR	10%	0.474 ± 0.023	0.465 ± 0.000	0.491 ± 0.004	0.500 ± 0.001	0.470 ± 0.015	0.465 ± 0.000
	30%	0.606 ± 0.009	0.465 ± 0.000	0.614 ± 0.011	0.608 ± 0.014	0.612 ± 0.012	0.465 ± 0.000
MNIST	10%	0.313 ± 0.028	0.267 ± 0.000	0.335 ± 0.013	0.330 ± 0.018	0.299 ± 0.013	0.267 ± 0.007
	30%	0.493 ± 0.019	0.267 ± 0.000	0.447 ± 0.011	0.453 ± 0.002	0.435 ± 0.009	0.284 ± 0.012
OPTI	10%	0.665 ± 0.022	0.643 ± 0.000	0.733 ± 0.019	0.692 ± 0.001	0.606 ± 0.015	0.640 ± 0.008
	30%	0.766 ± 0.010	0.643 ± 0.000	0.795 ± 0.009	0.800 ± 0.004	0.800 ± 0.004	0.650 ± 0.013

medoids notably improves with a lower percentage of labeled data, surpassing our LabeledPAM approach. The classification quality for other datasets is summarized in Table 2, showing only the 10% and 30% values due to space restrictions.

Next, we assess the clustering quality for the same dataset. Figure 1c shows that the unsupervised FasterPAM remains unaffected by the amount of labeled data. Surprisingly, LabeledPAM with the random initialization performs as well as or better, even in cases where it underperformed in classification, indicating accurate clusters but poor medoid choices.

Similar to the classification task, our approach with a proper initialization outperforms COP k-medoids across the given label percentages, though the margin decreases as more labeled data reduces the solution space. In Fig. 1d, the clustering quality on the MNIST dataset shows similar patterns between the results for the datasets as seen in the classification task. A comprehensive overview of the clustering quality for the other datasets is provided in Table 3.

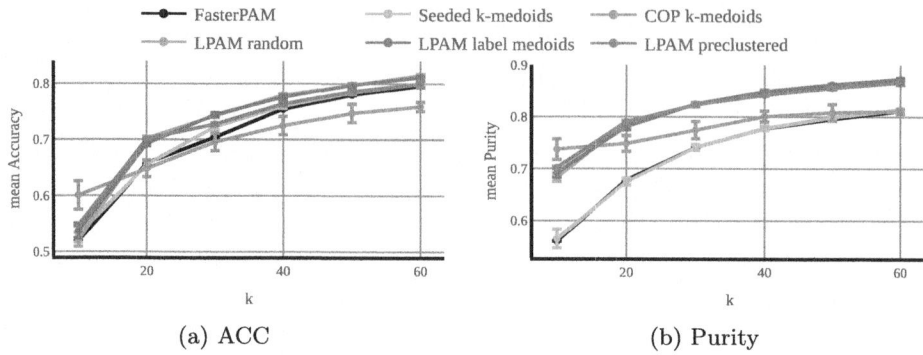

Fig. 2. Classification and clustering quality with increasing k on the MNIST dataset with 30% labels.

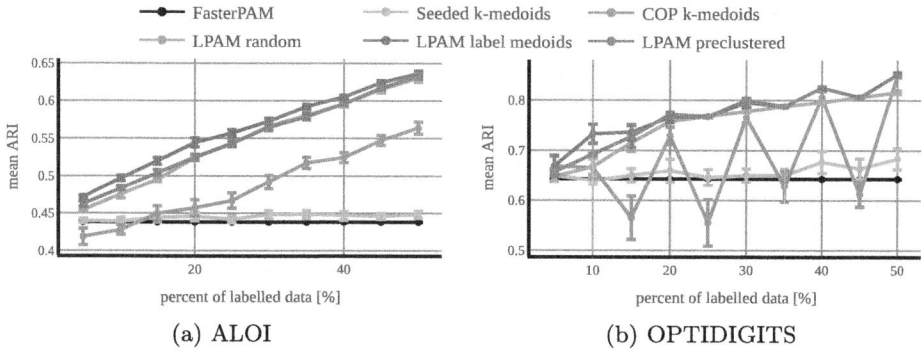

Fig. 3. Clustering performance when 20% of classes have their label withheld.

Overclustering: In Fig. 2, we explore the impact of an increasing k on classification and clustering accuracy on the MNIST dataset, which was problematic in previous experiments. Increasing the number of clusters beyond the number of labels can help uncover substructures and manage classes that have subtypes, while providing the optimization more freedom.

As shown in Fig. 2a, the accuracy increases noticeably for all algorithms. The COP k-medoids algorithm, which performed best in the previous experiment, exhibits the slowest and smallest accuracy improvement, even when the number of clusters is doubled (20 clusters). All other algorithms reach a higher accuracy, highlighting the inflexibility of COP k-medoids in relabeling the clusters.

The LabeledPAM variants show the earliest increase in accuracy and achieve the overall best performance. In Fig. 2b, we assess the purity because the mismatch between the number of classes and clusters would influence other metrics.

We observe that semi-supervised learning impacts clustering quality more than classification quality. The COP k-medoids algorithm does not benefit as much as the FasterPAM algorithm from an increased number of clusters. Additionally, FasterPAM outperforms COP k-medoids, despite its initial higher quality. Our LabeledPAM approach achieves the best cluster purity for larger k, except for the previously analyzed case of 10 clusters.

Missing labels: In our final research question, explored in Fig. 3, we investigate the impact of removing labels from 20% of the classes. This experiment showcases how individual algorithms can discern clusters containing merely unlabeled objects that semantically belong to different classes. Figure 3a and 3b show a slight improvement of Seeded k-medoids over FasterPAM as the percentage of labeled data increases, whereas the ARI of COP k-medoids increases linearly.

Notably, our approach outperforms each of the reference algorithms by 5–8% on the ALOI dataset and by 5–15% on the OPTIDIGITS dataset. These results confirm that our approach effectively shifts the medoids into better positions, thereby improving clustering performance. Consistent with other experiments, random initialization performs the worst, compared to other LPAM initializations. However, the differences are only of minor importance, especially on the OPTIDIGITS dataset. Noticeably, the LPAM methods also exhibit high stability in all experiments, with the 95% confidence intervals being much smaller than the confidence intervals of COP k-medoids.

Lastly, we also looked at the runtime of our approach in the basic experiments. The increase in quality, unfortunately, also increases the runtime by 100–400% compared to unlabeled FasterPAM.

5 Conclusions

We proposed and implemented LabeledPAM, an algorithm for clustering that combines k-medoids with principles of semi-supervised learning. Our algorithm is based on the FasterPAM algorithm, which is extended with cluster labeling using the cannot-link constraints.

Our extensive experimental evaluation reveals that LabeledPAM achieves the best average performance across various datasets and fractions of labeled data. Specifically, we found that in classification, our approach outperformed other reference algorithms by up to 5 percentage points on most datasets. In clustering, LabeledPAM outperforms FasterPAM, and Seeded k-medoids, and COP k-medoids consistently.

The third research question demonstrated that increasing the number of clusters is favorable for the clustering quality, by capturing diverse data shapes within a single label, with LabeledPAM performing the best. Additionally, our algorithm shows robustness, particularly when withholding some labels in the input, outperforming the other algorithms by 10–15%. The experimental evaluation shows that LabeledPAM frequently surpasses COP k-medoids thanks to its ability to shift medoids more effectively, thereby avoiding local minima.

For future work, we aim to reduce the amount of precomputed information required, as only a portion of the distance matrix is required during the algorithm's execution. Next, we intend to make our approach more robust towards adverse initialization to prevent early convergence to a local minimum. Lastly, we plan to investigate how the LabeledPAM algorithm can be used as an input to a regular kNN-classifier, reducing the number of training data needed and, therefore, improving its prediction run time.

Acknowledgments. Czech Science Foundation project No. GF23-07040K. Computational resources were provided by the e-INFRA CZ project, supported by the Ministry of Education, Youth and Sports of the Czech Republic, project No. ID:90254.

References

1. Alpaydin, E., Kaynak, C.: Optical Recognition of Handwritten Digits. UCI Machine Learning Repository (1998). https://doi.org/10.24432/C50P49
2. Balcan, M.F., Blum, A., Gupta, A.: Clustering under approximation stability. J. ACM **60**(2) (2013). https://doi.org/10.1145/2450142.2450144
3. Basu, S., Banerjee, A., Mooney, R.J.: Semi-supervised clustering by seeding. In: ICML, pp. 27–34 (2002)
4. Estivill-Castro, V., Murray, A.T.: Discovering associations in spatial data - an efficient medoid based approach. PAKDD **1394**, 110–121 (1998). https://doi.org/10.1007/3-540-64383-4_10
5. Estivill-Castro, V., Yang, J.: Fast and robust general purpose clustering algorithms. Data Min. Knowl. Discov. **8**(2), 127–150 (2004). https://doi.org/10.1023/B:DAMI.0000015869.08323.b3
6. Ezugwu, A.E., Ikotun, A.M., Oyelade, O.O., Abualigah, L., Agushaka, J.O., Eke, C.I., Akinyelu, A.A.: A comprehensive survey of clustering algorithms: state-of-the-art machine learning applications, taxonomy, challenges, and future research prospects. Eng. Appl. Artif. Intell. **110**, 104743 (2022). https://doi.org/10.1016/j.engappai.2022.104743
7. Hubert, L.J., Arabie, P.: Comparing partitions. J. Classif. **2**, 193–218 (1985)
8. Jiang, H., Ren, Z., Xuan, J., Wu, X.: Extracting elite pairwise constraints for clustering. Neurocomputing **99**, 124–133 (2013). https://doi.org/10.1016/j.neucom.2012.06.013
9. Kaufman, L., Rousseeuw, P.J.: Finding Groups in Data: An Introduction to Cluster Analysis. Wiley (2009). https://doi.org/10.1002/9780470316801
10. Krizhevsky, A.: Learning multiple layers of features from tiny images. Tech. rep., University of Toronto (2009)
11. Lange, T., Law, M.H., Jain, A.K., Buhmann, J.M.: Learning with constrained and unlabelled data. In: CVPR, pp. 731–738 (2005). https://doi.org/10.1109/CVPR.2005.210
12. LeCun, Y., Cortes, C., Burges, C.: MNIST handwritten digit database. http://yann.lecun.com/exdb/mnist (1998)
13. Lenssen, L., Schubert, E.: Sparse partitioning around medoids. In: Machine Learning Under Resource Constraints – Volume 1: Fundamentals, pp. 182–196. De Gruyter (2022). https://doi.org/10.1515/9783110785944-005
14. Leonard Kaufman, P.J.R.: Partitioning Around Medoids (Program PAM), pp. 68–125. Wiley (1990) (chapter 2)
15. Radford, A., Kim, J.W., Hallacy, C., Ramesh, A., Goh, G., Agarwal, S., Sastry, G., Askell, A., Mishkin, P., Clark, J., Krueger, G., Sutskever, I.: Learning transferable visual models from natural language supervision. In: ICML, pp. 8748–8763 (2021)

16. Reynolds, A.P., Richards, G., de la Iglesia, B., Rayward-Smith, V.J.: Clustering rules: a comparison of partitioning and hierarchical clustering algorithms. J. Math. Model. Algorith. **5**, 475–504 (2006). https://doi.org/10.1007/s10852-005-9022-1
17. Schubert, E.: Automatic indexing for similarity search in ELKI. In: Similarity Search and Applications. SISAP (2022). https://doi.org/10.1007/978-3-031-17849-8_16
18. Schubert, E., Rousseeuw, P.J.: Fast and eager k-medoids clustering: O(k) runtime improvement of the PAM, CLARA, and CLARANS algorithms. Inf. Syst. **101**, 101804 (2021). https://doi.org/10.1016/j.is.2021.101804
19. Schubert, E., Zimek, A.: ELKI Multi-View Clustering Data Sets Based on the Amsterdam Library of Object Images (ALOI). Zenodo (2010)
20. Van Engelen, J.E., Hoos, H.H.: A survey on semi-supervised learning. Mach. Learn. **109**(2), 373–440 (2020). https://doi.org/10.1007/s10994-019-05855-6
21. Wagstaff, K., Cardie, C., Rogers, S., Schrödl, S.: Constrained k-means clustering with background knowledge. In: ICML, pp. 577–584 (2001)

Hierarchical Clustering Without Pairwise Distances by Incremental Similarity Search

Erich Schubert(✉)

TU Dortmund University, Dortmund, Germany
erich.schubert@tu-dortmund.de

Abstract. Hierarchical clustering is a popular classic technique for cluster analysis, in particular, because it is easy to understand and explain. The key limitation of hierarchical agglomerative clustering is its run time: the standard algorithm runs in cubic time, and improved methods use at least quadratic time. We propose novel strategies for accelerating hierarchical clustering using incremental similarity search. Using a priority search on a vantage-point tree, we often find the next merge without computing all pairwise distances. We propose two strategies based on heaps of searches for single linkage and a third strategy based on the nearest-neighbor chain algorithm for Ward, centroid, and median linkage, other linkages are not supported efficiently (yet). Experimentally, we demonstrate 2 to 10-fold speedups on real data sets and show that subquadratic scalability is possible although it can not be guaranteed.

Keywords: Cluster analysis · Hierarchical Agglomerative Clustering · Index acceleration

1 Introduction

Hierarchical clustering is a key technique in data analysis for uncovering unknown structures within datasets. It can either be performed agglomerative, which merges smaller clusters into larger ones, or divisive, which splits larger clusters into smaller ones. Despite its applicability in biology, bioinformatics, image analysis, and market segmentation, hierarchical clustering is often limited by its computational complexity, which poses challenges when dealing with large datasets. The result of hierarchical clustering, at least for smaller data sets, is often presented using a dendrogram, a tree structure, that has the objects on the x-axis, and the distance at which they are merged on the y-axis. Interesting subtrees of the dendrogram, not necessarily at the same height, can be selected as clusters. The primary computational challenge in hierarchical clustering arises from its quadratic or cubic time complexity in the number of samples. This inefficiency renders it impractical for large datasets, constraining its use in big data environments. In this article, we focus on the much more common agglomerative approach, as divisive strategies suffer from different problems. Most algorithms for hierarchical agglomerative clustering (HAC) operate on a distance matrix, which implies a lower bound of $O(N^2)$ for their memory and run-time complexity. The standard agglomerative clustering algorithm taught in textbooks, AGNES, even needs $O(N^3)$ time, improved algorithms such as SLINK [24], CLINK [9], and the nearest-neighbor chain algorithm [20] are not commonly discussed there. Although GPUs provide a highly parallel means to compute the pairwise distance matrix, this does not solve the scalability problem. Other strategies to accelerate HAC involve approximation, but we focus on exact methods here.

This paper introduces a new approach to accelerating hierarchical clustering with indexes, which can reduce computational overhead without compromising the integrity of the cluster hierarchy. The method leverages incremental nearest-neighbor search data structures to improve performance. Experimental results on various datasets show the effectiveness in handling larger data, enhancing the practical application of hierarchical clustering in data-intensive scenarios.

2 Related Work

Hierarchical agglomerative clustering (HAC) is likely the earliest clustering approach, as evidenced by the early term "Numerical Taxonomy". Many of these methods were pioneered in biological sciences in the 50s and 60s [25–27].

2.1 Algorithms for Hierarchical Agglomerative Clustering

Hierarchical agglomerative clustering (HAC) methods will usually require a pairwise distance matrix to be computed and stored, hence they at least need $O(N^2)$ time and memory for clustering N objects. The standard algorithm discussed below requires $O(N^3)$ time. For the special case of single-linkage, $O(N)$ memory is sufficient when using a variant of Prim's algorithm known as SLINK [24]. For complete linkage, the linear-memory approximation CLINK [9] tends to produce worse results and the results depend on the ordering of points [9]. In Algorithm 1 we give the basic idea of hierarchical clustering: find the best merge, greedily execute the merge, and then update the distance matrix. The loop of line 3 will be executed exactly $N-1$ times until everything is merged into a single cluster. Searching for the best merge in line 4 by brute force requires $O(N^2)$ operations (we will discuss better strategies below). Merging clusters in line 6 is $O(N)$ or $O(1)$. Updating the dissimilarity matrix in line 7 can take considerable effort: A naive implementation that recomputes the entire matrix each iteration performs $O(N^2)$ work here while using the Lance-Williams recurrences allows computing this in just $O(N)$ from the previous values in the matrix. The resulting overall complexity of the standard algorithm is $O(N^3)$.

Algorithm 1: Abstract Hierarchical Agglomerative Clustering

1 $D \leftarrow$ pairwise dissimilarity matrix
2 $C \leftarrow \{\{x_1\}, \ldots, \{x_N\}\}$ singleton clusters
3 **while** $|C| > 1$ **do** // $O(N)$ times
4      $i, j \leftarrow \arg\min D$ // find the best merge -- $O(N^2)$
5   add (i, j, d_{ij}) to the dendrogram
6      $C \leftarrow$ merge clusters i and j into i, remove j // $O(N)$ or $O(1)$
7      $D \leftarrow$ update column and row i of D // $O(N^2)$ or $O(N)$

Table 1. Selected linkage strategies

Names	Definition (for distances)
Single [25], Minimum	$d(A, B) := \min_{a \in A} \min_{b \in B} d(a, b)$
Complete [12, 14, 28], Maximum	$d(A, B) := \max_{a \in A} \max_{b \in B} d(a, b)$
(Group) Average [26], UPGMA	$d(A, B) := \frac{1}{|A| \cdot |B|} \sum_{a \in A} \sum_{b \in B} d(a, b)$
Weighted Average [26], WPGMA, McQuitty [17]	$d(A \cup A', B) := \frac{1}{2}\left(d(A, B) + d(A', B)\right)$
Centroid [10], UPGMC	$d^2(A, B) := \|\mu_A - \mu_B\|_2^2$
Median [10], WPGMC	$d^2(A \cup A', B) := \left\|\frac{1}{2}(\mu_A + \mu_{A'}) - \mu_B\right\|_2^2$
Ward [30]	$d^2(A, B) := \frac{2|A||B|}{|A \cup B|} \cdot \|\mu_A - \mu_B\|_2^2$
Less common linkages:	
Mini-Max [2]	$d(A, B) := \min_{r \in A \cup B} \max_{p \in A \cup B} d(r, p)$
Hausdorff [3]	$d(A, B) := \max\{\max_{a \in A} \min_{b \in B} d(a, b), \max_{b \in B} \min_{a \in A} d(a, b)\}$
Medoid [11, 18]	$d(A, B) := d(\text{medoid}(A), \text{medoid}(B))$
Min-sum Medoid [21]	$d(A, B) := \min_{m \in A \cup B} \sum_{p \in A \cup B} d(m, p)$
Min-sum-increase [21]	$d(A, B) := \min_{m \in A \cup B} \sum_{p \in A \cup B} d(m, p) - \min_{m \in A} \sum_{p \in A} d(m, p) - \min_{m \in B} \sum_{p \in B} d(m, p)$

2.2 Linkage Strategies

The update in line 7 depends on the linkage, which defines how the distance between two clusters is computed from the distances of their elements. Simple examples include the minimum distance (single linkage), the maximum distance (complete linkage), and the average distance (group average linkage). Table 1 gives an overview of linkage strategies and their definition in terms of two clusters A and B, except for the WPGMA and WPGMC methods, which depend on the

Table 2. Recursive definitions of selected linkages

Name	Recurrence (for distances)
Single	$d(A \cup A', B) := \min\{d(A,B), d(A',B)\}$
Complete	$d(A \cup A', B) := \max\{d(A,B), d(A',B)\}$
Average	$d(A \cup A', B) := \frac{|A|}{|A \cup A'|}d(A,B) + \frac{|A'|}{|A \cup A'|}d(A',B)$
McQuitty	$d(A \cup A', B) := \frac{1}{2}d(A,B) + \frac{1}{2}d(A',B)$
Centroid	$d^2(A \cup A', B) := \frac{|A|}{|A \cup A'|}d^2(A,B) + \frac{|A'|}{|A \cup A'|}d^2(A',B) - \frac{|A| \cdot |A'|}{|A \cup A'|^2}d^2(A,A')$
Median	$d^2(A \cup A', B) := \frac{1}{2}\left(d^2(A,B) + d^2(A',B) - \frac{1}{2}d^2(A,A')\right)$
Ward	$d^2(A \cup A', B) := \frac{|A|+|B|}{|A \cup A' \cup B|}d^2(A,B) + \frac{|A'|+|B|}{|A \cup A' \cup B|}d^2(A',B) - \frac{|B|}{|A \cup A' \cup B|}d^2(A,A')$

previous clusters A and A'. Because computing these distances can cost up to $O(N)$ time each, fast algorithms usually compute this efficiently using Lance-Williams recurrences. For the top group of common linkages in Table 1, such recurrences exist; for the bottom group, it is unclear if such efficient updates are even possible. Table 2 gives the recursive definition of these linkages. It is easy to see that the required values (previous distances and cluster sizes) are already known and hence these equations can be computed in $O(1)$ each. For Centroid (UPGMC), Median (WPGMC), and Ward linkage, the proof of these equations depends on properties of least squares optimization, and they hence should only be used with (squared) Euclidean distances.

2.3 Accelerated Matrix-Based Algorithms

Because the bottleneck is the arg min operation in line 4 (when using Lance-Williams updates), many techniques have been proposed for this. Several authors have proposed to organize distances in heaps, which reduces the runtime to $O(N^2 \log N)$, but with fairly high constant factors (because the heap is of size $O(N^2)$, and needs up to $O(N)$ updates each step). An interesting alternative, which we can recommend because of its simplicity, is found in a small note of the book by Anderberg [1], who suggested caching the location of the minimum in each row. According to his analysis, this will typically reduce the effort to N^2, but the theoretical worst-case remains $O(N^3)$. In our experiments with real data, this practical result is much more useful than the theoretical analysis, and this low-overhead approach often outperforms heap-based approaches, as previously observed by Kriegel et al. [13]. Müllner [19] further improves this algorithm using a priority heap on the cached values of Anderberg. Nevertheless, the relatively simple idea of organizing the distance matrix in a heap forms a central insight for our new algorithms discussed below.

2.4 Nearest-Neighbor Chain Algorithm

The second important idea for our approaches is the nearest-neighbor chain algorithm of Murtagh [20]. This algorithm is based on the observation that we can merge two clusters A and B immediately – even if a smaller distance would exist elsewhere in the data set – when the "nearest neighbor" of A is B and vice versa (nearest refers to the current set of clusters, not the original data set) if the linkage satisfies the reducibility property [6]. Irreducible linkages can cause inversions in the dendrogram, where the linkage of a parent node is smaller than that of its children. The Centroid and Median methods can produce such inversions. In such cases, the NNChain algorithms will not fail but may return a different result than the standard algorithm, but usually of a similar quality. Note that some variability exists even within the standard algorithm when distances are identical, because the next merge may not be uniquely defined.

The NNChain algorithm searches the nearest neighbor $O(N)$ times, hence it uses at least $O(N^2)$ time with linear scans. For the linkages based on least squares optimization (Centroid, Median, and Ward), we can use the cluster centers μ instead of the data points, to compute the distances. This allows the implementation of the nearest-neighbor chain algorithm with only linear memory. At its core, this algorithm uses a chain of points, $c_1, c_2, \ldots, c_k$, such that c_{i+1} is the nearest neighbor cluster of c_i. A random unlinked point c_1 can be used as the starting point. When $c_k = c_{k-2}$ we have found mutual nearest neighbors and we merge c_{k-1} and c_k. We then cut the chain after c_{k-3} and search the (possibly different from $c_{k-2} = c_k$, which has just changed) new nearest neighbor of this cluster. When implemented with linear memory, this approach needs about three times as many distance computations as when computing a pairwise distance matrix: For each of the $N-1$ merges, the chain needs to be extended by three steps, which requires about $N/2$ distance computations on average. If memory permits, it is hence preferable to use a pairwise distance matrix. We will denote the linear-memory version as *LinNNC* below.

2.5 Dual-Tree Join Algorithms

One alternate approach that we, unfortunately, do not further explain in this work is based on dual-tree joins. Such algorithms have been proposed for Euclidean minimum spanning trees [15] (and hence can also solve the single-link problem), have later been extended to metric ball trees [8] and have successfully been used to accelerate HDBSCAN* [7,16]. Similar algorithms have previously been proposed in database systems as k-nearest-neighbor joins [5] using R-trees. While these methods do not use *incremental* nearest-neighbor search (the focus of this work), they are based on nearest neighbors and will, when applicable, likely yield similar performance. Our new algorithms based on incremental search are easier to implement.

2.6 Incremental Nearest-Neighbor Search

An API for incremental nearest-neighbor search has been proposed recently by Schubert [22] under the name "incremental priority search". We will use this API (and implementation) in our new algorithms. In classic nearest-neighbor search, we have to either give the search radius ("range search") or the number of nearest neighbors ("kNN search"). In incremental nearest neighbor search, we obtain an iterator that yields the next nearest neighbor instead. This allows us to decide later how many neighbors we need, and to stop or pause the search when we have found enough. This becomes useful if we want to find nearest neighbors, but have to filter and hence skip some results. But it makes sense to relax the requirements for such a searcher by, instead of requiring the nearest neighbors to be returned in strictly increasing order, allowing the search to yield neighbor candidates (*currentNeighbor* below) only in the approximately correct order as long as the search can also provide us with a useful current lower bound (*lowerBound* below) on the distance to the remainder of the data set. *advanceToNextCandidate()* advances the search one step, and *computeExactDistance()* computes the exact distance if the index has not yet done so for navigation. Many metric and spatial search indexes can support such queries by relying on such bounds to prune the search space. For example, a tree using ball covers such as the VP-tree [29,31] or the cover-tree [4] may enumerate the points in a leaf based on an estimated distance; based on the radius of the tree it can guarantee lower and upper bounds to each point. For example in DBSCAN-clustering, it is sufficient to know whether a point is within the search radius, for which an upper bound to the distance is sufficient. When used carefully, this allows us to avoid unnecessary distance computations.

3 Accelerated Hierarchical Clustering

In this section, we will introduce three new algorithms for hierarchical clustering that all use the incremental nearest neighbor search [22] and hence are of particular interest to the SISAP community.

3.1 Heap-of-Searchers Single-Linkage (HSSL) Algorithm

The first algorithm is a rather straightforward use of this API for single-linkage. As noted before, one method to implement single-linkage clustering is to store all pairwise distances in a heap, then repeatedly poll the shortest edge and merge these clusters (unless they are already connected). This is a heap-based version of Kruskal's algorithm for minimum spanning trees. For $|E|=O(N^2)$ edges, the complexity of this algorithm is $O(N^2 \log N)$. Heaps are typically recursive data structures that consist of smaller heaps, where each part satisfies has the smallest (resp. largest) element at the top. Furthermore, nearest-neighbor searchers typically involve a heap to store the current best candidates. Instead of materializing all pairwise distances into a heap, it is sufficient to have a heap that contains only the current best neighbor of each node at any time. After merging these clusters, we update the heap with the next nearest neighbor of both of these points. Ideally, we would *hope* that we can find the next nearest neighbor in amortized $O(\log N)$, update the heap in $O(\log N)$, and as we have to repeat this $O(N)$ times obtain an $O(N \log N)$ algorithm. Unfortunately, such a guarantee is impossible even for fairly low-dimensional Euclidean data. In the worst

Algorithm 2: HSSL: Heap-of-Searchers Single-Linkage

```
 1  U ← new union find data structure of size N
 2  M ← primary min-heap of size N
 3  foreach x_i in the data set do                    /* Build initial heaps */
 4  |   P[i] ← nearest neighbor searcher for x_i
 5  |   H[i] ← new min-heap for refined candidates
 6  |   while H[i].peekKey > P[i].lowerBound do
 7  |   |   H[i].insert(P[i].computeExactDistance(), P[i].currentNeighbor)
 8  |   |   P[i].advanceToNextCandidate()
 9  |   M.insert(H[i].peekKey, i)
10  while not everything merged do                    /* Hierarchical clustering */
11  |   d, i ← M.poll()
12  |   d, j ← H_i.poll()                             // same distance d
13  |   C_i, C_j ← U.find(i), U.find(j)
14  |   if C_i ≠ C_j then                             // not yet connected
15  |   |   add (d, i, j) to the dendrogram
16  |   |   C_i ← U.union(i,j)
17  |   while H[i].peekKey > P[i].lowerBound do  /* Refill the heap of i */
18  |   |   if U.find(P_i.currentNeighbor) ≠ C_i then // not yet connected
19  |   |   |   H[i].insert(P[i].computeExactDistance(), P[i].currentNeighbor)
20  |   |   P[i].advance()
21  |   M.insert(H[i].peekKey, i)
```

case, the algorithm remains in $O(N^2)$ because nearest-neighbor search is not guaranteed to be possible in sub-linear time in the general case.

Nevertheless, we implement and explore such a strategy using incremental nearest-neighbor searchers instead of heaps. For the HSSL algorithm, we run a priority search for every point in the data set in parallel (not with parallel processing, but we keep the search state to allow continuing), and put the searchers with the distances to the nearest neighbor in a heap. We then repeatedly poll the heap, perform the corresponding merge (unless the clusters are already merged), and then search for the next nearest neighbor using the corresponding incremental searcher. As our searchers may return results only in approximately increasing order, each uses a min heap of the discovered neighbors, and polls the priority search until the lower bound of the remainder is farther away than the current best result. Before computing the exact distance to a new candidate, we check whether they are linked using a union-find data structure with path-halving. If they are already connected, we do not need to compute their distance and can proceed to the next candidate immediately. In Algorithm 2 we give a pseudocode.

3.2 Restarting Search Single-Linkage (RSSL) Algorithm

The drawback of the above algorithm is that it is fairly memory intensive. While it typically does consume less memory than the full distance matrix, its worst-

Algorithm 3: RSSL: Restarting Search Single-Linkage

```
 1  U ← new union find data structure of size N
 2  M ← primary min-heap of size N
 3  P ← nearest neighbor searcher
 4  foreach xᵢ in the data set do              /* Find initial neighbors */
 5  |   d_B, B ← ∞, ⊥
 6  |   P.search(xᵢ)
 7  |   while d_B > P.lowerBound do
 8  |   |   d' ← Pᵢ.computeExactDistance()
 9  |   |   if d' < d_B then  d_B, B ← d', P.currentNeighbor
10  |   |   Pᵢ.advanceToNextCandidate()
11  |   M.insert(d_B, i, B)
12  while not everything merged do             /* Hierarchical clustering */
13  |   d, i, j ← M.poll()
14  |   Cᵢ, Cⱼ ← U.find(i), U.find(j)
15  |   if Cᵢ ≠ Cⱼ then                        // not yet connected
16  |   |   add (d, i, j) to the dendrogram
17  |   |   Cᵢ ← U.union(i,j)
    |   /* Find the next nearest neighbor of i               */
18  |   d_B, B ← ∞, ⊥
19  |   P.search(xᵢ).setLowerBound(d)
20  |   while d_B > P.lowerBound do            /* Refill the heap of i */
21  |   |   if U.find(P.currentNeighbor) ≠ Cᵢ then  // not yet connected
22  |   |   |   d' ← Pᵢ.computeExactDistance()
23  |   |   |   if d' < d_B then  d_B, B ← d', P.currentNeighbor
24  |   |   Pᵢ.advanceToNextCandidate()
25  |   M.insert(d_B, i, B)
```

case memory is even larger (albeit still $O(N^2)$). For larger data sets, it may become necessary to reduce the memory usage. If we only retain a single best candidate for each point, we reduce the memory requirements to just $O(N)$, at the cost of having to restart searches and hence perform more distance computations. Fortunately, we can also implement the ability to skip candidates based on a lower bound in the priority search. To initialize this algorithm, we first identify the 1-nearest neighbor of each point. Then when processing an edge, we search the next nearest unlinked neighbor, skipping over points that are guaranteed to be closer to the query, or that are already linked (closer points should already be, but the lower bound may allow the index search to skip entire subtrees). In Algorithm 3, we give a pseudocode for this second approach.

The above two algorithms could only solve the single-linkage problem. Accelerating other linkages is far from trivial. For example, the group average linkage depends on all distances between points, and hence can in the general case not be answered in subquadratic time. Supporting complete linkage also requires further research, as it depends on the maximum distance to the nearest cluster, which may require efficient ways to find the farthest neighbors to sets of query points. There has been occasional research in indexing for farthest neighbors, but this is currently not available in the ELKI framework we use. However, there is a class of linkages that we can support with a different, yet effective, strategy.

3.3 Incremental Nearest-Neighbor Chain (INNC) Algorithm

For geometric linkages (linkages that are based on the arithmetic mean of a set of points, i.e., Centroid, Median, and Ward linkage), we can adapt the LinNNC algorithm discussed above and give a pseudocode in Algorithm 4. These linkages must be used with squared Euclidean distances. While many indexes require a metric (which squared Euclidean is not), we can alternatively use an index for Euclidean distance, too. Again, the approach is fairly simple at its core, but instead of starting the search at an original data point, we begin searching at the current center of the cluster. We then exploit that these linkages tend to produce convex clusters, and in many cases, the nearest point will be enough to identify the closest center. With some simple bound adjustments, we can limit our search, and we can skip over any cluster we have previously tested. To find the necessary bounds, we study the distances in Table 1. The simplest case is Centroid linkage. A cluster with no point closer than the current best center can be skipped, and we can stop searching at the distance of the current best cluster center. For Median linkage, we observe that we know one center exactly, and for the other centers, the nearest cluster member is a lower bound. We hence can also stop searching at the distance of the current best cluster center. Ward linkage has an additional weighting term that depends on the cluster sizes. The size of the current cluster is known, and we can use the size of the largest different cluster to obtain an upper bound (this could be further tightened), because

$$2\frac{|A||B|}{|A|+|B|} \geq 2\frac{|A||B|_{\min}}{|A|+|B|_{\max}} \geq \frac{2|A|}{|A|+|B|_{\max}} \quad \Rightarrow \quad \frac{|A|+|B|_{\max}}{2|A|}d^2(A,B) \geq \|\mu_A - \mu_B\|_2^2.$$

4 Experiments

As the results will be equivalent to existing approaches except for differences due to data ordering with tied distances, we will not include a clustering quality evaluation, but instead focus only on the number of distance computations and the runtime. We implemented the algorithms in Java using the ELKI framework [22], as it contains the only implementation of incremental neighbor search that we are aware of. We use the VP-tree [29,31] with a fairly large leaf size of 100, which outperformed the cover-tree [4]. We also include the well-known hdbscan Python package which uses dual-tree joins [16], even though it is implemented in Cython. However, Java is not the optimum choice for this kind of algorithm, and we believe that a more memory-efficient implementation could further improve the runtime; the number of distance computations can serve as a proxy for the speedups possible.

Algorithm 4: INNC: Incremental Nearest-Neighbor Chain

```
 1  U ← new union find data structure of size N
 2  P ← nearest neighbor searcher
 3  S ← empty chain
 4  μ[] ← cluster center storage, initially xᵢ
 5  while not everything merged do              /* Hierarchical clustering */
 6      if chain S is empty then S.append(some unlinked point)
 7      Cᵢ ← S[|S| − 1]                          // last element in chain
        // Find the next nearest neighbor of i
 8      visited[] ← empty boolean array
 9      visited[Cᵢ] ← true
10      d_B, B, τ ← ∞, ⊥, ∞
11      if |S|¿1 then                            // prefer the back-edge on tied distances
12          B ← S[|S| − 2]
13          d_B ← distance Cᵢ to B
14          visited[B] ← true
15          τ ← compute threshold from d_B      // linkage dependent!
16      P.search(Cᵢ)                             // start search at current cluster center
17      while τ > P.lowerBound do                /* find nearby clusters */
18          C′ ← U.find(P.currentNeighbor)
19          if not visited[C′] then              // not yet connected
20              d′ ← Pᵢ.computeExactDistance()
21              if d′ < d_B then
22                  d_B, B ← d′, P.currentNeighbor
23                  τ ← compute threshold from d′   // linkage dependent!
24              visited[C′] ← true
25          Pᵢ.advance()
26      if |S| > 2 and B == S[|S| − 2] then     /* Back-edge found, merge */
27          C_m ← U.merge(Cᵢ, B)
28          μ[C_m] ← merged cluster centers      // linkage dependent!
29          S.truncate(|S| − 2)                  // remove modified elements from chain
30      else
31          S.append(B)                          // grow the nearest-neighbor chain
```

4.1 Data Sets and Parameterization

As we are interested in particular in the metric case, the first data set we use are Tweett coordinates in the city of Istanbul, using spherical distance (while the data could locally be mapped to Euclidean space, we opted to work with a spherical metric instead) and single linkage. The second data set contains HSV color histograms of ALOI, in the $7 \times 2 \times 2$ version [23], using squared Euclidean distance for geometric linkages such as Ward.

We will limit the data set size for the classic AGNES algorithm (which runs in $O(N^3)$) because of its non-competitive run-time. For the Anderberg and NNChain algorithms, the data set size is limited by the memory requirements of the distance matrix array of 2^{31} due to Java limitations.

4.2 Single Linkage

Figure 1 shows the run times of the single-linkage algorithms on the Istanbul data set. Because of the scale differences, we use a log-log plot. For AGNES we observe the expected $O(N^3)$ scaling, while SLINK, Anderberg, and NNChain scale approximately $O(N^2)$. While Anderberg cannot guarantee this run-time, it performs marginally faster than NNChain in our implementation because it is simpler. SLINK (closely related to Prim's algorithm) benefits from using only linear memory, a cache-friendly memory access pattern, and used to be the best algorithm for single-link in ELKI. For small data sets (less than 5000 points), it continues to be the best choice, but for larger data sets our new algorithms are faster. The more memory-intensive HSSL was faster and able to process the entire data set within the available memory of 32 GB. Both new algorithms suggest empirical sub-quadratic scalability in similar to the dual-tree join using the Cython `hdbscan` package (which runs out of memory at 500.000 instances), although we cannot guarantee this. For all algorithms, the run-time is dominated by the distance computation cost. In Fig. 2, we show the number of distance computations for the new algorithms, which need orders of magnitude fewer

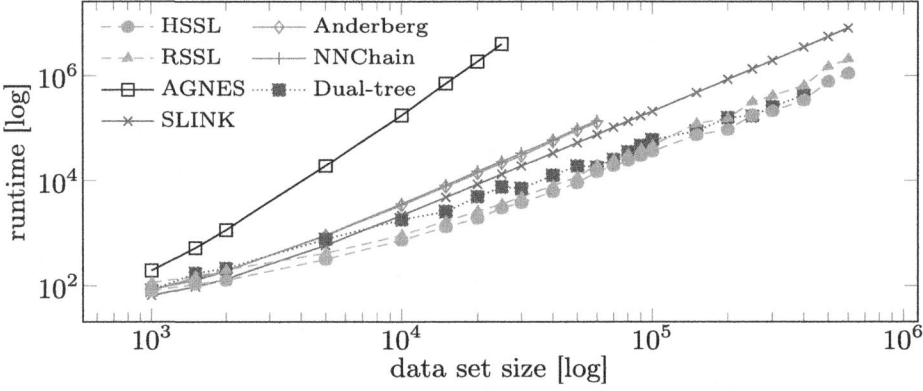

Fig. 1. Run time on the Istanbul data, single linkage, log-log plot

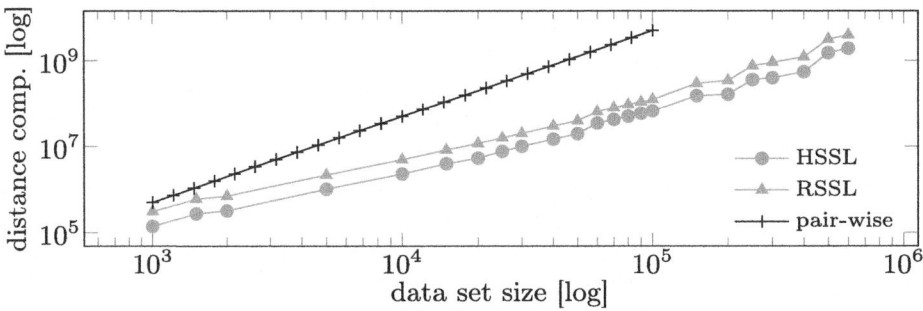

Fig. 2. Distance computations on the Istanbul data, single linkage, log-log plot

distance computations than all others, which use a pair-wise distance matrix: $N(N-1)/2$. For HSSL and RSSL, this includes the distance computations to build the index, which could be amortized across multiple runs.

4.3 Geometric Linkage

Single-linkage has severe limitations for noisy data sets. A popular alternative is Ward linkage, while Centroid and Median linkage are less common. We study these because their geometric properties allow for an efficient algorithm, but we are primarily interested in Ward, the most complex and important case. We like to emphasize that Ward inherently assumes distances to be least-squares, and should not be used with arbitrary distances. Mathematically, it tries to find the smallest increase in total variance in each step, similar to k-means.

Figure 3 shows the run times of the algorithms for geometric linkages on the ALOI $7 \times 2 \times 2$ data set. Again AGNES shows the expected $O(N^3)$ run-time, while Anderberg and the nearest-neighbor chain algorithm exhibit quadratic behavior. The linear-memory version of the nearest-neighbor chain algorithm is,

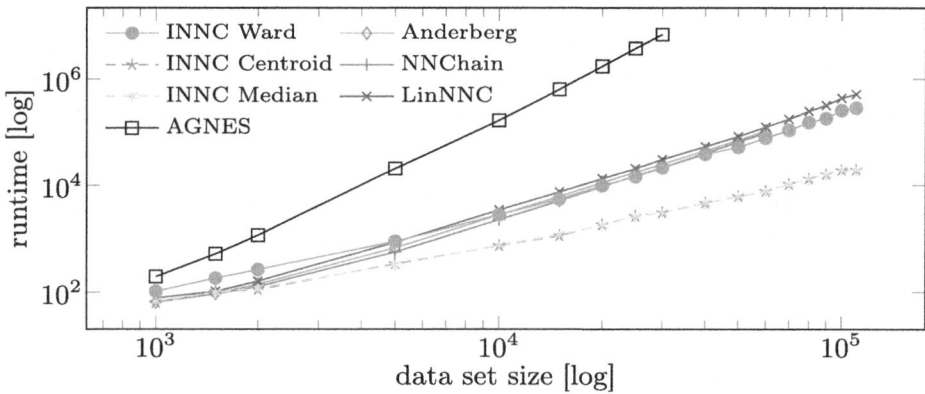

Fig. 3. Run time on the ALOI data, geometric linkages, log-log plot

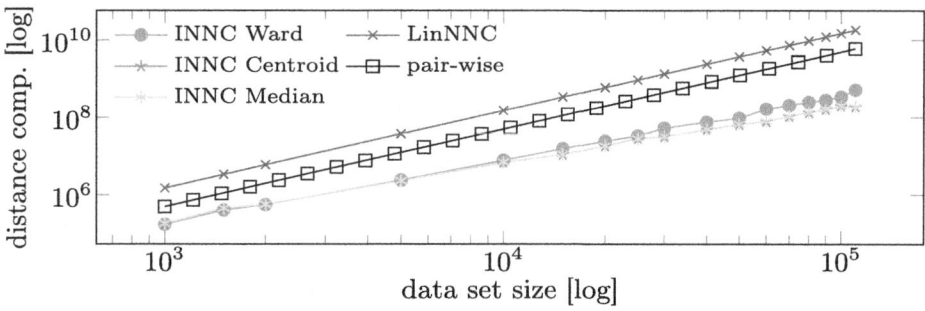

Fig. 4. Distance computations on the ALOI data, geometric linkages, log-log plot

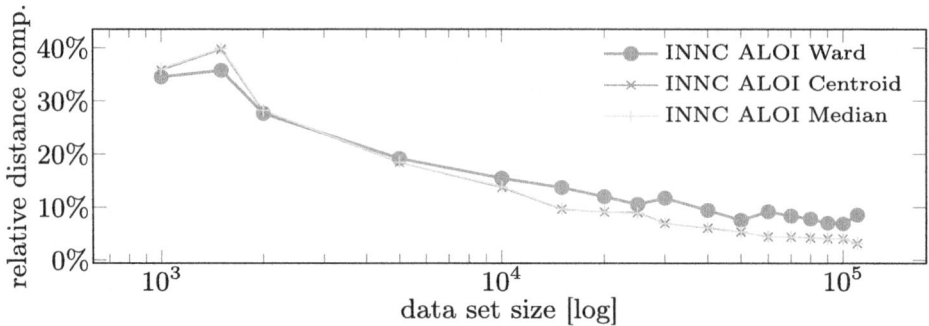

Fig. 5. Relative distance computations compared to pairwise distances, log-linear plot

as expected, almost three times slower than the matrix version, but continues to work on the full data set. The new INNC algorithm becomes the fastest at around 30,000 instances, but only 2× faster for Ward linkage at the full data set; likely due to Java overheads. For Centroid and Median linkage, the new algorithm scales much better because of the simpler geometry.

Figure 4 gives the distance computations needed. As expected, LinNNC needs approximately three times as many distance computations as a full distance matrix. The new HSSL algorithm needs 1–2 orders of magnitude fewer distance computations. In Fig. 5, we look at the quotient of the distance functions compared to a distance matrix (we omit the constant line of LinNNC at 300%). We also include Centroid and Median linkage, which exhibit similar behavior, but because the bounds for these are tighter, they are faster than Ward linkage. As they use the same bounds, the number of distance computations is almost identical for both. For the full 110,250 instances of the ALOI data set, we need less than 10% of the distance computations with the new algorithm. A more low-level implementation may improve the run-time further and bring it closer to the difference in distance computations.

5 Conclusion

In this article, we proposed three new algorithms for hierarchical clustering that can (but are not guaranteed to) achieve sub-quadratic run-time for hierarchical clustering. Currently, only single-linkage, Ward, centroid, and medoid linkage are supported, and other linkages (such as the important average linkage, but also complete linkage) will be harder to accelerate with a similar strategy because these approaches rely on particular geometric properties. We have empirically shown that the algorithms can achieve sub-quadratic run-time on real data sets, and with only linear memory requirements allow clustering of much larger data sets than before. The new algorithms only need a fraction of the distance computations of a pair-wise distance matrix, as required by all classic algorithms.

References

1. Anderberg, M.R.: Hierarchical Clustering Methods, chap. 6. Probability and Mathematical Statistics. Academic Press (1973). https://doi.org/10.1016/B978-0-12-057650-0.50012-0
2. Ao, S.I., et al.: CLUSTAG: hierarchical clustering and graph methods for selecting tag SNPs. Bioinformatics **21**(8) (2005). https://doi.org/10.1093/bioinformatics/bti201
3. Basalto, N., Bellotti, R., De Carlo, F., Facchi, P., Pantaleo, E., Pascazio, S.: Hausdorff clustering of financial time series. Physica A Stat. Mech. Appl. **379**(2) (2007). https://doi.org/10.1016/j.physa.2007.01.011
4. Beygelzimer, A., Kakade, S.M., Langford, J.: Cover trees for nearest neighbor. In: Proc. ICML (2006). https://doi.org/10.1145/1143844.1143857
5. Böhm, C., Krebs, F.: Supporting KDD applications by the k-nearest neighbor join. In: Proc. Database and Expert Systems Applications, DEXA (2003). https://doi.org/10.1007/978-3-540-45227-0_50
6. Bruynooghe, M.: Classification ascendante hiérarchique des grands ensembles de données: un algorithme rapide fondé sur la construction des voisinages réductibles. Cahiers de l'analyse des données **3**(1), 7–33 (1978)
7. Campello, R.J.G.B., Moulavi, D., Sander, J.: Density-based clustering based on hierarchical density estimates. In: Proc. PAKDD (2013). https://doi.org/10.1007/978-3-642-37456-2_14
8. Curtin, R.R.: Faster dual-tree traversal for nearest neighbor search. In: Proc. Similarity Search and Applications, SISAP (2015). https://doi.org/10.1007/978-3-319-25087-8_7
9. Defays, D.: An efficient algorithm for the complete link cluster method. Comput. J. **20**(4) (1977). https://doi.org/10.1093/comjnl/20.4.364
10. Gower, J.C.: A comparison of some methods of cluster analysis. Biometrics **23**(4) (1967). https://doi.org/10.2307/2528417
11. Herr, D., Han, Q., Lohmann, S., Ertl, T.: Visual clutter reduction through hierarchy-based projection of high-dimensional labeled data. In: Graphics Interface Conference (2016). https://doi.org/10.20380/GI2016.14
12. Johnson, S.C.: Hierarchical clustering schemes. Psychometrika **32**(3) (1967). https://doi.org/10.1007/BF02289588
13. Kriegel, H., Schubert, E., Zimek, A.: The (black) art of runtime evaluation: are we comparing algorithms or implementations? Knowl. Inf. Syst. **52**(2) (2017). https://doi.org/10.1007/s10115-016-1004-2
14. Macnaughton-Smith, P.: Some statistical and other numerical techniques for classifying individuals. Tech. Rep. Home Office Res. Rpt. No. 6, HMSO (1965)
15. March, W.B., Ram, P., Gray, A.G.: Fast Euclidean minimum spanning tree: algorithm, analysis, and applications. In: Proc. SIGKDD (2010). https://doi.org/10.1145/1835804.1835882
16. McInnes, L., Healy, J.: Accelerated hierarchical density based clustering. In: Proc. ICDM Workshops (2017). https://doi.org/10.1109/ICDMW.2017.12
17. McQuitty, L.L.: Elementary linkage analysis for isolating orthogonal and oblique types and typal relevancies. Educ. Psychol. Meas. **17**(2) (1957). https://doi.org/10.1177/001316445701700204
18. Miyamoto, S., Kaizu, Y., Endo, Y.: Hierarchical and non-hierarchical medoid clustering using asymmetric similarity measures. In: SCIS/ISIS (2016). https://doi.org/10.1109/SCIS-ISIS.2016.0091

19. Müllner, D.: Modern hierarchical, agglomerative clustering algorithms. CoRR **abs/1109.2378** (2011)
20. Murtagh, F.: A survey of recent advances in hierarchical clustering algorithms. Comput. J. **26**(4) (1983). https://doi.org/10.1093/comjnl/26.4.354
21. Schubert, E.: HACAM: hierarchical agglomerative clustering around medoids - and its limitations. In: Proc. Lernen, Wissen, Daten, Analysen. LWDA (2021)
22. Schubert, E.: Automatic indexing for similarity search in ELKI. In: Proc. Similarity Search and Applications, SISAP (2022). https://doi.org/10.1007/978-3-031-17849-8_16
23. Schubert, E., Zimek, A.: ELKI Multi-view Clustering Data Sets Based on the Amsterdam Library of Object Images (ALOI) (2010). https://doi.org/10.5281/zenodo.6355684
24. Sibson, R.: SLINK: an optimally efficient algorithm for the single-link cluster method. Comput. J. **16**(1) (1973). https://doi.org/10.1093/comjnl/16.1.30
25. Sneath, P.H.A.: The application of computers to taxonomy. Microbiology **17** (1957). https://doi.org/10.1099/00221287-17-1-201
26. Sokal, R.R., Michener, C.D.: A statistical method for evaluating systematic relationship. Univ. Kansas Sci. Bull. **38** (1958)
27. Sokal, R.R., Sneath, P.H.A.: Principles of Numerical Taxonomy. Books in Biology. W. H. Freeman (1963)
28. Sørensen, T.: A method of establishing groups of equal amplitude in plant sociology based on similarity of species and its application to analyses of the vegetation on Danish commons. Kongelige Danske Videnskabernes Selskab **5**(4) (1948)
29. Uhlmann, J.K.: Satisfying general proximity/similarity queries with metric trees. Inf. Process. Lett. **40**(4) (1991). https://doi.org/10.1016/0020-0190(91)90074-R
30. Ward, J.H.: Hierarchical grouping to optimize an objective function. J. Am. Stat. Assoc. **58**(301) (1963). https://doi.org/10.1080/01621459.1963.10500845
31. Yianilos, P.N.: Data structures and algorithms for nearest neighbor search in general metric spaces. In: Proc. Symposium on Discrete Algorithms (1993)

Indexing Challenge

Overview of the SISAP 2024 Indexing Challenge

Eric S. Tellez[1] (✉), Martin Aumüller[2], and Vladimir Mic[3]

[1] INFOTEC, IxM CONACyT, Mexico
eric.tellez@ieee.org
[2] ITU Copenhagen, Copenhagen, Denmark
maau@itu.dk
[3] Aarhus University, Aarhus, Denmark
v.mic@cs.au.dk

Abstract. The SISAP 2024 Indexing Challenge invited replicable and competitive approximate similarity search solutions for datasets of up to 100 million real-valued vectors. Participants are evaluated on the search performance of their implementations under quality constraints. Using a subset of the deep features of a neural network model provided by the LAION-5B dataset, the challenge posed three tasks, each with its unique focus:

- **Task 1, Unrestricted indexing**: Conduct a classical approximate nearest neighbors search, ensuring an average recall of at least 0.8 for 30-NN queries.
- **Task 2, Memory-constrained indexing with reranking**: Conduct nearest neighbors search in a low-memory setting where the dataset collection is only accessible on disk, ensuring the same quality as in Task 1.
- **Task 3, Memory-constrained indexing without reranking**: Conduct nearest neighbor search in a setting where the dataset cannot be accessed at search stage, ensuring an average recall of at least 0.4 for 30-NN queries.

The present paper describes the details of the challenge, the evaluation system that was developed with it, and gives an overview of the submitted solutions.

Keywords: Approximate nearest neighbor search · Indexing and searching pipelines · Experimental comparison of search methods

1 Introduction

Large Language Models (LLMs) led to a fundamental change in how computers can help humans to solve complex tasks. However, once model training has finished, the corpus of knowledge of the model remains static. A crucial component for the applicability of these systems is thus that knowledge can be added post-hoc. To do so, systems can retrieve semantically matching, additional knowledge from large knowledge corpora [11]. This semantic search is made possible through deep-learning based vector embedding techniques such as CLIP [7], in which a piece of text or an image, among others, can be represented as a real-valued vector in a high-dimensional space.

Given a large collection of high-dimensional objects in a metric space, the most common search operation is to find the nearest neighbors of a given point in the space. An *exact* metric search structure will either detorate to a linear scan of the whole data collection, or use too much space in these high-dimensional spaces [3]. This means that the search algorithms should become approximate, i.e., *inexact*, to let users trade between speed and quality. Furthermore, there is a compromise with other practical constraints like preprocessing and construction time, as well as the whole memory usage, leading to various indexing solutions, each with merits and drawbacks.

The SISAP Indexing Challenge[1] seeks to identify efficient similarity search algorithms that strike a balance between accuracy and practical constraints like construction time, search time, and memory consumption. To facilitate this, we devised a 100M vectors test using the LAION deep features English subset [12]. Additionally, there are two query sets: public and private, each comprising 10K vectors. The public queries were made available during the call for papers, while the private ones were revealed post-submission and were used for the evaluation.

Dataset. This challenge is the second iteration of the SISAP indexing challenge; see [17] for an overview of the first iteration. As in this previous iteration, we used the CLIP image embeddings from the LAION dataset [12] as basis for the challenge. We selected a 100-million vector subset excluding *Not Safe for Work* (NSFW) elements. Note that the LAION dataset has been removed from several mirrors due to illegal content access on their metadata; therefore, we remove metadata access and take synthetic decisions to construct queries that will be detailed in §2.2.[2] The dataset contains 768-dimension vectors in the unit sphere such that the cosine can be used as a similarity function without needing vector normalization. More details on the dataset can be found on the challenge's companion site and the overview of the previous challenge [17].

Related Challenges and Benchmarks. Given its wide application-range, numerous methods for approximate nearest neighbor search have been developed in the last decade. The ann-benchmarks project [1] provides a standard benchmarking environment for million-scale approximate nearest neighbor search with a large collection of implementations. Alternatively, Li et al. [8] provide an alternative experimental survey and Matsui [9] provides an alternative benchmarking environment. For larger data collections, the 2021 NeurIPS challenge by Simhardi et al. [13] focused on billion-scale approximate nearest neighbor search, and the 2023 NeurIPS challenge[3] focused on different search under different constraints such as the presence of metadata filters, or a streaming setting.

---

[1] Official site of the challenge https://sisap-challenges.github.io/.
[2] Our datasets are made of image embeddings avoiding any access to metadata.
[3] https://big-ann-benchmarks.com/neurips23.html.

2 Challenge Details, Evaluation Setup, and Task Description

The Indexing Challenge focuses on approximate k nearest neighbor search. Given a collection of d-dimensional vectors $S \subseteq \mathbb{R}^d$, each task comes in two phases: First, given S, the index construction phase asks to layout the data to answer nearest neighbor searches quickly. Traditionally, layouts can be based on building a graph, a tree, or partitioning the space using a quantizer such as clustering-based or hashing-based approaches, see [2] for a recent survey. In the second phase, the query workload $Y \subseteq \mathbb{R}^d$ is provided, and the task is to find the 30 nearest neighbors for each query in Y. The challenge tasks simulate scenarios with different needs in terms of quality, speed, and memory.

The reproducibility is crucial; therefore, submissions must have an operational Github Action (GHA) workflow.[4] Teams crafted their solutions by meticulously setting and benchmarking hyperparameters for each task and detailing their choices in their GHA entry point.

2.1 Indexing Tasks

All tasks revolved around retrieving the approximate $k = 30$ nearest neighbors. We expected teams to construct indexes that efficiently solve queries and excel under each task's specific conditions and metrics. We first review the evaluation hardware, and then describe the three different tasks.

Hardware Similarly to the previous SISAP Challenge 2023, we limit the index construction time in each task, making the hardware parameters important for participants. The evaluation was going to be conducted on 2x Intel(R) Xeon(R) CPU E5-2690 V4 CPUs (28 cores, 56 hyperthreads) workstation with 512GiB of RAM. The original dataset resided on a 1TB SSD. Despite the same hardware used a year ago, current tasks focus more on a search with limited hardware. These hardware limits were enforced through the use of Docker containers.

Task 1: Unrestricted Indexing Task 1 provides continuity to the SISAP Challenge 2023. Inspired by Task A from the previous year, it allows teams to use all resources on the evaluation computer to build an index for the respective datasets and carry out search operations as quickly as possible under quality thresholds. Unlike in the previous iteration, we (1) limit the index building time to 12 h instead of 24 h, (2) specify that the index does not have to be serialized on a disk, and (3) set a threshold for an average recall of at least 0.8. The solutions meeting these requirements using a private query set are sorted by their search time.

Task 2: Memory-Constrained Indexing with Reranking Task 2 is motivated by limited hardware situations, which is expected, for instance, when sharing machines in cloud services or whenever very large datasets are handled. This task considers the main memory to be the main bottleneck. Participants are

---

[4] Github Actions is a continuous integration platform that enables continuous testing of repositories within virtual machines.

given just eight virtual CPUs and 32 GiB of RAM. The wall clock time for index construction is 24 h. The search quality still targets an average recall of at least 0.8, as in Task 1. For 100 million vectors of 768 dimensions using 4 bytes per coordinate, storing the vectors requires more than 300 GiB of RAM. Therefore, the dataset cannot be stored uncompressed/unembedded, resulting in an accuracy loss. To reach the target quality, participants are expected to refine the candidates in a final phase, as the search dataset is available on SSD.

Task 3: Memory-Constrained Indexing without Reranking Task 3 considers persistent storage to be the main bottleneck. Participants are asked to develop a memory-efficient index that does not use re-ranking. Participants must build an index in up to 12 h using all available CPU cores and with at most 64 GiB of RAM. In the search phase, the original vectors cannot be used. The minimum recall to be considered in the final ranking is 0.4.

2.2 Public and Private Query Workloads

The indexing challenge provides two query workloads, each with 10 thousand query vectors; both query sets were extracted from different subsets not seen in any of the datasets given for indexing. The public query set was provided with the call for participation; it is intended to be used by participating teams to fine-tune their submissions and provide a thorough analysis of their system in their report. The private one is used to evaluate all systems and produce this manuscript's analysis.

This 2024 edition uses the private query set of the previous year (2023) as the public query set. We computed gold standards for k nearest neighbor queries, which are the foundation for calculating the recall score in the final results. Both the query set and the gold standard are accessible from the SISAP Indexing Challenge site.[5]

To make the evaluation more interesting, this year's private query set was chosen slightly differently than last year: Both query sets were selected from different LAION bundles[6]. For the public query set, we removed near-duplicate objects/queries, considering as *near-duplicate* any object inside a radius of 0.15[7]; for the private query set, we remove near-duplicates with a radius of 0.2. This slight change is reflected in variations on the expected quality that show the robustness of each system solution; note that the challenge rules allow up to 30 search hyperparameter probes that are expected to help overcome this query distribution change. Additionally, we removed any query retrieving a zero-distant nearest neighbor and those with distance values tying at 29th and 30th neighbors. The private query set was made available after the evaluation results were presented to the participants.

---

[5] Gold standards incorporate results up to $k = 1000$.
[6] The LAION dataset was available in bundles of close to one million vectors.
[7] Using the cosine distance, e.g., $1 - cosine(q_i, q_j)$; also note that we used an approximate algorithm to remove near duplicates.

Table 1. Quantile values of the 30th neighborhood, i.e., $1-cosine$; note that Q_0 means for minimum and Q_1 for maximum values.

Query set	Q_0	$Q_{0.1}$	$Q_{0.2}$	$Q_{0.3}$	$Q_{0.4}$	$Q_{0.5}$	$Q_{0.6}$	$Q_{0.7}$	$Q_{0.8}$	$Q_{0.9}$	Q_1
Public	0.012	0.08	0.105	0.127	0.149	0.171	0.196	0.224	0.26	0.313	0.604
Private	0.009	0.146	0.168	0.187	0.205	0.224	0.246	0.271	0.303	0.348	0.593

Table 1 shows the distance distribution of the 30th nearest neighbor for the two different query sets. We can notice the effect of the near duplicate radius by observing larger distances in the private query set for all quantiles (0.1 to 0.9) except for the extremes.

3 Solutions Overview

This section describes the set of solutions to the SISAP 2024 Indexing Challenge. The solutions were implemented in C++ and Rust programming languages using Python as a wrapper. Our baselines used the Julia programming language. As detailed below, system solutions used different index structures, graph-based indexes, combinatorial and numerical optimization, and neural networks to learn how to project the dataset. Their submissions were carefully crafted to exploit the dataset characteristics and running setups.

3.1 Baselines

We present two baselines: **BL-SearchGraph** and **BL-Bruteforce**, based on on the Julia's package `SimilaritySearch.jl`. The BL-SearchGraph uses the SearchGraph index; see [15,16]. This index is a graph-based index similar to the HNSW, but instead of a hierarchy, it uses a small sample of disjoint neighbors to get fast navigation. In contrast to HNSW, it uses variable-size neighborhoods with an upper bound size defined as $M = O(\log i)$. It uses Beam Search (BS) as a search algorithm; that is, it is based on storing candidates in a priority list of maximum size (beam size) and also controls what is inserted into the beam using a parameter $0 < \Delta < 2$; the result set is populated during the navigation, and the search finishes when the result set does not improve and the beam is empty. It supports single-pass automatic index optimization for a given quality score. As a baseline, it was constructed with 0.95 as objective recall and a neighborhood size of $M = \log_{1.2} i$ for Task 1. On the other hand, we use $M = \log_2 i$ for memory-limited tasks, i.e., 2 and 3. The database was also projected to reduce its memory footprint, using 96-dimensional PCA for Task 2 and 128-dimensional PCA for Task 3. Note that for tasks 2 and 3, the objective recall is achieved in the projected space and not in the original metric database. During the search stage, we varied the self-optimized Δ parameter in the range $\Delta/1.05^2 \leq \Delta' < 2$ growing exponentially in a 1.05 factor.

The BL-Bruteforce is a parallelized exhaustive evaluation of all objects with queries. Task 1 is approached as a direct search over the 768-dimensional vectors using the cosine distance. In contrast, Task 3 uses the Euclidean distance over a 128-dimensional PCA projection of the dataset to fulfill the requirement of not touching the original vectors during the search stage. Task 2 is not tackled with this baseline.

3.2 Teams Solutions

Four teams submitted a candidate for evaluation; one team (HIOB) targeted all three tasks, two teams (DEGLIB and LMI) solved two tasks (Task 1 and 3),[8] and the remaining team (HSP) focused on efficient retrieval in the standard setting (Task 1). Teams used their own implementations or tuned well-known approximate nearest neighbor search libraries. We encouraged participants to use multithreading or multiprocessing in the construction and searching stages to take advantage of the hardware. We give a short overview over the approaches next and refer to the individual papers for more details.

DEGLIB [6]. The solution uses the Continuous Refining Exploratory Graph [5], which is a graph index using monotone metric approximation properties but also ensuring three properties: (i) no redundant paths, (ii) good starting points, (iii) compression of the metric database (vectors under the cosine similarity). The submission is tuned to take advantage of low-recall requirements and work with low-memory limits. Task A projects the LAION 768-dimensional vectors into 512-dimensional vectors using their compression technique, while Task C compresses to dimension 64, ensuring working in a low-memory regime.

HIOB [19]. This approach continues the research on HIOB binary sketching presented at SISAP 2023 [18]. The current approach addresses the theoretical limits of Hamming spaces used as a proxy for the memory-effective search in complex spaces. The proposed index uses the HIOB sketches and groups their bits into integers to efficiently find solution candidates. Considering the available main memory size, the authors identify proper parameters for the sketching and grouping: the length of sketches is 384 bits, and sketches are indexed using 57 (overlapping) groups of bits, each covering 12 randomly selected bits of sketches. The authors observed a speed-up of about 17 times in comparison with the brute force search in Hamming space.

LMI [10]. The Learned Metric Index (LMI) is an updated version of the 2023 submission [14] to the indexing challenge. LMI clusters the dataset using spherical k-means and learns the cluster labels using an MLP. To cope with larger datasets that do not fit into memory, TruncatedSVD is employed to embed the vectors into a lower-dimensional space. Further improvements to the previous pipeline are query parallelism, improved sampling schemes during index creation, and better memory management.

---

[8] LMI submitted a solution for Task 2, but was wall-clocked after 36hrs of construction.

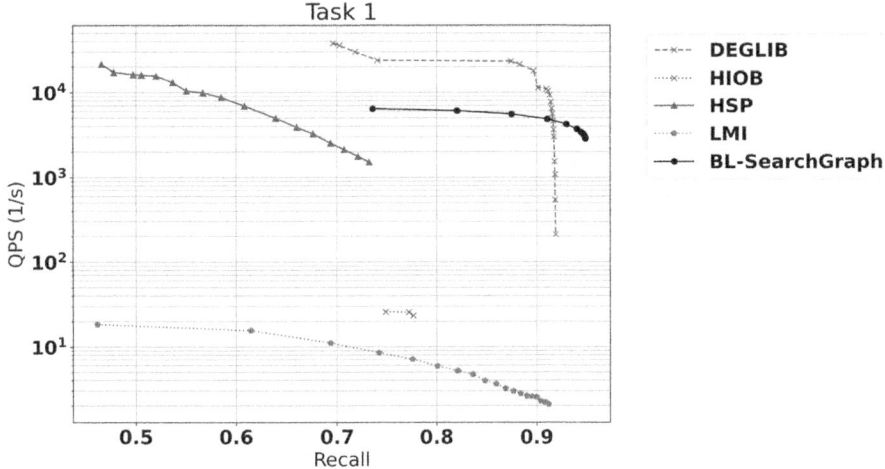

Fig. 1. Recall vs query throughput in our unrestricted indexing task under our private query set using 30 nearest neighbor queries. Up and to the right is better.

HSP [4]. This submission proposes the construction of an approximate half-space proximal graph (HSP) with local monotonicity properties. Instead of applying the HSP test to all possible neighbors of a node v, which results in a construction time of $O(N^2)$ on a dataset collection with N vectors, it instead first finds an approximate neighborhood $R(v)$ and applies the test only locally. A hierarchy of HSP graphs is built recursively through sampling to enable fast construction time.

4 Results and Discussions

Figure 1 presents the results of the evaluation of Task 1 for our 10K private query set. For fixed search hyperparameters, it related the achieved quality (in terms of the average recall over all 10K queries) to the observed throughput measuring queries per second. As mentioned in §2, all teams built a single index and provided a list of at most 30 different search hyperparameters. Disregarding the baseline, and focusing on the quality in terms of the achieved recall, teams DEGLIB and LMI surpass the 0.8 recall limit specified by the task. At the same time, HIOB kept just behind the lower line, presumably due to the differences between public and private query sets and a possible overfitting of the hyperparameter setting. Team HSP did not provide search parameters that would provide a quality that exceeds the recall threshold. As clear from the plot, team DEGLIB's solution provides the best recall-throughput trade-off. It provides a speed-up of almost a factor 3 over the baseline SearchGraph at around .8 recall.

Figure 2 compares search indexes in the memory-constrained task with reranking using the private query set. Only HIOB and LMI teams submitted system solutions for this task, but LMI's run was terminated due to exceeding

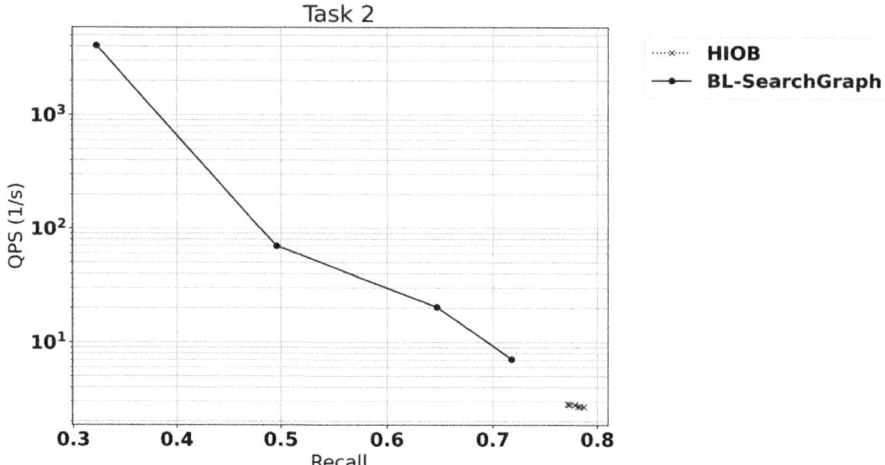

Fig. 2. Recall vs query throughput for the memory constrained task with reranking.

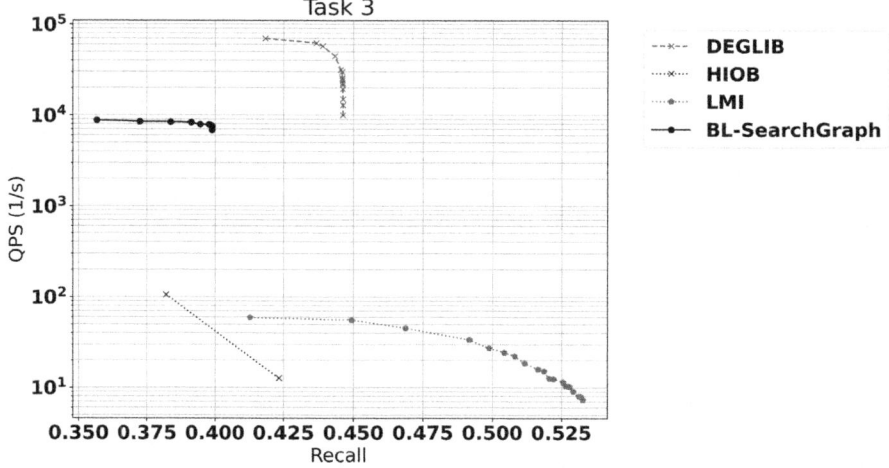

Fig. 3. Recall vs query throughput for our memory-restricted task without reranking, i.e., Task 3.

24 h of wall-clock time. The figure shows that neither HIOB nor our baselines achieved the requested average recall of 0.8. Even when undesired, this is a consequence of how private queries differ from public ones.

Finally, our memory-constrained task without reranking is evaluated in Table 3. Please remember that the challenge asks for fastest searching system solutions achieving a recall of at least 0.4. Here, we observe that DEGLIB, HIOB, and LMI improves the recall requirement; all of them improving on the baseline consisting of 128-dimensional PCA projection with Search Graph. Note that

Table 2. Summary of the evaluation process on the 100M database and the 10K private query set for Task 1, 2, and 3. Bold marks indicate best performing solutions under the challenge specifications; the strikethrough mark indicates that the recall limit was achieved with numerical round by the left, and ignored from the ranking. Build times are given in hours and minutes and query time in seconds.

Team	Task 1				Task 2				Task 3			
	Build time	Query Time	Recall	Rank	Build Time	Query Time	Recall	Rank	Build Time	Query Time	Recall	Rank
DEGLIB	2h 44m	**0.4 s**	0.87	1	–	–	–	–	1h 33m	**0.14 s**	0.42	1
HIOB	13 m	425.9 s	0.78	–	33 m	3,712.3 s	0.79	–	33 m	789.31 s	0.42	3
HSP	11h 00m	6.6 s	0.73	–	–	–	–	–	–	–	–	–
LMI	2h 07m	1,691.6 s	0.80	3	>24h	–	–	–	3h 08m	169.04 s	0.41	2
BL-SearchGraph	9h 20m	1.6 s	0.82	2	4h 17m	1,401.7 s	0.72	–	3h 21m	1.46 s	~~0.40~~	–
BL-Bruteforce	–	∼3×10⁴ s	1.0	4	–	–	–	–	–	3638.1 s	0.41	4

highest average recall is achieved by LMI while the best search performance is again achieved by DEGLIB; as in Task 1, the performance gain is remarkable.

Summary and Final Ranking. Table 2 summarizes the challenge results, showing the index construction time, the shortest search time that exceeded the recall requirement, and the actual recall achieved within that time. We also include the best runs for solutions that do not meet the requirements yet cannot get a rank position. We can observe that DEGLIB stands out in both Task 1 and Task 3, achieving the best performances. Task 2 is officially declared unmatched, but in practice, HIOB is better in this modality. HIOB also has a pretty low construction time, which is remarkable for the database size (100 million vectors). On the other hand, the LMI achieves high recall values in low-memory environments, but it holds high search times yet uses moderate building times. The HSP team also achieves high query throughput but does not satisfy the recall requirements.

5 Conclusions

The SISAP 2024 Indexing Challenge offered three tasks to assess indexing methods' performance under various operational constraints. Participants proposed approaches that effectively balanced search speed, result quality, memory usage, and indexing time when dealing with high-dimensional and large-scale metric datasets. The challenge inspired innovative methods, and the solutions exemplify carefully engineered data structures, efficient search algorithms, and feature projection techniques.

We are confident that the challenge contributes to future research and development efforts in approximate nearest neighbor searches. While the methods submitted and tested represent a significant step forward, there is still room for improvement in the fields of faster indexing and searching algorithms, more robust algorithms, and tackle the secondary memory challenge.

Acknowledgments. We thank Edgar Chavez for useful discussion during the planning phase of the challenge. The second co-author received funding from the Innovation Fund Denmark for the project DIREC (9142-00001B).

References

1. Aumüller, M., Bernhardsson, E., Faithfull, A.J.: ANN-benchmarks: a benchmarking tool for approximate nearest neighbor algorithms. Inf. Syst. **87** (2020)
2. Aumüller, M., Ceccarello, M.: Recent approaches and trends in approximate nearest neighbor search, with remarks on benchmarking. IEEE Data Eng. Bull. **46**(3), 89–105 (2023)
3. Chávez, E., Navarro, G., Baeza-Yates, R., Marroquín, J.L.: Searching in metric spaces. ACM Comput. Surv. **33**(3), 273–321 (2001). https://doi.org/10.1145/502807.502808
4. Foster, C., Chavez, E., Kimia, B.: Top-down construction of locally monotonic graphs for similarity search. In: SISAP. Lecture Notes in Computer Science. Springer (2024)
5. Hezel, N., Barthel, K.U., Schall, K., Jung, K.: An exploration graph with continuous refinement for efficient multimedia retrieval. In: Proceedings of the 2024 International Conference on Multimedia Retrieval. ICMR '24, pp. 657–665. Association for Computing Machinery, New York, NY, USA (2024). https://doi.org/10.1145/3652583.3658117
6. Hezel, N., Schilling, B., Barthel, K., Schall, K., Jung, K.: Adapting the exploration graph for high throughput in low recall regimes. In: SISAP. Lecture Notes in Computer Science. Springer (2024)
7. Khandelwal, U., Levy, O., Jurafsky, D., Zettlemoyer, L., Lewis, M.: Generalization through memorization: Nearest neighbor language models. In: ICLR. OpenReview.net (2020)
8. Li, W., Zhang, Y., Sun, Y., Wang, W., Li, M., Zhang, W., Lin, X.: Approximate nearest neighbor search on high dimensional data - experiments, analyses, and improvement. IEEE Trans. Knowl. Data Eng. **32**(8), 1475–1488 (2020)
9. Matsui, Y.: https://github.com/matsui528/annbench
10. Procházka, D., Slanináková, T., Čerňanský, J., Ol'ha, J., Antol, M., Dohnal, V.: Scaling learned metric index to 100m datasets. In: SISAP. Lecture Notes in Computer Science. Springer (2024)
11. Radford, A., Kim, J.W., Hallacy, C., Ramesh, A., Goh, G., Agarwal, S., Sastry, G., Askell, A., Mishkin, P., Clark, J., Krueger, G., Sutskever, I.: Learning transferable visual models from natural language supervision. In: ICML. Proceedings of Machine Learning Research, vol. 139, pp. 8748–8763. PMLR (2021)
12. Schuhmann, C., Beaumont, R., Vencu, R., Gordon, C., Wightman, R., Cherti, M., Coombes, T., Katta, A., Mullis, C., Wortsman, M., et al.: Laion-5b: an open large-scale dataset for training next generation image-text models. Adv. Neural Inf. Process. Syst. **35**, 25278–25294 (2022)
13. Simhadri, H.V., Williams, G., Aumüller, M., Douze, M., Babenko, A., Baranchuk, D., Chen, Q., Hosseini, L., Krishnaswamy, R., Srinivasa, G., Subramanya, S.J., Wang, J.: Results of the neurips'21 challenge on billion-scale approximate nearest neighbor search. In: NeurIPS (Competition and Demos). Proceedings of Machine Learning Research, vol. 176, pp. 177–189. PMLR (2021)

14. Slanináková, T., Procházka, D., Antol, M., Olha, J., Dohnal, V.: Sisap 2023 indexing challenge - learned metric index. In: Proceedings of the 16th International Conference on Similarity Search and Applications (2023)
15. Tellez, E.S., Ruiz, G.: Similarity search on neighbor's graphs with automatic pareto optimal performance and minimum expected quality setups based on hyperparameter optimization. CoRR **abs/2201.07917** (2022), https://arxiv.org/abs/2201.07917
16. Tellez, E.S., Ruiz, G.: Similaritysearch.jl: autotuned nearest neighbor indexes for Julia. J. Open Source Softw. **7**(75), 4442 (2022)
17. Tellez, E.S., Aumüller, M., Chávez, E.: Overview of the SISAP 2023 indexing challenge. In: SISAP. Lecture Notes in Computer Science, vol. 14289, pp. 255–264. Springer (2023)
18. Thordsen, E., Schubert, E.: An alternating optimization scheme for binary sketches for cosine similarity search. In: Similarity Search and Applications: 16th International Conference, SISAP 2023, A Coruña Spain, October 9–11, Proceedings. Springer (2023)
19. Thordsen, E., Schubert, E.: Grouping sketches to index high-dimensional data in a resource limited setting. In: SISAP. Lecture Notes in Computer Science. Springer (2024)

Scaling Learned Metric Index to 100M Datasets

David Procházka[1(✉)], Terézia Slanináková[1,2], Jozef Čerňanský[1], Jaroslav Olha[1], Matej Antol[1,2], and Vlastislav Dohnal[1]

[1] Faculty of Informatics, Masaryk University,
Botanická 68a, 602 00 Brno, Czech Republic
{davidprochazka,slaninakova,524756,olha}@mail.muni.cz, antol@muni.cz,
dohnal@fi.muni.cz
[2] Institute of Computer Science, Masaryk University,
Šumavská 15, 602 00 Brno, Czech Republic

Abstract. Learned indexing of high-dimensional data is an indexing approach that is still in the process of proving its viability – the Learned Metric Index (LMI) stands as one of the pioneering methods in this regard. Earlier implementation of LMI [Slanináková et al., SISAP 2023] primarily served as experimental prototype, operating under unrealistic assumptions, such as the availability of unlimited main memory or unbounded index construction time. Recently, however, LMI made the leap towards practical applicability on real-world datasets when it was successfully deployed to efficiently index 214 million protein structures for near-instantaneous retrieval [Procházka et al., Nucleic Acids Research 2024]. This paper details the key improvements that enabled this transition, including the introduction of parallel query processing (with the possibility of GPU acceleration), adaptive memory usage, pre-construction of memory buckets for contiguous access, a shift from k-means to spherical k-means clustering, and faster index construction through fewer epochs and the use of smaller training samples. LMI is now capable of handling 100M datasets and supports both in-memory and on-disk indexing, marking several important steps towards practical viability of AI-enhanced indexes for high-dimensional complex data in real-world settings.

Keywords: learned metric index · high-dimensional data · memory efficiency · on-disk index · approximate nearest neighbor search · similarity search

1 Introduction

Indexing dense high-dimensional embeddings, extracted from complex data such as images, videos, and human motion, is a crucial part of modern similarity data management. The increasing volume of data, along with higher vector dimensionality and diverse modalities, necessitates the development of efficient and effective indexing methods. At the same time, enhancing indexes with artificial intelligence is becoming a prominent trend, though its practical viability in high-dimensional data remains to be fully established. To date, the Learned Metric Index (LMI) had mostly been implemented as an experimental prototype, assuming the availability of sufficiently large main memory. Even though LMI successfully indexed 214 million protein structures [11], there has been no systematic exploration of how to scale the index to large datasets.

We present a novel implementation of LMI that supports indexing data on disk, with the option of dimensionality reduction to compact the data. This advancement represents a significant step towards establishing LMI as a comprehensive solution for large-scale similarity search in production environments. Our implementation includes the option to offload the entire search computation to a GPU and tune the system for varying numbers of computational cores. In this paper, we assume that the dataset is available in advance, and that reasonable pre-processing time is allowed before any queries are made.

In the following sections, we detail the new architecture of LMI as it applies to the tasks defined in the SISAP 2024 Indexing Challenge, showcasing how the updated architecture performs in practice.

1.1 Task Description

The goal of the SISAP 2024 Indexing Challenge is to build such an index for 100 million 768-dimensional CLIP embeddings.[1] The search involves finding 30 nearest neighbors for 10,000 simultaneously issued queries in three distinct tasks, described as follows. **Task 1**: Given 12 h to construct an index, achieve at least 0.8 recall in the least amount of time. **Task 2**: Given 24 h to construct the index, achieve at least 0.8 recall in the least amount of time while using 8 CPU cores and 32 GB RAM. The dataset is accessible as read-only and stored on disk. **Task 3**: Given 12 h to construct the index, achieve at least 0.4 recall in the least amount of time while using 64 GB of RAM. The dataset cannot be accessed during the search.

2 Learned Metric Index

The Learned Metric Index [2], similar to other approaches [3,5], extends the learned indexing paradigm from the originally proposed one-dimensional [7] to high-dimensional data. During construction, the dataset is divided into non-overlapping partitions using a k-means algorithm, with each cluster represented by a centroid. A machine learning model that solves a supervised classification task is trained on vector label pairs, where the label for a vector is given by its nearest centroid. After the model is trained, the buckets are formed by performing inference on the entire dataset and placing each vector in the most probable bucket. During the search, the model produces a probability distribution, which

---

[1] https://sisap-challenges.github.io/2024/tasks/

determines the order of the visited buckets. Typically, a cut-off is imposed to consider only the most similar categories for the final k-nearest neighbor (k-NN) search.

LMI [2] was first implemented in Python and open-sourced within [13]. In 2023, it organized and searched 10M CLIP embeddings in the SISAP 2023 Indexing Challenge [12]. More recently, a new search engine for retrieval of protein structures from the AlphaFold Protein Structure Database [14], named AlphaFind, successfully used LMI as the backbone of its search and indexing operations, achieving yet unmatched querying speed. In this work, we detail a redesigned implementation of LMI capable of working with 100M CLIP embeddings and exploiting multiple CPU cores, while respecting strict RAM limits.

2.1 Advancements

The "LMI 2023" implementation [12] was not focused on indexing datasets beyond the 10M subset. After careful analysis, we have identified the following opportunities to scale LMI for larger datasets. First, we introduced parallelism in query processing; LMI can now utilize more than one CPU core in the search phase. Second, we incorporated the ability to adapt the index to a limited amount of main memory, which is also required for Tasks 2 and 3, refer to Sect. 1.1. The previous design where buckets were essentially scattered throughout the dataset, has been superseded by a more structured approach with explicit bucket construction. Finally, model training has been significantly streamlined by employing a more suitable clustering algorithm, reducing the number of learning epochs, and training on a sample of the data rather than the whole dataset.

Parallelism of query processing. We introduced inter-query parallelism for the k-NN search in the search phase. However, an inherent limitation of Python is that multi-threaded code is restricted to running on a single thread due to the Global Interpreter Lock (GIL). To bypass Python's GIL, we employed an exhaustive k-NN search from the FAISS library [4]. In addition, we rely heavily on PyTorch [10], which has the potential to offload parallelism to GPU[2].

Adapting to the available main memory. When the size of a dataset is larger than the available main memory, we use a dimensionality reduction technique to obtain vectors with smaller dimensionality, such that the whole reduced dataset fits into the main memory. In particular, each object is transformed with *TruncatedSVD* [6] into a lower-dimensional vector and stored in in-memory buckets to approximate the original vector stored on disk. In contrast to the Johnson-Lindenstrauss lemma, TruncatedSVD is a data-dependent dimensionality reduction technique. When applied to centered input data, its output is equivalent to that of principal component analysis.

---

[2] GPU was not provided for the SISAP 2024 Indexing Challenge.

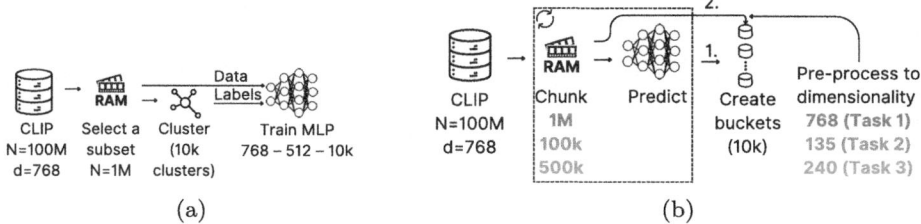

Fig. 1. Index construction.

Introduction of a bucket construction mechanism. LMI 2023 represented the entire dataset in a single large matrix and stored a bucket identifier next to each vector. This resulted in the buckets being scattered throughout the dataset, necessitating reconstruction during the search process. The current approach involves pre-allocating a matrix per bucket and populating them with data, thereby achieving predictable memory requirements and faster search.

Improving index construction. For partitioning, we replaced k-means with spherical k-means, which is better suited for approximate nearest neighbor search with inner product as the similarity function [15]. To achieve faster index construction, we select a random 1% sample of the dataset to train the model on, as was explored in [2]. We also reduced the number of epochs from hundreds to the low tens, since using fewer epochs does not degrade search performance [5,8]. Following the rationale that a sufficiently complex model can learn arbitrary partitioning, learned indexing techniques also show a preference for a single model instead of a hierarchy [8] – we adopted this recommendation in our index.

2.2 Pipeline

Index construction (Fig. 1) starts by selecting a random data subset, and clustering it via spherical k-means to obtain a label for each object. A model (multi-layer perceptron) is trained on the data and this partitioning (Fig. 1a). The model then iteratively processes the dataset in data chunks. For each chunk, it returns predictions, which indicate the placement of objects in the buckets. To maximize memory efficiency, we employ a two-pass approach: (i) we determine the size of each bucket to pre-allocate it; (ii) each object is assigned to the respective bucket. The chunk size and the use of dimensionality reduction for the data in buckets varies depending on the task (Fig. 1b).

The search pipeline (Fig. 2) begins by jointly obtaining predictions for all queries and identifying a number of their most similar buckets (*nprobe* parameter). Then, k-NN search is executed in parallel for each query and the *nprobe* buckets. The resulting vectors of c most similar objects[3] are then combined for

---

[3] c is equal to k in Tasks 1 and 3, and is set to 1,000 in Task 2.

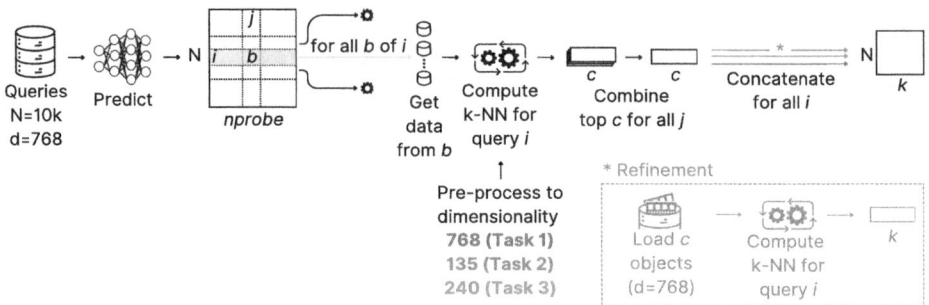

Fig. 2. The search process.

individual queries and later concatenated into the final result matrix. Task 2 additionally introduces a refinement step, where the full 768-dimensional vectors for c candidates are retrieved from the disk. The final answer is based on the k-NN search over such vectors.

3 Experiments

We have implemented the presented LMI architecture in Python. We provide the code[4] with predefined seeds and fixed package versions in two execution environments, Docker and Conda.[5]

3.1 Experimental Settings

We experimentally evaluated all tasks on an AMD EPYC 7532 CPU. Apart from the constraints imposed by the challenge, Tasks 1 and 3 were limited to 32 CPU cores, Task 1 ran with 200 GB of RAM (but only about 164 GB were allocated), and during the evaluation of Task 2, the dataset was stored on Micron 7300 SSD. In practice, the computation time of Tasks 1 and 3 could be lower, since the dataset was accessed over the network during the experiments.

To select appropriate parameters for the index and to replace the exhaustive grid search, we used Optuna [1,9], a hyperparameter optimization framework. The goal was to achieve the highest possible accuracy, i.e., how well the model predictions match the underlying clustering. The optimization was done on the 300K subset provided by the organizers of the SISAP 2024 Indexing Challenge, and the following configuration was obtained: a two-layer MLP, with a hidden layer of size 512, trained for 15 epochs, with the learning rate set to an initial value of 0.00098 and progressively adjusted by the Adam optimizer.

---

[4] https://github.com/Coda-Research-Group/LearnedMetricIndex/tree/paper-sisap24-indexing-challenge
[5] PyTorch does not guarantee completely reproducible results across platforms, see https://pytorch.org/docs/stable/notes/randomness.html.

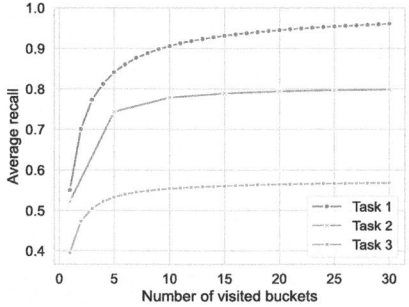 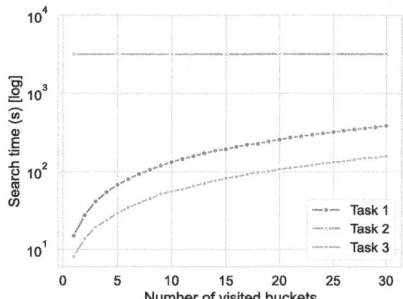

(a) The impact of the number of visited buckets on average recall (higher and leftward positions indicate better performance).

(b) The impact of the number of visited buckets on the search time (lower and leftward positions indicate better performance).

Fig. 3. Relationships between the number of visited buckets and average recall/overall search time.

Training was done using a sample of 1,000,000 vectors and the number of buckets was set to 10,000. To match the memory requirements, LMI used chunk sizes of 100,000 and 500,000 in Tasks 2 and 3, respectively. Task 2 reduced the vectors to 135 dimensions, while Task 3 reduced the vectors to 240 dimensions. The number of candidates for refinement in Task 2 was set to 1,000.

3.2 Retrieval Evaluation

Figure 3 shows the results of the evaluation on the public set of 10,000 queries. On average, more than half of the nearest neighbors are found after visiting one bucket in Tasks 1 and 2 (Fig. 3a). In Task 3, the LMI is slightly below the required threshold of 40% (with 39.46%) when visiting a single bucket.

In Tasks 1 and 3, the search time scales linearly with the number of buckets visited, while in Task 2, the search time increases only slightly (Fig. 3b). The difference between visiting one bucket and 100 buckets is about one and a half minutes, but this value is negligible compared to the disk latency of retrieving 1,000 candidates for refinement.

In Task 1, the recall of LMI increases rapidly with the first few buckets visited, exceeding 80% recall with only 4 buckets, and 96% recall with 30 buckets. LMI in Task 2 achieves nearly 75% recall by visiting two buckets, and essentially plateaus at 80% after 30 inspected buckets. In Task 3, the required recall of 40% is quickly reached by visiting two buckets, while the configuration converges to around 57% recall. Note that the maximum possible recall in Tasks 2 and 3 is limited by the quality of the dimensionality reduction technique and the size of the main memory, and cannot get to 100%. The time required to train the dimensionality reduction technique on the sample was approximately 11 min for Task 2 and less than 17 min for Task 3. The total time to transform the entire dataset was less than nine minutes.

Table 1. Performance of the fastest configurations reaching the required recall.

Task	Build time	Stop condition	Query time	Avg. candidates	Recall
1	1 h 36 m	4	55.2 s	53,948	81.3%
2	2 h 41 m	40	53.7 m	499,670	80.0%
3	2 h 12 m	2	13.7 s	27,997	47.3%

In all tasks, over 400 buckets contain less than a hundred objects. The largest and only bucket with over 100,000 objects contains over 165,000 vectors, more than 16 times the average bucket size.

Best-performing setups. We list the best-performing setups that meet the task requirements in Table 1. All configurations were built in less than three hours, while the pure overhead of LMI without considering dimensionality reduction was less than 2.5 h. Applying dimensionality reduction techniques to all queries took less than 100 ms for each task. The search throughput of Task 1 reached over 181 queries per second (QPS), over 3 QPS for Task 2[6], and 729 QPS for Task 3.

Performance in Task 2. The suboptimal performance of LMI in Task 2 is due to the large number of candidates that need to be refined. Since our engineering efforts were primarily focused on optimizing CPU utilization, the implementation did not take full advantage of techniques for handling I/O-bound workloads. As a result, we observed up to 78 s of disk latency when requesting the 1,000 candidates for refinement. This highlights some of the intricacies involved in redesigning a structure for on-disk indexing. In a heavily I/O-bound use case, merely preventing memory overflow might not be sufficient; it could be necessary to redesign core aspects of the search process itself.

4 Conclusion

We presented a new implementation of the Learned Metric Index, rebuilt from the ground up to enhance scalability. This redesign enabled the LMI to achieve a search throughput of 181 queries per second (QPS) on raw 768-dimensional vectors and 729 QPS in memory-constrained settings, while also leveraging multiple cores for parallel search and processing. This represents a notable progress towards a fully optimized indexing structure – nevertheless, the results of this challenge reveal many untapped opportunities for performance optimization. Future research will concentrate on several key areas: refining methods for indexing data stored on disk, enhancing LMI to improve its dynamicity while tackling

---

[6] A rather large stop condition of 40 bucket was needed to achieve the required recall. It is not displayed in the graph.

related challenges in machine learning models, and investigating the theoretical foundations and limitations of high-dimensional learned indexes. Finally, the implementation of LMI in a compiled language should enhance the overall utilization of hardware resources.

Acknowledgments. Supported by Czech Science Foundation project No. GF23-07040K. Computational resources were provided by the e-INFRA CZ project (ID:90254), supported by the Ministry of Education, Youth and Sports of the Czech Republic and the ELIXIR-CZ project (ID:90255), part of the international ELIXIR infrastructure.

References

1. Akiba, T., Sano, S., Yanase, T., Ohta, T., Koyama, M.: Optuna: a next-generation hyperparameter optimization framework. In: Proceedings of the 25th ACM SIGKDD (2019)
2. Antol, M., Ol'ha, J., Slaninákova, T., Dohnal, V.: Learned metric index-proposition of learned indexing for unstructured data. Inf. Syst. **100**, 101774 (2021)
3. Dong, Y., Indyk, P., Razenshteyn, I.P., Wagner, T.: Learning space partitions for nearest neighbor search. In: ICLR (2020)
4. Douze, M., et al.: The Faiss library (2024). https://arxiv.org/abs/2401.08281
5. Gupta, G., Medini, T., Shrivastava, A., Smola, A.J.: Bliss: a billion scale index using iterative re-partitioning. In: Proceedings of the 28th ACM SIGKDD, pp. 486–495 (2022)
6. Halko, N., Martinsson, P.G., Tropp, J.A.: Finding structure with randomness: probabilistic algorithms for constructing approximate matrix decompositions. SIAM Rev. **53**(2), 217–288 (2011)
7. Kraska, T., Beutel, A., Chi, E.H., Dean, J., Polyzotis, N.: The case for learned index structures. In: Proceedings of the SIGMOD 2018, pp. 489–504 (2018)
8. Li, L., Han, A., Cui, X., Wu, B.: Flex: a fast and light-weight learned index for KNN search in high-dimensional space. Inf. Sci. **669**, 120546 (2024)
9. Oguri, Y., Matsui, Y.: General and practical tuning method for off-the-shelf graph-based index: SISAP indexing challenge report by team UTokyo. In: International Conference on Similarity Search and Applications, pp. 273–281. Springer, Berlin (2023)
10. Paszke, A., et al.: Pytorch: an imperative style, high-performance deep learning library. In: Advances in Neural Information Processing Systems, vol. 32 (2019)
11. Procházka, D., et al.: AlphaFind: discover structure similarity across the proteome in AlphaFold DB. Nucleic Acids Res. **52**(W1), W182–W186 (2024)
12. Slaninákova, T., Procházka, D., Antol, M., Olha, J., Dohnal, V.: SISAP 2023 indexing challenge—learned metric index. In: International Conference on Similarity Search and Applications, pp. 282–290. Springer, Berlin (2023)
13. Slaninákova, T., et al.: Reproducible experiments with learned metric index framework. Inf. Syst. **118**, 102255 (2023)
14. Varadi, M., et al.: AlphaFold protein structure database: massively expanding the structural coverage of protein-sequence space with high-accuracy models. Nucleic Acids Res. **50**(D1), D439–D444 (2022)
15. Vecchiato, T., Lucchese, C., Nardini, F.M., Bruch, S.: A learning-to-rank formulation of clustering-based approximate nearest neighbor search. In: Proceedings of the 47th ACM SIGIR, pp. 2261–2265 (2024)

Grouping Sketches to Index High-Dimensional Data in a Resource-Limited Setting

Erik Thordsen[✉] and Erich Schubert

TU Dortmund University, Data Mining 44227, Dortmund, Germany
erik.thordsen@tu-dortmund.de

Abstract. Indexing very large and high-dimensional data in resource-limited settings is a challenging task. While sketching techniques can reduce the data size, the resulting sketches – if useful for indexing – are often (almost) uncorrelated. That makes it hard to build an index structure for the sketches. In this work, we propose to group the bits of the sketches and use these groups as hash values. We derive expected recall values and show that the recall of the index can be controlled by the number of groups and bits per group. The parameter choice is a trade-off between recall, speed, and memory consumption. In a practical setting, the approach is combined with Stochastic HIOB sketches and achieves a speedup of near 100 times over a linear scan of the sketches.

Keywords: Locality Sensitive Hashing · Binary Sketches · High-Dimensional Data · Indexing · Resource-limited

1 Introduction

This paper is a submission to the SISAP Indexing Challenge 2024 and focuses on Task 3 of the challenge. The goal is to create an index structure for very large high-dimensional data in a resource-limited setting. The data is a subset of approximately 100M vectors of LAION5B [7] with 768 dimensions. The target machine is a single node with 64 GB of RAM and 8 cores, similar to a modern desktop environment. As the data set consists of roughly 300 GB 32-bit floats, it is not possible to store the data in memory. Aside from being unable to load the entire data set during training, submissions are not permitted to access the data set during query time. It is, thus, mandatory to achieve a data compression ratio of at least 1:5 and one must exceed a recall on the 30-nearest neighbors of 40% to be considered for the challenge. The recall is measured as the fraction of true nearest neighbors that are found in the 30-nearest neighbors returned by the index. In our submission, we focus on the data compression ratio and intend to use the remaining memory for an auxiliary index structure. In the literature, multiple data compression techniques have been proposed, such as Principal Component Analysis (PCA) [6], Random Projections (RP), Locality Sensitive

Hashing (LSH) [2], or Binary Sketches – a special case of LSH where each hashing function has a binary output. Since we intend to build an index structure, we rule PCA and RP out, as we could at best retain 77 dimensions, if we were to allocate half our memory to the compressed representation. We propose to combine the remaining LSH and Binary Sketches into a joined approach, where we base our hashing functions on the binary sketches produced by Stochastic HIOB [9]. We use sequences of the sketches as hashes and the Hamming distance for the refinement of the queries. This allows to at best achieve the 30@30-recall of the binary sketches. In this work we will first give an overview over the Stochastic HIOB approach to binary sketching. We will then give an overview over the theoretical properties of the proposed LSH index assuming completely random bit strings. Afterwards we provide empirical evidence on the efficacy of the approach, considerations on parameter choice, and implementation details. This paper is then concluded with a summary and an outlook on future work.

2 Stochastic HIOB

HIOB [9] is an alternating optimization scheme to improve the quality of binary sketches in terms of k-nearest neighbor recall. It is based on hyperplanes as decision boundaries. The assignments – 0 for "below" or 1 for "above" – of any single hyperplane are called a *bit*. Starting from any initialization, the hyperplanes are iteratively rotated to minimize the correlation between the bits. To achieve this, the number of equal and opposite values of a bit pair are translated into a rotation angle. One of the hyperplanes is then rotated away or towards the normal angle of the other hyperplane. Stochastic HIOB is an extension of HIOB, which uses a varying noised subsample of the full data set during the optimization to massively reduce the computational cost. For the challenge data set, training the Stochastic HIOB with 384 bits finished in about 10 min on a reference system with 14 cores. This bit length was the smallest multiple of 64 that reliably provided a recall above 40%. Our focus in this work is to improve on the search time of the sketches, which previously required a linear scan.

3 From Sketches to Grouped LSH

The sketches produced by Stochastic HIOB are negligibly correlated. This allows for a good recall in nearest neighbor queries, but also makes it hard to create any kind of (exact) index structure for the sketches. All small and fast to build index structures – considering the constraints of the challenge – fail to perform faster than a linear scan. Other indices such as HNSW [4] may be capable of indexing the sketches with sufficient recall and speed, but are significantly too large. However, if we know the approximate number of queried neighbors, we can exploit some probabilistic features of the sketches. We propose to create g groups of b bits each. The bits of each group are chosen at random, thus the bits of different groups may overlap. Each group of bits produces a hash value for each sketch simply by concatenating its selected bits and interpreting them as an integer. Using these hash values, we can create a hash table for each group and similar to classical LSH, concatenate the matches of all groups to find the nearest neighbors. The number of groups g and the number of bits per group b can be chosen to trade off between recall, speed, and memory consumption. Additionally, we can further concatenate the results from all hash values at a maximum Hamming distance of m to the query hash. This improves the recall, albeit at the cost of speed. We will first inspect the relevant probabilities and then discuss the expected properties based on the parameter choice.

We assume uniform random bit strings of length B, which is the worst case for the recall of the grouped LSH index. In practice, we observed that the recall can be significantly better than expected from the theoretical derivation, even if the bit strings are almost uncorrelated. If two bit strings s_1 and s_2 have a Hamming distance of d, the probability, that the substring created from b random positions in s_1 and s_2 have a maximum Hamming distance of $m \leq b$, is given by

$$p_1(B,d,b,m) = \sum_{i=0}^{m} \frac{\binom{B-d}{b-i}\binom{d}{i}}{\binom{B}{b}} \stackrel{m=0}{=} \frac{\binom{B-d}{b}}{\binom{B}{b}}. \tag{1}$$

Consequentially, the probability that any of the g substring pairs of length b in s_1 and s_2 differs by at most m bits is given by

$$p_g(B,d,b,m) = 1 - (1 - p_1(B,d,b,m))^g. \tag{2}$$

On the other hand, the probability that two uniformly random bit strings have a Hamming distance of d is given by

$$p_d(B,d) = \frac{\binom{B}{d}}{2^B}. \tag{3}$$

From here on we will denote the corresponding cumulative mass function of some probability mass function p_x as P_x and vice versa. All probabilities are cumulated over the Hamming distance d. The cumulative mass can be calculated by summation of individual probability masses. The individual probability masses are simply the differences between the cumulative mass at d and $d-1$, except for the first value, which is equal to the cumulative mass at $d=0$. The cumulative mass function of the k-th order statistic, i.e., the k-th smallest Hamming distance between N pairs of random bit strings of length B, follows as

$$P_{d_k}(B,N,k,d) = \sum_{i=k}^{N} \binom{N}{i} P_d(B,d)^i (1-P_d(B,d))^{N-i} \tag{4}$$

$$= \sum_{i=0}^{k-1} \binom{N}{i} P_d(B,d)^i (1-P_d(B,d))^{N-i}. \tag{5}$$

Since the k-th nearest neighbor is at least as far away as the $(k-1)$-th nearest neighbor, the probabilities for the order statistics are not independent. As a naive relaxation, we can, however, assume them to be independent, whereby the probability mass function of the k-nearest neighbor distances is approximately

$$p_{knn}(B,N,k,d) \approx \frac{1}{k}\sum_{i=1}^{k} p_{d_k}(B,N,i,d). \tag{6}$$

The naive relaxation is quite close to the true values for the sizes of B relevant to us ($B \gg 1$). The k-nearest neighbor distances in a data set are not independent either, but that relaxation also does not affect the results much. An exemplary comparison to empirical values to support these claims is displayed in Fig. 1. The expected recall of the grouped LSH index then follows from the collision probability p_g and the probability of observing certain neighbor distances p_{knn} as

$$r(B,N,k,g,b) = \sum_{d=0}^{B} p_g(B,d,b) p_{knn}(B,N,k,d). \tag{7}$$

The ideal "compression ratio", i.e. the number of Hamming distance computations required divided by N, is $(g/2^b)\sum_{i=0}^{m}\binom{b}{i}$ as each group splits uniformly random bit strings into 2^b equally sized bins and $\binom{b}{i}$ hash values are at a Hamming distance of i to the query. Lastly, the memory footprint of the index is given by $\mathcal{O}(Ng)$ as we have to store references to all sketches in each hash table. The number of indices per hash table 2^b can be omitted as it should be $<N$. This naturally leads to the limits $\log_2(g) < b \ll B$ and $1 \leq g < \frac{M}{Nu}$, where M is the available memory in bytes and u is the number of bytes per reference. We further suggest $N/2^b \gg k$, i.e. $b < \log_2(N/k)$, to ensure that each bin in the hashing tables contains at least k elements. While it is obvious, that individually changing g, b, and m to increase the compression ratios leads to an improved recall at the price of higher computational cost, their interplay is not as clear. We can use the expected recall and compression ratio to find "optimal" parameters – not accounting for side effects like overhead or caching. Figure 2 displays the expected recall for different parameter choices as well as different memory limitations. Both the recall and compression ratios are approximately lower bounds for practical cases. Correlations in the bit strings can lead to a higher recall than expected from the theoretical derivation while the overhead due to the index structure and unbalanced hash bins can lead to higher computational cost than indicated by the compression ratio.

4 Application to the Challenge

In this section we will first discuss some implementation details of the Stochastic HIOB and the parameter choice for the sketches. We will then expand on the parameter choice and implementation of the grouped LSH and the practical performance of the combination.

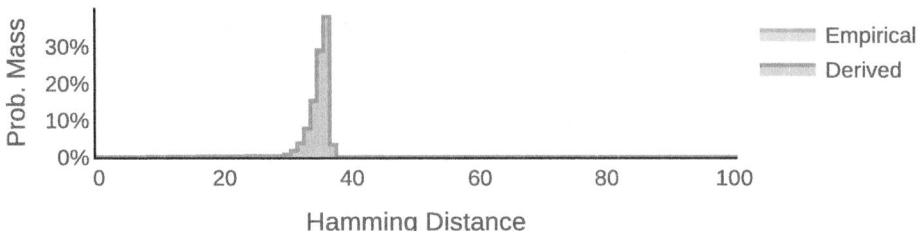

Fig. 1. Comparison of the derived naive probability mass function of the 30-nearest neighbor distances p_{knn} to the empirical distribution of nearest neighbor distances of 10 000 bit strings with 100 bits length. The distributions are almost identical.

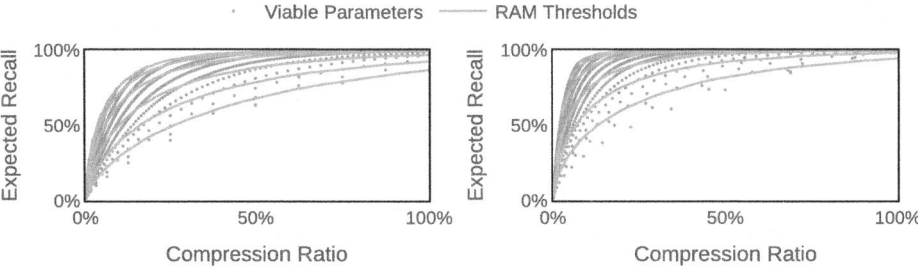

Fig. 2. Expected recall and compression ratio for various parameter choices of g and b. The left plot uses exact matches of the hash groups only ($m = 0$) and the right uses a maximum distance of $m = 2$. Each blue dot represents one integer parameter pair. The plots assume $N = 100M$ uniformly random bit strings of length $B = 384$ and 30-nearest neighbor queries. Below each curved line are parameters with a total RAM usage below 3, 10, 30, 100, 200, and 300 GB, respectively, from bottom to top. The required memory was computed assuming 64-bit references.

4.1 Stochastic HIOB

As the memory limitations do not allow to store the full data set even during training, the Stochastic HIOB approach is a natural choice. The sketches can be trained on a subsample of the data set and the full data set can afterwards be processed in chunks. We separately investigated the impact of the different parameters of Stochastic HIOB on the recall when using brute force search. As for the number of bits B, we found that 384 bits are sufficient to achieve a 30@30-recall of about 43%, which even under some measurement error exceeds the requirements of the challenge. The Stochastic HIOB was modified since its original publication to improve the scaling performance in terms of sketch length. The original squared dependency during training was replaced by a linear dependency by using Anderberg's update rule for dynamic matrix argmax [1]. Further, the initialization was replaced with Iterative Quantization [3] to improve the initial (and final) recall whenever B is smaller than the data dimensionality. The choice of sketch length led to a memory footprint of the sketches of approximately 38.4 Gbit or 4.8 GB. As for the other parameters of Stochastic HIOB, we obtained stable results for a training set size of $\geq$ 25K samples for 150K iterations, exchanging the stochastic sample every 1K iterations. In general, larger sample sizes as well as fewer iterations per sample lead to a better representation of the full data set. We, however, found that the recall did not significantly improve beyond the chosen values and thus kept the parameters with the least required IO. Each stochastic sample was noised with a standard deviation of 0.05, which produced good results in the 2023 Challenge [8]. The chunk size for IO operations was set to 300K samples, i.e. approximately 0.92 GB.

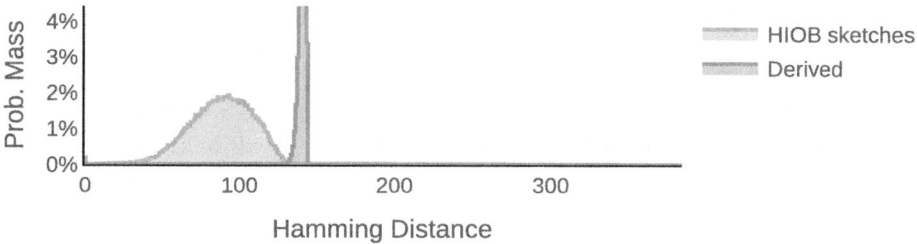

Fig. 3. Comparison of the derived naive probability mass function of the 30-nearest neighbor distances p_{knn} and the empirical distribution of nearest neighbor distances of 384 bit long HIOB sketches on the LAION 100M subset. The y-axis is clipped for better visibility.

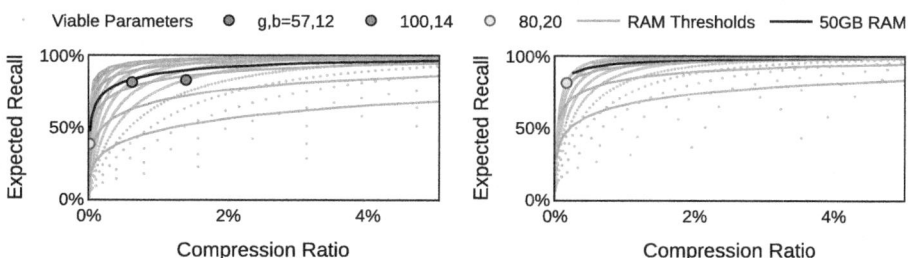

Fig. 4. Expected recall and compression ratios for various parameter choices of g and b. The left plot uses exact hash matches ($m = 0$) and the right uses a maximum distance of $m = 1$. Each blue dot represents one integer parameter pair. The plot used $N = 100M$ HIOB sketches of length $B = 384$ computed on the LAION subset and the 30-nearest neighbor distances for the sketches of the public query set. Below each curved line are parameters with a total RAM usage below 3, 10, 30, 50 (black), 100, 200, and 300 GB, respectively, from bottom to top. The required memory was computed assuming 32-bit references. Note, that the x-axis is cropped to end at 5% for improved visibility.

4.2 Grouped LSH

With about 6 GB of memory for the sketches and IO operations, and about 4 GB for the operating system, we were left with a little less than 54 GB for the index structure. While HIOB minimizes the pairwise correlation of the bits, it does not consider larger groups of bits. We thus observed that the 30-nearest neighbor distances are much closer than expected from the naive relaxation as displayed in Fig. 3. We assume, that certain parts of the data's inherent structure is translated from the original space into the sketch space. Replacing the theoretical distribution of the nearest neighbor distances with the empirical distribution, we found that the practical recall of the grouped LSH index is significantly higher than the recall derived for uniformly random bit strings. The recall for different parameter choices is displayed in Fig. 4. Intuitively, the overall recall could be assumed to be the product of the recall in Euclidean space when approximating with the sketches and the recall in Hamming space when approximating the sketches with the grouped LSH. Thus, to not drop below the required recall of 40%, given a recall purely from sketches of 43.6%, we would need a recall of at least 91.7% in the grouped LSH. However, the measured recall when choosing parameters with approximately 90% grouped LSH recall was still close to 43%. We assume, that the "near misses" of both approximations tend to cancel out: A 28th neighbor in Euclidean space may end up being the 32nd neighbor in Hamming space. Its Hamming distance to the query is almost as small as the 30 neighbors returned by the sketches. It thus has a high likelihood of being a "false positive" of the grouped LSH. This effect may be even more pronounced as we are in the $< 50\%$ recall regime. We initially submitted $g = 57, b = 12$ to the challenge, which uses a bit under 50 GB of memory for 64-bit references. Later improvements in the implementation allowed us to use $g = 100, b = 14$ for 32-bit references and $g = 80, b = 20$ for $m = 1$. These parameter pairs are highlighted in Fig. 4. All chosen parameters use less than 50 GB memory and provide a grouped LSH recall of above 80%. The overall recall of all parameters was between 40.6 and 40.8%. We specifically chose them not right on the pareto front to allow for an increased margin of error in the final query set of the challenge. To build the grouped LSH index, we used a list-of-lists based structure. The bits selected for each group were produced with a permutation-based approach, that first used all bits once before repeating any. We intended to obtain very diverse and uncorrelated groups of bits in that fashion. We then iterated over the bit groups in parallel and over the sketches sequentially and added the indices of the sketches to the lists corresponding to their hash value. The hash values were computed by interpreting the bits of the group as an integer. The hash values at a Hamming distance of $m = 1$ were precomputed and stored as arrays. For querying, we iterated over the queries in parallel and over the groups and matching bins sequentially. We used a max heap of size kg to keep track of the (kg)-nearest neighbors. As any sketch index can occur in more than one matching bin, we chose to keep all (up to g) duplicates at this stage. We then removed duplicates from the heap and collected the k-nearest neighbors. On a reference system with 14 cores, a linear scan of the sketches took about 36 min for 10K queries or 205 ms per query on average. The grouped LSH index took about 2 min for the same queries or 12 ms per query on average for $(57, 12)$ and 65 s in total/6.5 ms per query on average for $(100, 14)$ and 22 s in total/2.2 ms per query on average for $(80, 20)$ and $m = 1$. We thus achieved a speedup of at most 99×, which is well below the theoretical maximal speedup of $((80/2^{20}) \sum_{i=0}^{1} \binom{20}{i}))^{-1} \approx 624\times$. The remaining factor of ≈ 6 is likely due to the overhead induced by the index and the non-sequential access of the sketches. Building the grouped LSH indices took less than 2.5 min each, making both building and querying in total faster than a linear scan.

5 Conclusion

In this work we explored an approach to index the (almost) uncorrelated binary sketches produced by Stochastic HIOB. The grouped LSH index structure we proposed is based on the idea of using groups of randomly selected bits as hash. The structure is very fast to build and has a controllable recall, which can be optimized if given a rough estimate of the nearest neighbor distance distribution and a memory limit. We found that the recall of the grouped LSH index can be significantly higher in practical settings than expected from the theoretical derivation and that "double errors" of Stochastic HIOB and the grouped LSH allow for a better recall than expected. Although well suited for joint operation with Stochastic HIOB, the grouped LSH index can also be used with other binary data. It is very fast to build and easy to implement efficiently. It is, though, somewhat wasteful in regards to memory consumption, which is its main drawback. We have shown that non-exact hash matches allow to circumvent that issue in part and can further improve the speed. Further, using smaller reference types can reduce the memory footprint, but only as long as the total number of samples can be represented. In future work, the impact of unbalanced bits could be investigated. Sketches with unbalanced bits have been shown to provide a higher recall in nearest neighbor queries [5], though they would also imply unbalanced hash bins. Whether the better recall outperforms potential issues due to imbalanced hash bins remains to be investigated.

References

1. Anderberg, M.R.: Cluster Analysis for Applications. Probability and Mathematical Statistics: A Series of Monographs and Textbooks. Academic Press (1973). https://doi.org/10.1016/C2013-0-06161-0
2. Gionis, A., Indyk, P., Motwani, R.: Similarity search in high dimensions via hashing. In: Very Large Data Bases, VLDB, pp. 518–529 (1999)
3. Gong, Y., Lazebnik, S.: Iterative quantization: a procrustean approach to learning binary codes. In: CVPR, 2011, pp. 817–824 (2011). https://doi.org/10.1109/cvpr.2011.5995432
4. Malkov, Y.A., Yashunin, D.A.: Efficient and robust approximate nearest neighbor search using hierarchical navigable small world graphs. IEEE Trans. Pattern Anal. Mach. Intell. **42**(4), 824–836 (2018). https://doi.org/10.1109/TPAMI.2018.2889473

5. Mic, V., Novak, D., Zezula, P.: Sketches with unbalanced bits for similarity search. In: Similarity Search and Applications (2017). https://doi.org/10.1007/978-3-319-68474-1_4
6. Pearson, K.: On lines and planes of closest fit to systems of points in space. Philos. Mag. Ser. **1**(2), 559–572 (1901). https://doi.org/10.1080/14786440109462720
7. Schuhmann, C., Beaumont, R., Vencu, R., Gordon, C., Wightman, R., Cherti, M., Coombes, T., Katta, A., Mullis, C., Wortsman, M., Schramowski, P., Kundurthy, S., Crowson, K., Schmidt, L., Kaczmarczyk, R., Jitsev, J.: LAION-5B: an open large-scale dataset for training next generation image-text models. In: NeurIPS (2022)
8. Tellez, E.S., Aumüller, M., Chávez, E.: Overview of the SISAP 2023 indexing challenge. In: International Conference on Similarity Search and Applications, SISAP, pp. 255–264 (2023). https://doi.org/10.1007/978-3-031-46994-7_21
9. Thordsen, E., Schubert, E.: An alternating optimization scheme for binary sketches for cosine similarity search. In: Similarity Search and Applications (2023). https://doi.org/10.1007/978-3-031-46994-7_4

Adapting the Exploration Graph for High Throughput in Low Recall Regimes

Nico Hezel[(✉)], Bruno Schilling, Kai Uwe Barthel, Konstantin Schall, and Klaus Jung

Visual Computing Group, HTW Berlin 12459, Berlin, Germany
{hezel,bruno.schilling,barthel,konstantin.schall,
klaus.jung}@htw-berlin.de

Abstract. Nearest neighbor search is vital for modern search systems, particularly in high-dimensional spaces. This paper addresses the 2024 SISAP Indexing Challenge, which involves searching a dataset of 100 million 768-dimensional feature vectors under low recall and memory constraints. We explore the trade-offs between navigation and exploration graphs, where the latter is typically better suited for high recall scenarios. To tackle the challenge, we combine the state-of-the-art continuous refining Exploration Graph (crEG) with feature compression techniques. Although compression reduces overall recall accuracy, it significantly improves search speed. Given the challenge's focus on low recall, this trade-off enables crEG to perform efficiently, making it competitive even against navigation graphs traditionally favored in such settings.

Keywords: Approximate Nearest Neighbor Search · Proximity Graph

1 Introduction

Nearest neighbor search is essential in modern search systems, including recommender systems and similar item searches. Items are compared using a metric based on high-dimensional feature vectors. Factors influencing comparison effort include the distance function, dimensionality, and data type of feature vectors. When datasets grow too large or comparisons become too costly, approximate nearest neighbor search algorithms (ANNS) are employed.

The SISAP Indexing Challenge, an annual event, benchmarks various ANNS approaches [11]. The 2024 edition features a dataset of 100 million feature vectors, each with 768 dimensions in float16 format, sourced from the LAION-5B dataset [9]. The dataset alone requires 150 GB of memory. The challenge comprises three tasks, of which we aim to solve two. Submissions are evaluated on a machine with 512 GB RAM and two Intel E5-2690 V4 CPUs with 28 cores.

---

https://visual-computing.com/.

© The Author(s), under exclusive license to Springer Nature Switzerland AG 2024
E. Chávez et al. (Eds.): SISAP 2024, LNCS 15268, pp. 283–290, 2024.
https://doi.org/10.1007/978-3-031-75823-2_24

Task 1 is limited only by the available hardware and a 12-h time limit to construct the search index. The goal is to search 10,000 queries as quickly as possible, with at least 80% of relevant results found. Each search must return the top 30 results from the entire dataset (100M).

Task 3 further limits the available memory to 64 GB RAM. Here, only 40% of relevant results need to be found, focusing on handling limited hardware resources. To meet Task 3's requirements, the original data must be compressed or quantized to fit within the limited memory resources.

In previous years, data was often indexed and searched using graph-based approaches [3,7] and compressed with methods such as PCA or learned feature space transformations [1]. Other teams utilized bit-feature vectors and Hamming distance [12].

The greedy search algorithm for the graph-based approaches, as investigated by Boguna [2], involves two phases: the "navigation" phase and the "exploration" phase. During navigation, the search algorithm starts at one or multiple seed vertices and approaches the query quickly using long edges. The exploration phase begins when no closer vertex can be found, requiring backtracking to explore the near neighborhood to find the best k closest vertices to the query. The exploration phase becomes more critical when high recall results are required, whereas navigation is more important for short result lists and small recall values.

Last year, a modified HNSW graph, known for its good navigation properties, won the SISAP Indexing Challenge [3]. This year, the same data is used, but the result set size has increased from $k = 10$ to $k = 30$. We aim to investigate whether a modified exploration graph is better suited than HNSW for this year's challenge.

2 Related Work

Theoretical proximity graphs, such as k-Nearest Neighbor Graphs (kNNG) [8] and Monotonic Relative Neighborhood Graphs (MRNG) [4], are known for their computational expense. Monotonic graphs possess the property that a greedy search algorithm will move closer to the query with each step, ensuring efficient navigation. However, due to the high computational cost of constructing these graphs, many approaches only approximate them.

The Hierarchical Navigable Small World (HNSW) graph [6] is widely regarded as a state-of-the-art approach for many tasks, offering short search times through its hierarchical structure. The structure enables rapid navigation using long edges in the upper layers, though it is less suited for detailed exploration in the lower layers. In contrast, the continuous refining Exploration Graph (crEG) [5] excels in exploration, particularly in high recall scenarios, where it significantly outperforms HNSW. It approximates both Monotonic Relative Neighborhood Graphs (MRNG) and k-Nearest Neighbor Graphs (kNNG), combining their strengths to enhance search efficiency.

In the context of the SISAP Indexing Challenge, recent works have demonstrated the effectiveness of graph-based approaches. Cole Forster [3] and Oguri

[7] achieved notable results in the 2023 challenge by selecting and modifying two well-suited navigation graphs: HNSW and NSG [4]. These graphs were particularly effective in meeting the challenge's emphasis on low recall rates and short result lists, leading to high search speeds.

3 Contribution

The continuous refining Exploration Graph (crEG) [5] has demonstrated state-of-the-art performance on datasets up to 1 million data points within the high recall regime. This paper extends the evaluation of crEG to datasets containing up to 100 million items, focusing on lower recall requirements. To enhance its navigation efficiency and overall search speed for the specific tasks of the challenge, we have implemented three key improvements:

- **Elimination of Redundant Paths:** To improve navigation efficiency, unnecessary alternative edges in crEG were removed, significantly reducing the total number of edges.
- **Optimal Starting Points:** Cluster centers in the feature space were identified and used as starting points during the graph search to reduce the navigation effort.
- **Feature Vector Compression:** Feature vectors were compressed using a neural network and quantized. The compression network's loss function enforces the preservation of the relative rankings between data points in the reduced-dimensional space.

4 Elimination of Redundant Paths

During the construction phase, the Exploration Graph aims to incorporate edges present in a Monotonic Relative Neighborhood Graph (MRNG). Edges appearing in this graph are termed "'MRNG conformant". Given the computational expense of exact conformity calculation, the crEG paper approximates it using Algorithm 1. For a graph G, an edge between vertices $v1$ and $v2$ is checked for MRNG conformity by ensuring $v1$ has no neighbor u also connected to $v2$, where the connections $v1$ to u and u to $v2$ are each shorter than the direct connection between $v1$ and $v2$.

Algorithm 1 checkMRNG$(G, v1, v2)$

Require: graph G, vertex $v1 \in V$ of G, vertex $v2 \in V$ of G
Ensure: an edge between $v1$ and $v2$ is MRNG conformant
1: **for all** $u \in N(G, v1) \cap N(G, v2)$ **do**
2:     **if** $\delta(v1, v2) > \max(w_{v1,u}, w_{v2,u})$ **then**
3:         **return** false
4: **return** true

Since EG employs undirected edges, adding more vertices can result in the loss of this conformity. Simultaneously, reducing the number of edges can decrease the number of distance calculations required during a search and improve the overall performance. Therefore, we remove all non-conformant edges in the fully constructed graph using the following algorithm:

1. Convert the undirected edges of the graph into two directed edges.
2. Check the RNG conformity 1 for each edge.
3. Remove the edge if it is not RNG-conformant.

The graph remains connected because each edge removal is verified to ensure an alternative path exists. Transforming and thinning the edges causes the EG to lose its dynamic properties and preventing it from further expansion. However, this is not an issue for the challenge, as it involves a static dataset.

5 Smart Entries

The Exploration Graph (EG) is optimized for exploration, leading to suboptimal properties during the navigation phase. To address this inefficiency, multiple starting points ("Smart Entries") can be employed in the graph search to avoiding excessively long search paths.

Given data points are not uniformly distributed in the feature space and exhibit certain structures, some entry points are more advantageous than others. Optimal search entry points should be close to many other data points while being sufficiently distant from each other. To identify these points, Mini-batch k-Means [10] is utilized. This variant of the k-Means algorithm processes data in small batches rather than using the entire dataset. After the cluster centers are determined, the closest vertex in the graph for each center is identified and used as an entry point for future searches.

6 Compression

The search time is primarily determined by the number of distance calculations, and the time required for each calculation increases linearly with the dimensionality of the feature vectors. To accelerate this process and reduce storage requirements, one can either binarize the values or reduce the dimensionality of the feature vectors. However, it's important to note that aggressive quantization, such as binarization, often leads to a significant loss in accuracy.

Feature vectors can be compressed into a lower-dimensional space using techniques like PCA. However, this compression often significantly alters the relative distances between vectors, which will negatively impact retrieval accuracy. Therefore, it is crucial to reduce dimensions in a way that prioritizes preserving the relative rankings of similar feature vectors, rather than the distances between them. If the relative rankings can be maintained, the exploration and navigation properties of graphs will remain largely unaffected.

Our compression approach utilizes a fully connected neural network with three layers to map the original feature vectors to a reduced-dimensional space. Let $X = \{x_1, x_2, ..., x_n\}$ be the original high-dimensional vectors and $Y = \{y_1, y_2, ..., y_n\}$ the compressed lower dimensional vectors. d_{ij} denotes the euclidean distance between two vectors x_i and x_j. The normalized rank r_{ij} of a vector x_j relative to a vector x_i is defined as

$$r_{ij} = \frac{d_{ij} - \min_i}{\max_i - \min_i} \in [0,1] \qquad (1)$$

with $\quad \min_i = \min\{d_{ij}\}_{j=1}^n \quad$ and $\quad \max_i = \max\{d_{ij}\}_{j=1}^n \quad$ for $j \neq i \qquad (2)$

For two vectors x_j and x_k their rank difference Δr_{ijk} with respect to x_i (boosted by $\alpha > 1$) is defined as

$$\Delta r_{ijk} = \tanh(\alpha(r_{ij} - r_{ik})) \in (-1,1) \quad \text{for} \quad j \neq i, k \neq i, k > j \qquad (3)$$

Additionally we define the similarity sim_{ij} of two vectors x_i and x_j as

$$\text{sim}_{ij} = \frac{s_{ij}}{\sum_{k, k \neq i} s_{ik}} \in (0,1) \quad \text{for} \quad j \neq i \qquad (4)$$

with $\quad s_{ij} = -0.5 * \tanh(\beta(r_{ij} - 0.5)) + 0.5 \qquad (5)$

β again is a boosting factor > 1 squashing the similarity value to 1 for similar vectors and to 0 for dissimilar ones.

For the low-dimensional y vectors all these values are calculated in the same way, where a superscript indicates the corresponding value in the compressed domain. For example, d'_{ij} is the distance between y_i and y_j. For the loss J, the rank differences of the high-dimensional and the low-dimensional space are compared, using an additional weighting factor w_{ijk} depending on the similarity of the original vectors x_j and x_k to x_i.

$$w_{ijk} = \frac{\tilde{w}_{ijk}}{\sum_{\substack{l,m \\ l \neq i, m \neq i, m > l}} \tilde{w}_{ilm}}, \quad \tilde{w}_{ijk} = \max(\text{sim}_{ij}, \text{sim}_{ik}) \text{ for } j \neq i, k \neq i, k > j \quad (6)$$

$$J = \sum_{\substack{i,j,k \\ j \neq i, k \neq i, k > j}} w_{ijk}(\Delta r_{ijk} - \Delta r'_{ijk})^2 \qquad (7)$$

For the training, 8 batches are used in each iteration, whereby each batch consists of 200 vectors, which in turn consist of 5 random vectors plus their 39 nearest neighbors. The calculations of all terms like $\min_i$, $\min'_i$, Δr_{ijk}, $\Delta r'_{ijk}$, w_{ijk}, etc. are performed per batch. The boosting parameters were empirically evaluated and set to $\alpha = 10$ and $\beta = 8.4$.

7 Experiments

We evaluated the effectiveness of three proposed improvements on a 100M dataset and 10,000 public queries with corresponding ground truth data. All experiments were conducted on a server equipped with an AMD EPYC GENOA 9254 CPU (24 cores) and 768 GB of DDR5 RAM clocked at 4800 MHz.

For generating and testing crEG, the deglib library was employed[1]. Preliminary investigations on a 10M dataset identified the following effective hyperparameter configuration: edges-per-vertex set to 32, an extended search radius of 102% (extEps = 0.02), examination of the top 64 candidates (extK = 64) and use of construction scheme = D. Given the data is approximately L2-normalized, the inner product was used as metric. The graph edges were not further optimized, hence referred to as Exploration Graph (EG).

Using the identified hyperparameters, an EG was constructed for the 100M dataset to address Task 1, taking approximately one hour on the specified hardware. Its search efficiency compared to HNSW is shown in Fig. 1 (EG32). While crEG is faster than HNSW in the high recall range, HNSW is more efficient in the lower recall range.

Next, non-conformant MRNG edges were removed, and "Smart Entries" were used during the search, improving search speed by approximately 16% at a recall of 0.8. Although the achieved search efficiency still does not match HNSW, EG now outperforms HNSW at a recall above 0.85. This advantage is crucial, as compression and quantization degrade recall for all search algorithms equally. To achieve a recall of 0.8 after compression, a graph performing efficiently in the high recall range without compression is necessary.

To further accelerate searches, feature vectors were reduced to 512 dimensions using the compression network described in Sect. 6 and subsequently quantized to uint8. This network also transformed the feature space to allow the use of L2 distance, enabling efficient CPU instructions based on the uint8 data type. Additionally, memory requirements are reduced, allowing for better CPU cache utilization. The curve "EG32 + MRNG + smart + 512q" in Fig. 1 was generated by compressing all data points and constructing an EG with uint8 data and an L2 distance function. Due to the new metric, a higher search radius of 110% (extEps = 0.1) was used during construction. This configuration was submitted to fulfill Task 1 of the challenge.

Impact of compression on tasks 1 and 3: A crucial parameter influencing how well distances are preserved after compression is the desired output dimension. We chose 512D for Task 1 and 64D for Task 3. The significant data reduction for Task 3 was necessary to fulfill its memory requirements. The ability of a compression network to compress to these target dimensions depends on the network architecture. A more complex network architecture positively impacts compression quality but also increases the time required to compress the query feature vectors. Since the compression time is part of the search time, a good

---
[1] https://github.com/Visual-Computing/DynamicExplorationGraph

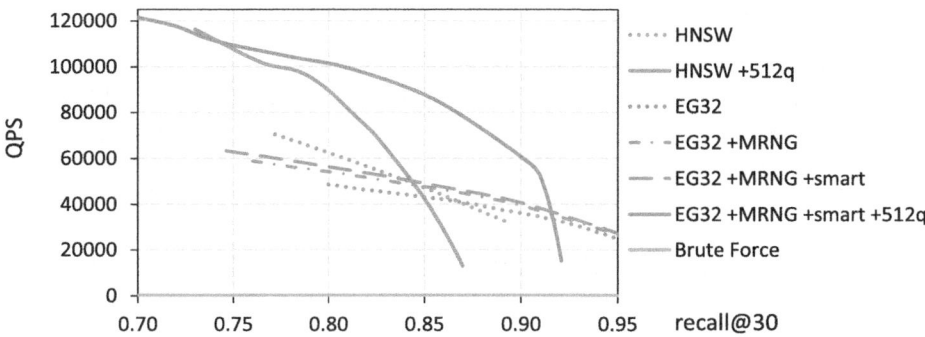

Fig. 1. Comparison of the EG with 32 edges per vertex and HNSW for Task 1. The proposed improvements significantly enhanced search efficiency.

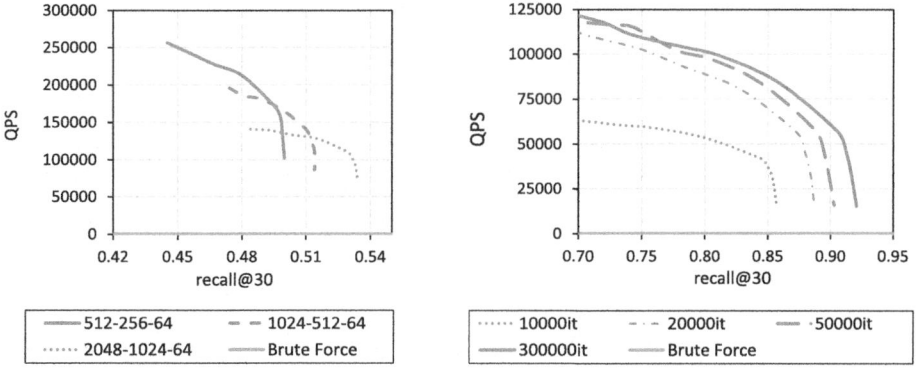

Fig. 2. A smaller architecture for the compression network delivers faster search times while still achieving recalls above 0.4 for Task 3 (left). Longer training times for the compression network improve search efficiency for Task 1 (right).

tradeoff must be found here. Figure 2 examines networks with three layers and varying numbers of neurons. Using smaller networks increases search throughput, but it comes at the cost of not achieving high recall ranges. For Task 3, the 512-256-64 network was used. Compression time during the search accounts for approximately 35% of the total search time. Further reduction was not made because the private test set used by the organizers of the competition is not known and we want to ensure a recall above 0.4 in the unknown test set.

During network training, it was observed that useful compressed feature vectors are generated after only a few iterations, but the training continues to improve search quality over time. Figure 2 shows that search efficiency increases rapidly with a recall of 0.8 up to 50,000 iterations, after which diminishing returns set in. In the higher recall range, improvements can still be observed after longer training durations.

8 Conclusion

In this paper, we extended the evaluation of the continuous refining Exploration Graph (crEG) to a 100 million-item dataset, focusing on lower recall requirements. We introduced three key improvements: elimination of redundant paths, utilization of optimal starting points, and compression of feature vectors. These enhancements significantly boost navigation efficiency, as demonstrated in our experiments. Notably, feature compression enables crEG to perform effectively in tasks where low recall value are sufficient, a domain typically dominated by navigation graphs.

References

1. Antol, M., Ol'ha, J., Slanináková, T., Dohnal, V.: Learned metric index—proposition of learned indexing for unstructured data. Inf. Syst. **100**, 101774 (2021)
2. Boguñá, M., Krioukov, D., Claffy, K.C.: Navigability of complex networks. Nat. Phys. **5**(1), 74–80 (2009)
3. Foster, C., Kimia, B.B.: Computational enhancements of HNSW targeted to very large datasets. In: Pedreira, O., Estivill-Castro, V. (eds.) SISAP. Lecture Notes in Computer Science, vol. 14289, pp. 291–299. Springer, Berlin (2023)
4. Fu, C., Xiang, C., Wang, C., Cai, D.: Fast approximate nearest neighbor search with the navigating spreading-out graph. Proc. VLDB Endow. **12**(5), 461–474 (2019)
5. Hezel, N., Barthel, K.U., Schall, K., Jung, K.: An exploration graph with continuous refinement for efficient multimedia retrieval. In: Gurrin, C., Kongkachandra, R., Schoeffmann, K., Dang-Nguyen, D.T., Rossetto, L., Satoh, S., Zhou, L. (eds.) ICMR, pp. 657–665. ACM (2024)
6. Malkov, Y.A., Yashunin, D.A.: Efficient and robust approximate nearest neighbor search using hierarchical navigable small world graphs. IEEE Trans. Pattern Anal. Mach. Intell. **42**(4), 824–836 (2020)
7. Oguri, Y., Matsui, Y.: General and practical tuning method for off-the-shelf graph-based index: SISAP indexing challenge report by team UTokyo. In: Pedreira, O., Estivill-Castro, V. (eds.) SISAP. Lecture Notes in Computer Science, vol. 14289, pp. 273–281. Springer, Berlin (2023)
8. Paredes, R., Chávez, E.: Using the k-nearest neighbor graph for proximity searching in metric spaces. In: Consens, M.P., Navarro, G. (eds.) SPIRE. Lecture Notes in Computer Science, vol. 3772, pp. 127–138. Springer, Berlin (2005)
9. Schuhmann, C., Beaumont, R., Vencu, R., Gordon, C., Wightman, R., Cherti, M., Coombes, T., Katta, A., Mullis, C., Wortsman, M., Schramowski, P., Kundurthy, S., Crowson, K., Schmidt, L., Kaczmarczyk, R., Jitsev, J.: Laion-5b: an open large-scale dataset for training next generation image-text models. In: Proceedings of the 36th International Conference on Neural Information Processing Systems, NIPS'22. Curran Associates Inc., Red Hook, NY, USA (2022)
10. Sculley, D.: Web-scale k-means clustering. In: Rappa, M., Jones, P., Freire, J., Chakrabarti, S. (eds.) WWW, pp. 1177–1178. ACM (2010)
11. Tellez, E.S., Aumüller, M., Chavez, E.: Overview of the SISAP 2023 indexing challenge. In: Pedreira, O., Estivill-Castro, V. (eds.) Similarity Search and Applications, pp. 255–264. Springer Nature, Cham (2023)
12. Thordsen, E., Schubert, E.: An alternating optimization scheme for binary sketches for cosine similarity search. In: Pedreira, O., Estivill-Castro, V. (eds.) SISAP. Lecture Notes in Computer Science, vol. 14289, pp. 41–55. Springer, Berlin (2023)

Top-Down Construction of Locally Monotonic Graphs for Similarity Search

Cole Foster[1]([✉]), Edgar Chávez[2]([✉]), and Benjamin Kimia[1]

[1] Brown University, Providence, RI 02912, USA
{cole_foster,benjamin_kimia}@brown.edu
[2] Centro de Investigación Científica y de Educación Superior de Ensenada, Ensenada, México
elchavez@cicese.edu.mx

Abstract. Similarity search is a fundamental task in applications such as recommender systems, image retrieval, and text retrieval. Graph-based indexes for similarity search traverse a graph constructed on the dataset to retrieve the query's neighbors, using edges to navigate to and explore the query's local neighborhood. Edge selection techniques are crucial for the performance of graph-based indexes, enhancing accuracy and efficiency by preventing local minima, reducing graph diameter, and improving sparsity. The Half-Space Proximal (HSP) Graph is an edge-minimal monotonic graph defined by a geometric edge selection which ensures a diverse, yet sparse set of edges. Unfortunately, the quadratic construction complexity of the HSP Graph renders it impractical for large-scale search scenarios. This work investigates an approximation of the HSP Graph that aims to preserve the monotonic property locally. By leveraging a hierarchical partitioning of the dataset, this work proposes a top-down, distributed graph construction which uses a coarse-scale graph on pivots to facilitate the construction of the layer below. This paper investigates the effectiveness of this approach as a submission to the SISAP 2024 Indexing Challenge.

Keywords: similarity search · hierarchy · graph · monotonic

1 Introduction

Large-scale collections of images, documents, social media posts, *etc.*, power modern applications such as image retrieval [23], retrieval-augmented generation [6], and recommender systems [28]. Underlying these applications are similarity search systems designed to efficiently retrieve relevant information while meeting the constraints posed by the application. Formally, similarity search is the task of searching a database S to collect the elements most similar to a query q. The search problem typically takes the form of the k-nearest neighbor (kNN) search, finding the k elements in S that maximize some similarity function or minimize some distance function to q.

The brute-force approach to similarity search involves computing the distance between the query and all elements of the dataset, but this $O(N)$ expense is infeasible under the real-time constraints posed by applications. A *search index* is a system that leverages an offline organization of the dataset to enable fast online search. Non-exhaustive search indexes, which avoid computing the distance between the query and all elements, can be categorized into *partitioning-based indexes* and *graph-based indexes*.

Partitioning-based indexes divide the dataset into groups and only examine a subset of groups rather than the entire dataset. These indexes improve accuracy at the expense of efficiency by increasing the number of groups searched. Metric-space partitioning-based indexes rely on proximity to *pivots*, representative members of the dataset, to organize elements into groups [5,9]. In a vector space, other modes of partitioning may be defined, *e.g.*, based on proximity to learned centroid vectors [15,17], random hyperplanes [25], or position along dimensions [4]. *Quantization-based indexes* [15,17] and *Hashing-based indexes* [1,24] both fall into this category, and further define vector compression techniques to dramatically reduce the memory footprint of the dataset at the cost of reduced accuracy.

Graph-based indexes construct a graph on all elements of the dataset, where each element acts as a node and edges encode the local relationships between nodes. These approaches are state-of-the-art for high-accuracy similarity search in high-dimensional spaces [27]. This success may be attributed to the fine-grain detail encoded by graphs, allowing for precise exploration of the query's local region, in contrast to the coarser exploration defined by partition-based indexes. Searching over a graph involves the local traversal of the edges of the graph, iteratively computing the query's distance to the neighbors of the current node and then choosing the next node to explore. A *greedy search* is a best-first strategy that chooses the next node to explore as the neighbor that decreases the distance to the query. This form of search is susceptible to local minima, which occurs when no neighbor decreases the distance to the query, causing early termination and incorrect results. A common approach to overcoming local minima is *beam search*, a generalization of greedy search which stores a "beam" of the closest encountered nodes and uses the beam for backtracking when faced with local minima. Increasing the beam size b can overcome local minima at the expense of longer search times.

The accuracy and efficiency of graph-based search is highly dependent on the graph itself. The most well-known graph structure is the kNN Graph, which connects each node to its k closest elements in $\mathcal{S}$ [7,13]. The kNN Graph is highly susceptible to local minima, and may even be disconnected, requiring high connectivity or excessive backtracking. More detailed edge-selection strategies consider the higher-order relationships among elements, aiming to reduce local minima and improve sparsity [14,18,20]. Many approaches also focus on minimizing graph diameter, which promotes shorter search paths and increases efficiency. This may be achieved through long-range links [16,20] or a hierarchical structure [18].

Monotonic Graphs: Graphs with the monotonic property [10] ensure a *monotonic path* between each pair of nodes in $x_1, x_n \in \mathcal{S}$, *i.e.*, a path $x_1, x_2, \ldots, x_n$ where $d(x_{i+1}, x_n) < d(x_i, x_n)$ for $i = 1, \ldots, n$. Such a path is traversable by greedy search, but does not guarantee greedy search for all queries [19]. Despite this, the principles of monotonicity are used for edge selection in many modern graph-based indexes [14,18,20], including the well-known Hierarchical Navigable Small World (HNSW) index [18].

The Half-Space Proximal (HSP) Graph [8,21] is an edge-minimal, monotonic graph defined by applying a geometric edge-selection algorithm, the *HSP Test* [8], on each node independently. The graph has recently received popularity under a different name, the "Monotonic Relative Neighborhood Graph (MRNG)" [14]. Unfortunately, the $O(N^2)$ construction complexity of the HSP Graph makes it infeasible for use on large-scale datasets: building the graph on a subset of 1M elements of the LAION dataset [22,27] takes 16 h on a 32-core CPU, and would require over 18 years for the full 100M dataset! There have been a few attempts to improve the efficiency of the HSP Test, which in turn, accelerates the entire graph construction. The Hierarchical HSP [12] uses a hierarchical partitioning of the dataset to improve efficiency while preserving exactness, but this approach is limited to lower-dimensional spaces and is not effective on the LAION dataset. The HSP neighbors of a node x can be approximated by constraining the neighbors to be within some local region $R(x)$ defined around x, rather than the entire dataset. In [26], HNSW [18] was used to retrieve the region $R(x)$ around the node x, and the HSP Test is applied on that region to find the neighbors of x.

Graph Construction: The effectiveness of a graph is due to its ability to capture the local relationships of elements, but this poses an interesting problem for its construction. Before a node can find its neighbors, it must first identify its local region, *i.e.*, a set of nearby candidate neighbors. Thus, the construction of a graph-based index requires the availability of existing index! A common approach is to define some starting graph structure that serves as the starting index, and gradually refine this index by incrementally updating the neighbors of each node. The starting graph structure may be defined as a random graph [11], a *k*NN graph [14], or even an empty graph [18]. For example, the Hierarchical Navigable Small World (HNSW) index [18] begins with an empty graph and incrementally adds each element to the graph. However, this incremental approach prevents a fully distributed graph construction, as the graph quality is dependent on the incremental refinement of the graph.

Overview: This submission[1] to the SISAP 2024 Indexing Challenge investigates the use of the Half-Space Proximal (HSP) Graph [8,21], a monotonic proximity graph, for similarity search. To address its quadratic construction time complexity and the inefficiency of the high-out degree, this work proposes to use a *locally monotonic graph*, a graph which preserves the monotonic property between nearby nodes. The approximate HSP Graph serves as the locally mono-

---

[1] source code: https://github.com/cole-foster/sisap-2024.git

tonic graph, constructed by constraining the HSP neighbors to a region of nearby elements. By leveraging a hierarchical partitioning of the dataset with a scaling factor s, the graph is built through a top-down construction process, where the graph on each layer guides the distributed construction of the graph on the layer below. Each node finds its neighbors independently, using the coarse-layer graph to locate the p closest pivots and finding the HSP neighbors from the union of those pivots' domains, Fig. 1. Finally, the hierarchical partitioning provides a fast entry-point into the bottom-level graph, serving as a starting point for beam search.

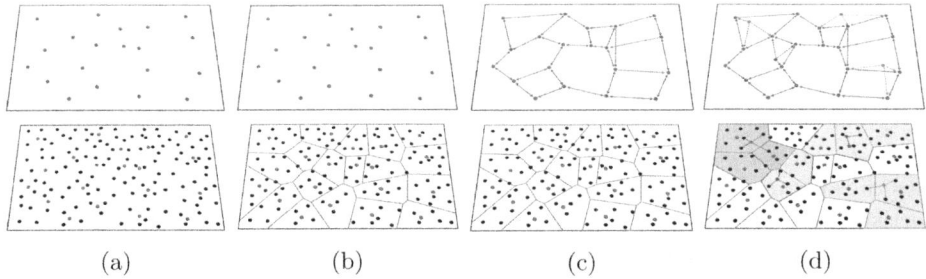

Fig. 1. The top-down, distributed graph construction. (a) Pivots are selected from the dataset according to the scaling factor s, (b) elements are assigned to the domain of their closest pivot, (c) a coarse-layer HSP Graph is constructed on the pivots, and (d) each element finds their approximate HSP neighbors independently, using the coarse graph to find the $p = 3$ closest pivots and performing the HSP Test on the union of their domains.

SISAP 2024 Indexing Challenge: The Similarity Search and Applications (SISAP) 2024 Indexing Challenge is a competition to perform similarity search on a dataset of 100M 768D vector embeddings of the LAION-5B dataset [22]. This submission competes in the first task of the competition, which allows for 512 GB RAM and 12 h construction time limit. The remaining tasks place extensive constraints on RAM, requiring dataset compression techniques which are not a focus of this work. The goal of the first task is to perform kNN Search ($k = 30$) with the greatest throughput, or queries-per-second (QPS), with over 80% recall. This paper accompanies our submission to this challenge under the team name "HSP".

2 Hierarchical, Approximate HSP Graph Construction

The Half-Space Proximal (HSP) Graph is a directed proximity graph defined by applying the *HSP Test* [8] to each node, resulting in a sparse and geometrically diverse graph. To find the neighbors of a single node $x \in \mathcal{S}$, the HSP Test takes an inside-out approach which iteratively chooses each neighbor y as the closest available element and then removes from consideration all other elements z satisfying $d(y, z) < d(x, z)$. This process continues until all elements $z \in \mathcal{S} \setminus x$ are either neighbors of x or satisfy $d(y, z) < d(x, z)$ for some neighbor y. The process resolves the condition of monotonicity for a single node, and when applied to each node, it results in a fully monotonic graph on $\mathcal{S}$. Note that the graph is edge-minimal; removing any edge results in the loss of monotonicity guarantees.

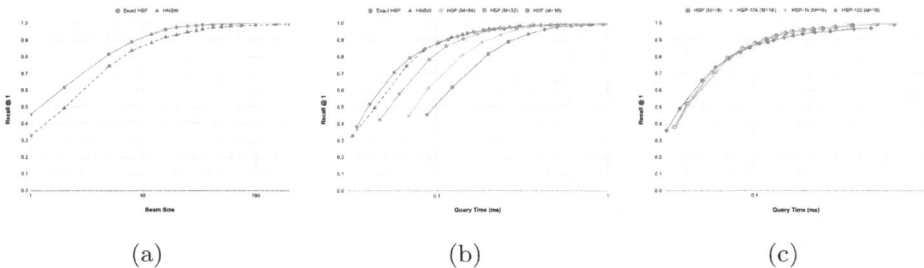

Fig. 2. (a) The recall of beam search on the exact HSP Graph compared to HNSW [18] on a 1M subset of the LAION dataset. (b) The search time vs. recall trade-off of HSP Graph and HNSW. Restricting the HSP Graph to a maximum of M neighbors improves efficiency. (c) Constraining the HSP neighbors to the $k = 10$k, 1k, 100 closest elements of each node provides little reduction in search performance.

Observe the performance of beam search on the exact HSP Graph for a subset of 1M elements of the LAION dataset, Fig. 2(a). For a fixed beam size, the HSP Graph provides an improved recall over HNSW [18], but the high out-degree (68.9) leads to long search times, Fig. 2(b). By constraining the maximum number of HSP neighbors, specifically, by only connecting each node to its M closest HSP neighbors, the search time is dramatically lowered with only a slight reduction in recall. From a practical perspective, there is no need for a monotonic path between all nodes in the dataset once the search is sufficiently close to the query.

This suggests that a fully monotonic graph may be unnecessary and monotonicity can simply be preserved locally, *i.e.*, with a *locally monotonic graph*. Such a graph may be built by using the HSP Test to provide monotonicity within the local region $R(x)$ of each node $x \in \mathcal{S}$. By defining the region $R(x)$ as the set of elements within a distance r of x, the HSP Test will define the approximate HSP Graph that guarantees a monotonic path for nodes within a distance r. Another option is to define $R(x)$ as the k closest elements to x, which loses the guarantees of exactness but has the convenience of constant region $R(x)$ size and independence from distance metric. Observe the search performance of this constrained HSP Graph on the LAION 1M subset when $R(x)$ is defined by the exact $k = 10$k, 1k, 100 closest elements in the dataset, Fig. 2(c). Constraining the HSP neighbors these k closest elements results in a near-identical identical performance to the unconstrained version, showing promise for locally-monotonic graph.

Algorithm 1 Top-Down Distributed Graph Construction

1: **Input:** s as scaling factor, n as number of domains to define the region.
2: **Output:** L as the number of layers, $\mathcal{P}_\ell$ as the set of pivots/elements in layer $\ell = 1 \ldots L$, $\mathcal{D}_\ell(p)$ as the set of layer $\ell+1$ pivots/elements in the domain of $p \in \mathcal{P}_\ell$, G_ℓ as the graph on layer $\mathcal{P}_\ell$.
3: Select the pivots $\mathcal{P}_\ell$, $\ell = 1 \ldots L$ according to scaling factor s, Fig. 1(a).
4: **for** $p \in \mathcal{P}_1$ **do** ▷ Build graph on $\mathcal{P}_1$ (parallel), Fig. 1(c)
5:    $G_1(p) \leftarrow$ the HSP neighbors of p within P_1
6: **for** $\ell = 2$ to L **do**
7:    **for** $p \in \mathcal{P}_\ell$ **do** ▷ Create partitioning of $\mathcal{P}_\ell$ (parallel), Fig. 1(b)
8:       $\bar{p} \leftarrow$ closest pivot in $\mathcal{P}_{\ell-1}$ by brute-force or graph-search on $G_{\ell-1}$.
9:       $\mathcal{D}_{\ell-1}(\bar{p})$.insert($p$)
10:    **for** $p \in \mathcal{P}_\ell$ **do** ▷ Build graph on $\mathcal{P}_\ell$ (parallel), Fig. 1(d)
11:       $v_{\bar{p}} \leftarrow n$ closest pivots in $\mathcal{P}_{\ell-1}$ by brute-force or graph-search on $G_{\ell-1}$.
12:       $R(p) \leftarrow$ union of $\mathcal{D}_{\ell-1}(\bar{p})$ for all $\bar{p} \in v_{\bar{p}}$.
13:       $G_\ell(p) \leftarrow$ the HSP neighbors of p within $R(p)$

Rather than defining $R(x)$ by an exact range search or kNN search, this region can be defined by search index for efficiency. A graph-based index can provide a high accuracy region, but defining a large region with a graph is costly and requires a full index to be constructed ahead of time. On the other hand, a partitioning-based index allows for a more coarse definition of $R(x)$: given s elements in each domain, $R(x)$ may be defined by the p closest domains. This work proposes a hybrid approach to use the two in tandem: N/s pivots selected from $\mathcal{S}$ form a partitioning of the dataset, where elements are assigned to the domain of their closest pivot, and a graph on the pivots guides x to find its p closest domains. The local region $R(x)$ is defined by the union of the domains of the p closest pivots, $|R(x)| \approx s \cdot p$. This concept of using a graph on pivots is similar to the use of HNSW on centroid vectors in quantization [3].

Top-Down Graph Construction: This work proposes a distributed graph construction approach which uses a partitioning and coarse-scale graph to guide the construction of the fine-scale graph. Given the coarse-scale graph built on the N/s pivots, each node x finds its HSP neighbors independently by: (i) using the coarse-scale graph to find the p closest pivots, (ii) defining the region $R(x)$ as the union of the domains of the pivots, and (iii) performing the HSP Test on $R(x)$ to find the neighbors of x. Since this process is independent for each x, the graph can be constructed in a distributed manner, maximizing the system's parallel capabilities.

Of course, the brute-force construction of the coarse-layer graph on N/s pivots has a high cost for large N. Similarly, the construction of this graph can also benefit from the guidance of a coarser-layer graph. This suggests a hierarchical, top-down graph construction, where the graph on each layer of this guides the construction of the graph on the layer below, Algorithm 1. The selection of the pivots on each layer may take on many different forms, *e.g.*, clustering [2]. This work chooses to probabilistically select the pivots in each layer according to the scaling factor s for its simplicity, its efficiency, and its applicability in a general metric space. While this graph construction is defined for the construction of the Approximate HSP Graph, it may easily be adapted to other graphs by defining a different edge selection technique on $R(x)$, *e.g.*, kNN Graph or the RNG Graph.

3 Experiments

Search over this graph defined by Algorithm 1 requires a starting node sufficiently close to the query, and then proceeds with a beam search over the graph to return the kNN of the query. While the entry-point may be defined in many ways, this work simply uses the hierarchical partitioning without backtracking, recursively exploring the domain of the closest pivot until the bottom layer is reached. Although closer entry-points may be found using the graph on each layer, the partitioning approach is faster and provides no reduction in recall.

The following results were run on a workstation with a 32-core Intel Xeon Gold 6242 CPU with 512 GB of RAM. For the LAION dataset, the inner product was converted into a distance measure using $d(x,y) = 1- <x,y>$. The Approximate HSP Graph was constructed a maximum of $M = 16$ neighbors. After graph construction, the hierarchical partitioning was rebuilt with $s = 16$ for consistency across graph construction configurations. This paper compares to HNSW, the state-of-the-art graph-based index which the competition in 2023 [27]. The construction parameters $M = 16$ and ef_construction $= 400$ were used for these results. At search time, the hyperparameter b controls the beam size of beam search, defining the trade-off between speed and accuracy. Accuracy is measured by recall@k, which indicates the percentage of correctly returned neighbors among the k returned elements. Speed is measured in query latency, the average time to perform a single query, and Queries-per-Second (QPS), the average rate at which queries can be processed. The queries are processed concurrently by the 32-core CPU to enhance throughput.

Table 1. The impact of hyperparameters on construction time and graph quality for a 1M subset of the LAION dataset.

s	p	Construction Time (s)	% Correct HSP Neighbors	Recall@30, b = 30
10	100	347.11	56.5%	0.813
20	50	223.43	53.1%	0.818
50	20	221.00	47.7%	0.799
100	10	198.52	42.1%	0.774
10	200	809.54	66.0%	0.830
20	100	486.53	63.7%	0.836
50	40	493.57	58.9%	0.825
100	20	392.33	54.0%	0.817

Importance of Hyperparameters: The goal of this submission is to provide a high-quality approximation of the HSP Graph. The hyperparameters s and p have an inverse impact on the construction time and graph quality, meaning a higher-quality graph takes longer to build, Table 1. The graph quality, measured by the percent of correct HSP neighbors, is correlated to the final recall of beam search on the graph. Under no time constraint, the highest quality graph may be achieved with a small s and large p.

Performance on the LAION Dataset: The results on the LAION 10M and 100M subset are shown in Fig. 3. For the 10M subset, the construction time constraint is not an issue. To show the impact of graph quality on performance, the parameters are chosen as a fixed $s = 10$ and $p = 100, 200, 300, 400$, resulting in construction times of 0.99, 1.38, 2.10, and 2.76 h, respectively. Interestingly, the performance of saturate as $p = 200$ despite an improvement in graph quality, suggesting that the maximum performance of the HSP Graph was reached. On the 100M subset, the 12 h construction time limit of the competition posed constraints on the hyperparameters. The configuration $s = 10, p = 100$ results in an index construction time of 33.5 h, which is well over the time limit. By fixing region size at $s \cdot p =$1k and varying s and p, the construction time can be reduced at the cost of graph quality: the configuration $s = 20, p = 50$ requires 14.7 h; $s = 50, p = 20$ requires 7.9 h; and $s = 100, p = 10$ requires 4.5 h. The parameters of $s = 100, p = 10$ were initially chosen for our submission, which resulted in a poor quality graph and subpar performance in the competition. Recent improvements to implementation further improve the graph quality, leading to the better performance here than in the competition.

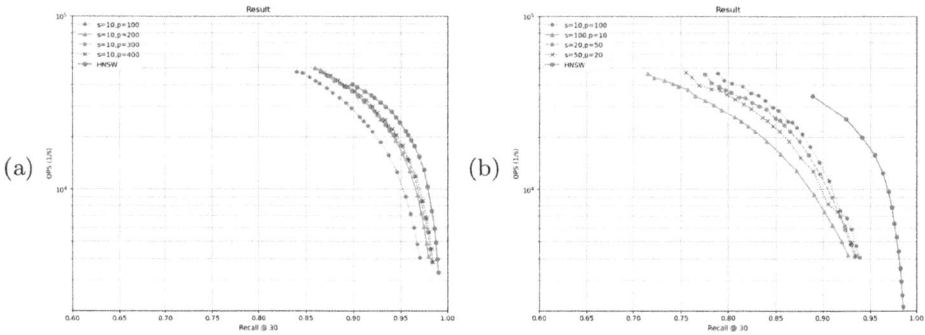

Fig. 3. Results of kNN search ($k = 30$) on the (a) a 10M subset and (b) a 100M subset of the LAION dataset. The parameters s and p influence graph quality and performance. Comparing to the state-of-the-art graph-based index HNSW [18].

4 Conclusion

This submission to the SISAP 2024 Indexing Challenge investigates the use of the Half-Space Proximal (HSP) Graph [8,21], a monotonic proximity graph, for similarity search. This work proposes to use a local variant of the Half-Space Proximal (HSP) Graph, which is a monotonic graph, as a locally monotonic graph. The construction leverages a hierarchical partitioning of the dataset, using a coarse-layer graph to guide the distributed construction of the graph on the layer below. The resulting graph quality is highly dependent on hyper-parameter selection, which further influences the index construction time. In future work, this trade-off can be avoided by fully leveraging the parallel potential of the distributed construction, using MPI to coordinate more CPUs or GPUs. It will also be important to test the generalization capabilities of this approach by evaluating it on a diverse array of datasets. Finally, this general graph construction approach can be applied to other types of graphs, including the RNG Graph and the kNN Graph, and can serve as a test-bed for examining the performance of different edge selection techniques.

Acknowledgments. We gratefully acknowledge the support of NSF award 1910530.

References

1. Andoni, A., Indyk, P.: Near-optimal hashing algorithms for approximate nearest neighbor in high dimensions. Comm. ACM **51**(1), 117–122 (2008)
2. Azizi, I., et al.: ELPIS: graph-based similarity search for scalable data science. VLDB **16**(6), 1548–1559 (2023)
3. Baranchuk, D., et al.: Revisiting the inverted indices for billion-scale approximate nearest neighbors. In: ECCV, pp. 202–216 (2018)
4. Bentley, J.L.: Multidimensional binary search trees used for associative searching. Comm. ACM **18**(9), 509–517 (1975)
5. Beygelzimer, A., et al.: Cover trees for nearest neighbor. In: ICML, pp. 97–104 (2006)
6. Borgeaud, S., et al.: Improving language models by retrieving from trillions of tokens. In: ICML, pp. 2206–2240 (2022)
7. Bratić, B., et al.: NN-descent on high-dimensional data. In: WIMS, pp. 1–8 (2018)
8. Chavez, E., et al.: Half-space proximal: a new local test for extracting a bounded dilation spanner of a unit disk graph. In: OPODIS, pp. 235–245. Springer, Berlin (2005)
9. Ciaccia, P., Patella, M., Zezula, P.: M-tree: an efficient access method for similarity search in metric spaces. In: VLDB, vol. 97, pp. 426–435 (1997)
10. Dearholt, D.W., et al.: Monotonic search networks for computer vision databases. In: ACSSC, vol. 2, pp. 548–553. IEEE (1988)
11. Dong, W., Moses, C., Li, K.: Efficient k-nearest neighbor graph construction for generic similarity measures. In: TheWebConf, pp. 577–586 (2011)
12. Foster, C., Chávez, E., Kimia, B.: Finding HSP neighbors via an exact, hierarchical approach. In: SISAP, pp. 3–18. Springer, Berlin (2023)
13. Fu, C., Cai, D.: EFANNA: an extremely fast approximate nearest neighbor search algorithm based on kNN graph (2016). arXiv preprint arXiv:1609.07228
14. Fu, C., et al.: Fast approximate nearest neighbor search with the navigating spreading-out graph. VLDB **12**(5), 461–474 (2019)
15. Guo, R., et al.: Accelerating large-scale inference with anisotropic vector quantization. In: ICML, pp. 3887–3896 (2020)
16. Jayaram, S., et al.: DiskANN: fast accurate billion-point nearest neighbor search on a single node. NeurIPS **32** (2019)
17. Johnson, J., et al.: Billion-scale similarity search with GPUs. Trans. Big Data **7**(3), 535–547 (2019)
18. Malkov, Y.A., Yashunin, D.A.: Efficient and robust approximate nearest neighbor search using hierarchical navigable small world graphs. IEEE **42**(4), 824–836 (2018)

19. Navarro, G.: Searching in metric spaces by spatial approximation. VLDB J. **11**(1), 28–46 (2002)
20. Peng, Y., et al.: Efficient approximate nearest neighbor search in multi-dimensional databases. ACM Manag. Data **1**(1), 1–27 (2023)
21. Ruiz, G., Chávez, E.: Proximal navigation graphs and t-spanners (2014). arXiv
22. Schuhmann, C., et al.: LAION-5B: an open large-scale dataset for training next generation image-text models. NeurIPS **35**, 25278–25294 (2022)
23. Shiau, R., et al.: Shop the look: building a large scale visual shopping system at Pinterest. In: SIGKDD, pp. 3203–3212 (2020)
24. Shrivastava, A., Li, P.: Asymmetric LSH (ALSH) for sublinear time maximum inner product search (MIPS). NeurIPS **27** (2014)
25. Spotify: Annoy (2023). https://github.com/spotify/annoy
26. Talamantes, A., Chavez, E.: Instance-based learning using the half-space proximal graph. Pattern Recogn. Lett. **156**, 88–95 (2022)
27. Tellez, E.S., Aumüller, M., Chavez, E.: Overview of the SISAP 2023 indexing challenge. In: SISAP, pp. 255–264. Springer, Berlin (2023)
28. Vemuri, H., et al.: Personalized retrieval over millions of items. In: SIGIR, pp. 1014–1022 (2023)

Correction to: A Dynamic Evaluation Metric for Feature Selection

Muhammad Rajabinasab, Anton D. Lautrup, Tobias Hyrup, and Arthur Zimek

---

Correction to:
Chapter 6 in: E. Chávez et al. (Eds.): SISAP 2024, LNCS 15268, pp. 65–72, 2024.
https://doi.org/10.1007/978-3-031-75823-2_6

In the originally published version of Chapter 6 an error was introduced in Equation 1, which has now been corrected as in below.

$$\text{FSDEM} = \frac{\int_a^b g(x)\,dx}{(b-a)} \tag{1}$$

The chapter has been updated with the change.

The updated version of this chapter can be found at
https://doi.org/10.1007/978-3-031-75823-2_6

Author Index

MISC
Čerňanský, Jozef 266
Černek, Andrej 18
Šikyňa, Matúš 126

A
Albrechts, Kevin 207
Amagata, Daichi 3
Antol, Matej 266
Aumüller, Martin 155, 170, 255

B
Bailey, James 111
Barthel, Kai Uwe 97, 283
Beecks, Christian 207
Budikova, Petra 18, 73, 223

C
Campello, Ricardo J. G. B. 111, 215
Carlini, Emanuele 140
Chávez, Edgar 291
Claydon, Ben 49, 57
Connor, Richard 49, 57, 140
Culemann, Wolf 34

D
Dearle, Alan 49, 57
Dohnal, Vlastislav 223, 266
Dostalova, Nicol 34

E
Erfani, Sarah Monazam 111

F
Foster, Cole 291

G
Gennaro, Claudio 140

H
Hara, Takahiro 3
Hezel, Nico 97, 283
Houle, Michael E. 111, 193
Huang, Hanxun 111
Hüwel, Jan David 207
Hyrup, Tobias 65

I
Iglesias, Félix 88

J
Jánošová, Miriama 73, 223
Johnsen, Malte Helin 155
Joukhadar, Zaher 111
Jung, Klaus 97, 283

K
Kimia, Benjamin 291

L
Lang, Andreas 223
Lautrup, Anton D. 65
Liarou, Margarita 185

M
Mahrík, Marek 126
Marchand-Maillet, Stephane 185
Marques, Henrique O. 215
Martínez, Conrado 88
Mic, Vladimir 126, 255
Molo, Mbasa Joaquim 140

O
Okkels, Camilla Birch 170
Olha, Jaroslav 266
Oria, Vincent 193

P
Procházka, David 266

R
Rajabinasab, Muhammad 65
Reyes, Victor 185
Röchner, Philipp 215
Rothlauf, Franz 215

S
Sabaei, Hamideh 193
Schall, Konstantin 97, 283
Schilling, Bruno 283
Schlake, Georg Stefan 207
Schubert, Erich 223, 238, 274
Sedmidubsky, Jan 18, 34, 73
Slanináková, Terézia 266
Svaricek, Roman 34

T
Tellez, Eric S. 255
Thordsen, Erik 274

U
Uemura, Reon 3

V
Vadicamo, Lucia 49, 140

Z
Zezula, Pavel 126
Zimek, Arthur 65, 170, 215
Zseby, Tanja 88

The manufacturer's authorised representative in the EU is Springer Nature Customer Service Centre GmbH, Europaplatz 3, 69115 Heidelberg, Germany. If you have any concerns regarding our products, please contact ProductSafety@springernature.com

Printed and bound by CPI Group (UK) Ltd, Croydon, CR0 4YY

25/03/2026

02078195-0012